I0050186

UNITED STATES DEPARTMENT OF THE INTERIOR
Ray Lyman Wilbur, Secretary

GEOLOGICAL SURVEY
W. C. Mendenhall, Director

Professional Paper 172

GOLD QUARTZ VEINS OF THE ALLEGHANY DISTRICT CALIFORNIA

BY

HENRY G. FERGUSON

AND

ROGER W. GANNETT

UNITED STATES
GOVERNMENT PRINTING OFFICE
WASHINGTON : 1932

For sale by the Superintendent of Documents, Washington, D. C.   -   -   -   -   -   -   Price $2.00 (Paper cover)

CONTENTS

	Page
Outline of the report	1
Part 1. Geology	3
Introduction	3
Scope of the report	3
Location and topography	3
Field work and acknowledgments	3
Previous work	5
Rock formations	5
"Bedrock series"	6
Blue Canyon formation	6
Tightner formation	6
Kanaka formation	7
Areal extent	7
Conglomerate member	8
Lower slate and greenstone member	10
Chert member	11
Upper slate and greenstone member	11
Sections	11
Relief quartzite	12
Cape Horn slate	12
Serpentine	13
Gabbro and diorite	14
Quartz diorite	15
Granite and aplite	16
Age of the "Bedrock series"	16
"Superjacent series"	17
Early gravel	17
Intervolcanic gravel	17
Andesitic breccia	18
Basalt	18
Pleistocene and Recent gravel	19
Geologic structure and history	19
Structure of the "Bedrock series"	19
Faulting	21
Tertiary history	22
Development of the present topography	23
Part 2. Ore deposits	25
History and production	25
The veins	28
Scope of descriptions	28
Distribution	29
Relation to country rock	31
Size and structure	33
Mineralogy	38
Summary	38
Chlorite stage	39
Quartz stage	40
Arsenopyrite and pyrite	40
Quartz	40
General features	40
Vacuoles	41
Plagioclase	44
Barite	44
Chalcedony	44
Carbonate and sericite	45
Apatite	45
Beidellite	45

	Page
Part 2. Ore deposits—Continued.	
Mineralogy—Continued.	
Carbonate stage	45
Character of mineralization	45
Carbonate	46
Mica (mariposite and sericite)	46
Chlorite	47
Talc	47
Carbon (graphite)	47
Palygorskite	48
Rutile	48
Quartz, chalcedony, and opal	48
Sulphides	49
Gold	50
Final stage	51
Oxidation	51
Ore shoots	52
Character	52
Productivity of veins of different types	53
Favorable structural features	56
Mineralogic indications	58
Probable persistence in depth	60
Origin of the deposits	60
Age of the veins	61
Depth of vein formation	61
Origin of the fissure systems	70
Pressure and temperature	72
Source of the minerals	73
Possibility of derivation from the wall rock	73
Minerals of the chlorite stage	74
Quartz and carbonate	74
Arsenopyrite	75
Mariposite	75
Chromite	75
Gold	75
Graphite	76
Plagioclase	77
Barite	77
Method of vein formation	77
Chlorite stage	77
Quartz stage	77
Possible recrystallization	77
Replacement	80
Fissure filling	80
Conclusions	84
Carbonate stage	85
Summary	86
Mine descriptions	87
Mines of Oregon Creek drainage basin	87
Brush Creek mine	87
Finan prospect	89
Kate Hardy mine	89
Hardy Kate prospect	91
Tomboy prospect	91
Eureka mine	92
Oak prospect	92

Part 2. Ore deposits—Continued.
   Mine descriptions—Continued.

	Page
Mines of Oregon Creek drainage basin—Con.	
Mugwump mine	92
Gold Bug prospect	94
Federal mine	94
South Fork mine	94
Amethyst mine	94
Diadem mine	95
North Fork mine	95
Mines of Kanaka Creek drainage basin	97
Kenton mine	97
Roye-Sum prospect	98
Oriental mine	98
Frances D prospect	101
Spoohn mine	101
General Sherman prospect	101
Wyoming group (Dead River)	102
Wonder prospect	102
Rainbow mine	102
Rainbow Extension mine	105
Sixteen to One mine	106
Ophir mine	110
Eclipse mine	110
Morning Glory mine	111
Extension of the Minnie D mine	112
Minnie D and Lucky Larry mines	113
Osceola mine	113
Panama prospect	115
Red Star mine	115
Rao prospect	117
Mayflower prospect	117

Part 2. Ore deposits—Continued.
   Mine descriptions—Continued.

	Page
Mines of Kanaka Creek drainage basin—Con.	
Mariposa mine	117
Central mine	117
Snowdrift prospect	118
Eldorado mine	118
Yellow Jacket (Colorado) mine	119
Eastern Cross group	120
Rising Sun mine	121
Iroquois prospect	122
Dreadnaught mine	122
Belmont prospect	122
Golden King mine	123
Docile mine	123
Continental mine	124
Mammoth Springs mine	125
Kinselbach (Acme) mine	125
Mines in the drainage basin of the Middle Fork of the Yuba River	125
Mountain View prospect	125
Maryland prospect	126
Oriflamme mine	126
Irelan mine	127
Arcade mine	128
Cooper prospect	129
Plumbago mine	129
German Bar mine	132
Gold Canyon mine	134
Independence mine	135
Future of the district	135
Index	137

ILLUSTRATIONS

                                                            Page

PLATE 1. Geologic map of the Alleghany district, California — In pocket.

2. *A*, Crumpled hornblende schist of Tightner formation between Morning Glory and Osceola mines; *B*, Slate of the Blue Canyon formation, near Mammoth Springs mine; *C*, Crumpled hornblende schist, siliceous facies of Tightner formation, valley of Kanaka Creek; *D*, View from Alleghany, looking down valley of Kanaka Creek — 10

3. Geologic map of the region surrounding the Alleghany district — 10

4. *A*, Hornblende schist with streaks of quartz, Tightner formation, near lower Osceola tunnel; *B*, Basal conglomerate (tillite?) of Kanaka formation near transformer house, Alleghany; *C*, Chert of Kanaka formation, South Fork of Yuba River near Washington; *D*, Detail showing contorted banding in chert, South Fork of Yuba River near Washington — 10

5. *A*, Relief quartzite, fold in quartzitic slate, Kanaka Creek near Rapps Ravine; *B*, Typical exposure of serpentine, nearly bare of vegetation, valley of Kanaka Creek; *C*, Eocene gravel at old hydraulic workings south of Alleghany; *D*, Flows of Pleistocene basalt capping Miocene andesitic breccia, Bald Mountain — 11

6. Map showing location of fault east of Alleghany with reference to the Mother Lode system — 26

7. *A*, North side of valley of Kanaka Creek looking toward Sixteen to One mill, showing shoulder below lava; *B*, Outcrop of vein on footwall side of Sixteen to One vein; *C*, Junction of Sixteen to One vein and vein along steep fault — 26

8. Principal fissures of the Alleghany district, projected to plane of 3,700-foot contour — In pocket.

9. Map of the Alleghany district, showing principal mine workings and mining claims — In pocket.

10. Structure sections through the eastern part of the Alleghany district — In pocket.

11. *A*, Vein with septa of altered country rock separating quartz strands, North Fork mine; *B*, Junction of main vein and fault vein, Sixteen to One mine; *C*, Altered but unsheared country rock separating quartz strands, Rainbow mine; *D*, Banding in vein due to shearing, Sixteen to One mine — 34

12. *A*, Quartz stringers in footwall of vein, Eldorado mine; *B*, "Crossover" stringers between hanging wall and footwall, Kate Hardy mine; *C*, Vein with sharp footwall due to faulting and frozen hanging wall with quartz stringers, Plumbago mine — 34

13. *A*, Quartz stringers in altered slate close to main vein, Rainbow mine; *B*, Lenses of quartz in crushed slate along wall of Clinton vein, Rainbow mine; *C*, Lenses of quartz and carbonate in crushed slate, Kate Hardy mine; *D*, Quartz stringers cutting Cape Horn slate, Brush Creek mine — 34

14. *A*, Breccia containing quartz fragments, cemented by fine-grained quartz, along hanging wall of vein, Eldorado mine; *B*, "Headcheese" breccia, Eldorado mine — 34

Page

PLATE 15. A, Vein with well-marked "walls," Plumbago mine; B. Walls of vein due to postmineral movement, Plumbago mine_____ 34

16. A, Footwall of vein marked by curving fault, Plumbago mine; B, Fault plane passing from footwall to hanging wall of vein, Sixteen to One mine_____ 34

17. A, Quartz displaced by minor fault, Eldorado mine; B, Wedge of quartz bounded by faults, Oriental mine____ 34

18. A, Specimen of ribbon quartz, German Bar mine; B, Ribbon quartz with bands parallel to walls, Sixteen to One mine; C, Ribbon quartz cut by transverse sheeting, Kate Hardy mine_____ 35

19. A, Junction of Oriental and Alta veins, Oriental mine; B, Vein with lower strand showing crinkly banding and strand above containing wall-rock inclusions, Eldorado mine_____ 42

20. A, Development of chlorite and albite in saussuritized gabbro; B, Grain of chromite in serpentine veined by antigorite; C, Vein quartz showing inclusions associated with septa of country rock; D, Vein quartz with abundant inclusions_____ 42

21. A, Arsenopyrite veined by quartz and crystalline against quartz, with gold replacing arsenopyrite; B, Arsenopyrite broken and veined by quartz, with gold for the most part replacing quartz_____ 42

22. A, Pyrite veined by quartz; B, Pyrite veined by quartz which shows clear border; C, Quartz grain showing crystal faces surrounded by quartz with allotriomorphic texture; D, Quartz crystal with core containing abundant vacuoles, surrounded by growth of relatively clear quartz; E, Same as D, with crossed nicols_____ 42

23. A, B, C, Vein quartz showing rosette of outward-pointing quartz crystals; D, Quartz crystals normal to face of arsenopyrite crystal_____ 42

24. A, Quartz crystals elongate normal to wall of vein; B, Quartz veinlet with cross-fiber structure; C, Same as B, with crossed nicols; D, Vein quartz showing strain twinning and microbrecciation; E, Quartz crystal showing vacuoles in zones parallel to prism with grain boundary due to later fracture; F, Same as E, with crossed nicols_ 42

25. A, Cloudy quartz cleared in zone of microbrecciation; B, Same as A, with crossed nicols; C, Change in grain close to wall of vein, probably due to later fracturing; D, Same as C, with crossed nicols_____ 42

26. A, Albite, dark from abundance of vacuole inclusions, intergrown with quartz; B, Same as A, with crossed nicols; C, Detail of A, showing contact of quartz and feldspar; D, Albite with zoning due to vacuoles_____ 42

27. A, Intergrown quartz and albite showing vacuole inclusions; B, Vacuoles with gas bubbles in quartz; C, Inclusions in quartz—clusters of minute inclusions and negative crystals with bubbles; D, Quartz crystal showing zonal structure due to arrangement of vacuoles_____ 42

28. A, Complex zoning in quartz due to vacuole inclusions; B, Same as A, with crossed nicols; C, Quartz grain showing crystal boundaries and core containing abundant inclusions; D, Quartz crystals elongate normal to wall of vein, crossed by lines of vacuoles; E, Same as D, with crossed nicols_____ 42

29. A, Lines of vacuoles crossing quartz grains; B, Linearly arranged vacuoles; C, Relation of vacuoles to strain twinning; D, Same as C, with crossed nicols_____ 42

30. A, Lines of vacuoles radiating from sulphide; B, Granite altered to fine-grained albite; C, Tubes in quartz radiating out from carbonate; D, Minute tubes in quartz_____ 43

31. A, Quartz crystal showing twinning in inner and outer zones; B, C, Vein of albite crossing older albite crystal___ 58

32. A, Zonal distribution of sericite in quartz crystal; B, Detail of A; C, Original albite of granite veined by quartz and later albite; D, Spherule of chalcedony in quartz_____ 58

33. A, B, Zoned carbonate replacing albitized granite; C, Quartz replaced by chlorite and two generations of carbonate; D, Early arsenopyrite veined by quartz and fringed by crystals of later arsenopyrite; E, Same as D, with crossed nicols_____ 58

34. A, Breccia of quartz fragments cemented by carbonate; B, Fault breccia containing quartz fragments with matrix of mariposite and ankerite; C, Microbrecciated quartz replaced by carbonate; D, Saussuritized gabbro replaced by chlorite and albite, with later development of carbonate and chlorite_____ 58

35. A, Mariposite in parallel arrangement in quartz; B, Rutile in oligoclase; C, Botryoidal talc replacing quartz; D, Detail of C, showing contact of talc and quartz_____ 58

36. A, Veinlet of later quartz with opal and chalcedony; B, Same as A, with crossed nicols; C, "Headcheese" breccia__ 58

37. A, Veinlet of clear quartz cutting strain-twinned crystal of older quartz; B, Same as A, with crossed nicols; C, Vug filled with opal and chalcedony; D, Veinlet containing quartz, chalcedony, and opal cutting opalized granite_____ 58

38. A, Veinlet of later quartz crossing area of microbrecciation; B, Same as A, with crossed nicols; C, Veinlet of later quartz with carbonate; D, Opal containing metallic mineral_____ 58

39. A, Later arsenopyrite fringing earlier; B, Gold in breccia with opal cement; C, Opal replacing albitized granite____ 58

40. A, Pyrite replaced by tetrahedrite; B, Jamesonite crystals in vug; C, Needles of jamesonite in quartz; D, Arsenopyrite along crenulations in quartz_____ 58

41. A, Gold and galena replacing quartz in contact with coarse arsenopyrite; B, Crushed arsenopyrite fringed by gold_____ 58

42. A, Gold replacing fractured quartz; B, Gold in quartz and carbonate; C, Specimen showing crinkly banding____ 59

43. Map of a portion of the Sierra Nevada showing major features of Tertiary drainage_____ 74

44. A, B, Crinkly banding in quartz with later microbrecciation; C, D, Crinkly banding with texture suggesting recrystallization_____ 74

45. A, Sericite band crossing single quartz grain; B, C, Folded septum of slate between quartz strands_____ 74

46. A, B, Crinkly banding crossing strain-twinned quartz; C, Zoned lamellar quartz crystals; D, Arsenopyrite veined by quartz with growth of quartz crystals outward from face; E, Fragments of arsenopyrite in quartz; F, Clustered arsenopyrite crystals veined by quartz_____ 74

47. Geologic map of Kate Hardy mine_____ 90

Page

PLATE 48. Sketch map of South Fork, Mugwump, and neighboring mines_____ 90
49. Geologic map of vicinity of Oriental shaft_____ 98
50. Geologic map of Oriental mine_____ 98
51. Geologic sections, Oriental mine_____ 98
52. Map of workings of Rainbow mine, Rainbow Extension mine, and Spiritualist tunnel_____ In pocket.
53. Map of Sixteen to One mine_____ 106
54. Map of workings of Morning Glory, Extension of Minnie D, Minnie D, Lucky Larry, and Eclipse mines_____ 106
55. Map of Yellow Jacket, Eldorado, and neighboring mines_____ 122
56. Map of workings on Eldorado vein, Eldorado mine_____ 122
57. Map of Irelan mine_____ 122
58. Map showing principal workings of Plumbago mine_____ 122

FIGURE   1. Map showing location of the Alleghany district, California_____ 4
2. Sketch map of outcrops along the Middle Fork of the Yuba River near Jackass Ravine_____ 8
3. Section along prospect tunnel south of Irelan mine_____ 14
4. Granite dike cutting schistose gabbro, French Ravine_____ 16
5. Gold produced from lode and placer mines of the Alleghany district, 1891–1927, and ore mined, 1903–1927____ 26
6. Details of junction of main vein and fault veins, Sixteen to One mine_____ 31
7. Map of western part of adit level, Oriental mine_____ 32
8. Section showing detail of junction of Sixteen to One and Ophir veins, Sixteen to One mine_____ 33
9. Section along vein in Eldorado mine_____ 34
10. Sketches showing thrusting in veins_____ 35
11. Sections across vein, Plumbago mine_____ 36
12. Plan and sections of a part of the Rainbow mine, showing relation of Rainbow and Clinton veins_____ 37
13. Section of Sixteen to One vein below drain tunnel_____ 38
14. Arsenopyrite veined by quartz_____ 40
15. Radiate needles of galena, probably replacing jamesonite_____ 50
16. Gold and galena in arsenopyrite_____ 50
17. Gold replacing quartz and arsenopyrite_____ 50
18. Sketch of Gold Canyon vein, showing occurrence of arsenopyrite and gold_____ 59
19. Sketches showing relation of crinkly banding to ribbon quartz_____ 78
20. Sketches showing possible folding of vein quartz_____ 79
21. American tunnel, Brush Creek mine_____ 88
22. Drifts from winze, Brush Creek mine_____ 88
23. Sketch map of Finan prospect_____ 89
24. Surface geology, Kate Hardy mine_____ 90
25. Map of lower tunnel, Tomboy prospect_____ 92
26. Sketch map of workings, Oak prospect, Bob Evans claim_____ 93
27. Plan of workings of North Fork mine and section along line A–B_____ 96
28. Sketch map of accessible workings of Kenton mine_____ 98
29. Map of General Sherman tunnel_____ 101
30. Plan and section of parts of the 900-foot, intermediate, and 1,000-foot levels, Sixteen to One mine_____ 108
31. Map of Ophir mine and adjoining workings of Sixteen to One mine_____ 110
32. Map of drain level, Ophir mine_____ 111
33. Map of upper tunnel, Extension of Minnie D mine_____ 113
34. Map of workings, Osceola mine_____ 114
35. Map of workings, Red Star mine_____ 116
36. Sketch map of Mayflower prospect_____ 117
37. Plan of Dreadnaught tunnel_____ 122
38. Map of workings, Docile mine_____ 124
39. Sketch map of lower tunnel, Mammoth Springs mine_____ 125
40. Sketch map of Mountain View tunnel_____ 126
41. Sketch map of Oriflamme tunnel_____ 127
42. Sketch map of upper tunnel, Arcade mine_____ 128
43. Sketch map of lower tunnel, Arcade mine_____ 129
44. Map of Plumbago mine_____ 130
45. Map of German Bar mine_____ 133
46. Map of Gold Canyon mine_____ 134

GOLD QUARTZ VEINS OF THE ALLEGHANY DISTRICT, CALIFORNIA

By Henry G. Ferguson and Roger W. Gannett

OUTLINE OF THE REPORT

The Alleghany district, in the southern part of Sierra County, Calif., has long been famous for the high-grade gold ore of its quartz veins. The oldest rocks of the district (pp. 6–17) are of sedimentary and volcanic origin and correspond to part of the Calaveras formation as mapped in the Colfax and Downieville folios of the Geologic Atlas of the United States. These rocks are divided into five formations, of which three—the Blue Canyon, Relief, and Cape Horn formations—follow the definitions laid down by Lindgren in the Colfax folio, and two—the Tightner and Kanaka formations—are new units required by the more detailed nature of the present study. It is thought possible that a conglomerate which forms the basal part of the Kanaka formation is of glacial origin. Intrusions of gabbro and more basic rocks, the latter now completely serpentinized, crop out over nearly half the area in which pre-Tertiary rocks are exposed. Small granitic dikes of later age than the basic intrusives are found in the western part of the district. Overlying and largely concealing the older rocks are auriferous gravel of Eocene and Miocene age, andesitic breccia of probable Miocen age, basalt flows of probable Pleistocene age, and minor amounts of Pleistocene and Recent gravel.

The metamorphism and deformation of the older rocks (pp. 19–22) is the result of two major epochs of diastrophism. The first of these occurred prior to the deposition of the Jurassic sediments that crop out east and west of Alleghany; the second, which involved rocks of Upper Jurassic age, was accompanied and followed by great intrusions, first of basic rocks and then, after an interval, of granitic rocks. A large fault with steep easterly dip borders the district on the east. The alinement of this fault with the similar fault bordering the Mother Lode region suggests that the two may be segments of a single major fault. The later history of the district involves successive wearing down and reelevation of the region, and, during Tertiary and early Pleistocene time, volcanic eruptions of considerable magnitude.

Lode mining in the Alleghany district (see pp. 25–28) began as early as 1853, and it is estimated that the total lode production from the veins of the district has been in excess of $20,000,000.

The principal veins (pp. 29–38) have northerly strikes with gentle easterly dips and follow minor reverse faults. Other veins strike in the same general direction but dip steeply west. These follow reverse faults that displace the eastward-dipping fissures. The age of the quartz filling is, however, the same in both series. During and after the filling of the veins there were continued movements along the walls.

The paragenesis of the vein minerals (pp. 38–51) indicates three stages of mineralization. The first caused the development of new minerals along the walls of the fissures and at least initiated serpentinization. The minerals formed in this stage do not appear to have required the introduction of notable amounts of new material but resulted from the recrystallization of the minerals of the wall rocks. The major feature of the second stage of mineralization was the deposition of the quartz. Accompanying the quartz were minor amounts of other minerals, chiefly arsenopyrite, pyrite, albite, oligoclase, and barite. The sulphides belonging to this stage crystallized prior to the quartz that surrounds them. They have sharp crystalline boundaries against the quartz and are veined by the quartz. Wall-rock alteration consisted largely in the recrystallization of the sodic feldspars and the replacement of the more calcic by albite. The third stage of mineralization was economically the most important, for the gold was introduced at this stage. Quantitatively its major feature was the replacement of the wall rock by carbonate, chiefly ankerite, and the micaceous minerals mariposite and sericite. At the same time the older quartz was fissured, and carbonate and mica, together with various sulphides, graphite, and gold, were deposited within the veins, chiefly by replacement of the quartz. To this stage also belong veinlets of quartz, chalcedony, and opal that cut the older quartz.

From a review of the geologic features that appear to control the deposition of the shoots of high-grade ore (pp. 52–60) it is considered that, inasmuch as the gold is later than the vein quartz, areas in which the veins are most fractured will on the whole be most favorable. Fracturing of a vein is largely dependent upon changes in strike and dip of the vein, and such changes are commonly found where the vein is close to serpentine masses, at junctions of veins, and where the vein has been faulted prior to the introduction of gold. Another influential factor in determining the position of the high-grade shoots is the presence of the early arsenopyrite, which has acted as a precipitant for the gold. As the arsenopyrite is found most abundantly in veins near the sepentine another reason is adduced for the observed close relations of high-grade ore and serpentine. Conditions are thought to be favorable for reasonable persistence of the high-grade ore in depth.

It is believed (see pp. 61–86) that at least 10,000 feet of material has been eroded from above the present outcrops of the veins, whose age of formation is placed at about the end of the Jurassic period. The eastward-dipping faults that are followed by the principal veins are thought to be auxiliary to the major fault along the eastern border of the district; those with steep westerly dip may owe their position to an elongate granite mass below the western part of the district. Early movement along the faults that carry the veins may have been due to the intrusions of granite. Pressure due to increase of volume during serpentinization caused irregularities in the original fissures and is competent to account for a part of the renewed movement during and after mineralization.

It is thought an unsafe generalization to assume a magmatic origin for all components of the vein minerals. Some of the elements composing several of these, including plagioclase, barite, ankerite, sericite, mariposite, and even gold, may have been derived from the wall rocks traversed by the ore-forming solutions. The present grain of the quartz is in part the result of recrystallization, and it is therefore doubtful how

1

far texture can be used as a criterion in considering methods of vein formation. None of the hypotheses of origin of quartz veins are found to be entirely free from objection. The minerals of the later stage were deposited chiefly as the result of replacement, both of the wall rock and of the quartz.

The detailed descriptions of the mines (pp. 87–135), though of interest chiefly to local operators, contain examples illustrating the generalizations of the main part of the report that will be of service to those who desire a thorough understanding of the geology of the ore deposits.

The report closes with a statement of opinion regarding the future of the district (pp. 135–136), in which a distinctly favorable view of the outlook for lode mining in the future is taken.

Part 1. GEOLOGY

INTRODUCTION

SCOPE OF THE REPORT

The Alleghany district (pl. 1) has been a producer of gold since the earliest days of California gold mining. Here, as in other California gold camps, placers yielded the bulk of the earlier production, but during the last 30 years lode deposits have been the source of nearly all the gold produced. The district differs from other California gold-quartz districts in that nearly all the production is obtained from small shoots of high-grade ore, which may yield several thousand dollars to the ton, and very little is obtained from the ordinary type of lower-grade ore. Because of this unusual concentration of the gold it has been the endeavor of the writer to describe in considerable detail all features of the geology that have any bearing on the localization of the high-grade shoots and to give his conclusions as to the determining causes.

As both field and office phases of an investigation of this kind are inevitably limited by lack of time, certain phases of the geology, in themselves of much interest, have perforce been neglected. The Tertiary rocks, being later than the veins, have received only the most scanty treatment. Also many problems of the bedrock geology, particularly those dealing with the metamorphism of the older sedimentary and volcanic rocks, have not received as full treatment as would have been desirable had unlimited time been available.

LOCATION AND TOPOGRAPHY

The mining camps of Alleghany and Forest are on the west flank of the Sierra Nevada (fig. 1) about 20 miles west of the crest. The Middle Fork of the Yuba River forms the southern boundary both of the district and of Sierra County. On the north the ridge north of Oregon Creek overlooks the deep valley of the North Fork of the Yuba River. The district is about 18 miles in a direct line from Nevada City, but owing to the deep valleys that cut the upland plateau the automobile road which connects the two places is 31 miles in length. The nearest railroad point is Nevada City, whence a narrow-gage road connects with the main line of the Southern Pacific at Colfax. The automobile route from San Francisco leaves the main highway at Auburn, about 50 miles by road from Alleghany. In winter freight by road between Alleghany and Nevada City costs about $20 a ton; in summer the rate is $15 a ton from Nevada City to Alleghany and $10 a ton in the reverse direction (1928).

The district has hot, rainless summers, though at this altitude (4,500 feet) the heat is far less intense than in the California Valley or in the towns of the lower foothills, such as Grass Valley. The winters are not cold—a temperature as low as 0° F. is almost unknown—but the snowfall is very heavy. It is not unusual for the roads to be blocked for short periods, and then the only egress from the district is by snowshoes. The mining companies, therefore, not later than September or October bring in sufficient supplies to last over the winter.

The topography is of the monotonous but impressive type that characterizes the west slope of the Sierra Nevada in this latitude. (See pl. 2, D.) The gently sloping lava-covered plateau that controls the upland topography has been trenched by deep canyons of the westward-flowing streams. Consequently the region is a succession of flat-topped ridges that have a very gentle westward slope, separated from one another by steep-walled canyons. The canyons of the major streams are cut to a depth of as much as 2,000 feet below the crests of the ridges. Such topography makes communication difficult in other directions than along the ridges. Roads that cross the grain of the country must zigzag wearily up and down the canyon walls, and as such road construction is expensive the grades, especially those of the minor roads, tend to be the maximum practicable for automobile traffic. Along the ridges the grades are easier, but during the rainless summers the roads in the soft andesite breccia, unless kept constantly in repair, become deeply rutted and thickly covered with dust, which the first rain in the fall turns to a slippery mud. Roads cut in the more resistant bedrock formations on the flanks of the ridges below the lava, such as the Foote road for some miles west of Alleghany, though much more expensive to construct, are free from this objectionable feature.

FIELD WORK AND ACKNOWLEDGMENTS

The senior author first visited the district in 1913 and as a result of two weeks spent in visits to the mines then in operation wrote a short description of the lode deposits.[1] At that time no attempt was made to study the general geology, but as a result of field and office study of the veins and ores certain theories were suggested to account for the formation of the

---

[1] Ferguson, H. G., Lode deposits of the Alleghany district, California: U. S. Geol. Survey Bull. 580, pp. 153–181, 1914.

veins and of the high-grade shoots. In the present report, as a result of much more detailed work, these theories are modified and elaborated.

The increasing productiveness of the district made it desirable that a more detailed study should be undertaken, and an accurate topographic map was made by W. B. Upton, jr., of the United States Geological

the field from July to October and Mr. Ferguson during parts of July and August. In September, 1928, after the completion of the first draft of the manuscript, Mr. Ferguson made another short visit to the district.

It had been the understanding that in the preparation of the report Mr. Gannett should have charge of

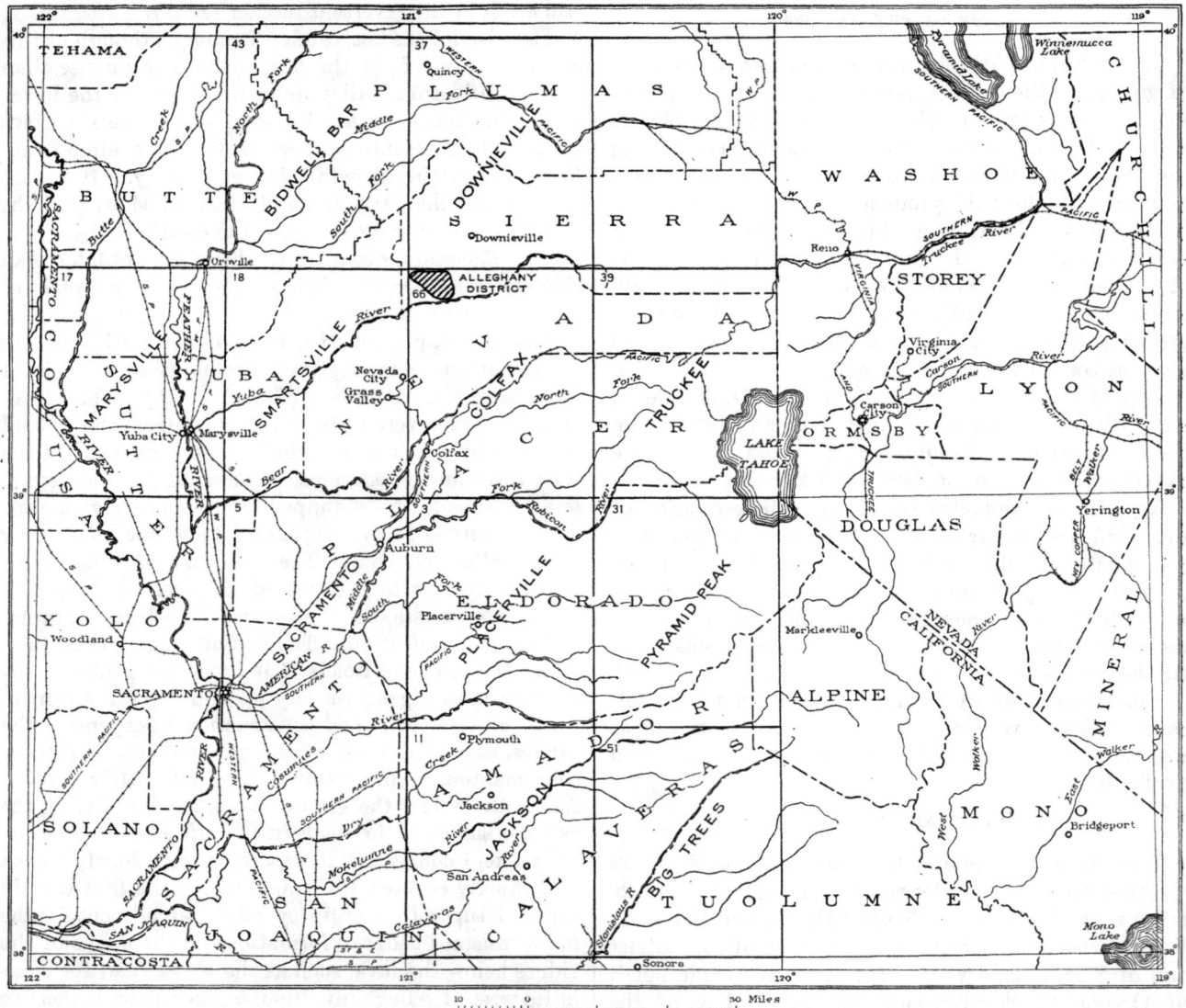

FIGURE 1.—Map showing location of the Alleghany district, California. Quadrangles described in geologic folios are indicated by numbers in upper left corner and names running diagonally

Survey, in the spring of 1924, and the senior author spent August and September of that year in the field. The complexity of the problem, as shown by the first season's field work, made it necessary to devote a second season to the investigation, and in this the Geological Survey was fortunate in procuring the services of Mr. Roger W. Gannett, who had already done detailed geologic work in the district for several of the mining companies. During 1925 Mr. Gannett was in

the section dealing with petrology and the detailed descriptions of the mines, and Mr. Ferguson should describe the general features of the geology and ore deposits and make the detailed mineralogic studies. Owing to the death of Mr. Gannett in October, 1925, just before his field work was to end, the work of preparation of the entire report fell to Mr. Ferguson. He takes this opportunity of acknowledging how deeply he is indebted for Mr. Gannett's excellent work in the field and his realization of how greatly this report suffers from lack of Mr. Gannett's assistance in the office. Specifically Mr. Gannett's contributions include a part of the geologic field mapping, including most of the detailed work in the vicinity of Alleghany; most of the detailed mine geology, including the essential portions of the descriptions of many of the mines and collection of data on the position of the high-grade shoots; the study of faulting along and in the veins, a subject on which he expected to prepare a special paper; and much of the general petrology, as a considerable amount of microscopic work was done in the field. The sketches showing details of the veins have been derived largely from his field notebooks, and most of the geologic mine maps are also his work. The text bearing on these subjects has been in part written on the basis of Mr. Gannett's notes, and the writer can not but feel that it is far less adequately presented than it would have been had Mr. Gannett lived to complete his section of the work.

During the authors' stay in Alleghany they received most cordial assistance from all those with whom they came into contact. Especial thanks are expressed to Mr. C. A. Bennett, Mr. T. J. Bradbury, Mr. A. F. Duggleby, Mr. W. H. J. Goldsworthy, and Mr. W. V. Wilson for the many courtesies they extended.

During the preparation of the report the writer has realized more clearly than ever before how largely a report, particularly one of this size, is the product of the organization rather than exclusively of the individuals whose names appear on the title-page. The publication could not have reached its present form without the hearty cooperation of his fellow workers on the Geological Survey, of both the scientific and the clerical staffs.

Thanks are due to all members of the section of metalliferous deposits for much friendly and searching criticism, the more valuable because of its informality, and in particular to T. B. Nolan, whose familiarity with the mines and geology of the Mother Lode made his comments of especial value; to C. S. Ross for assistance in petrologic work; to M. N. Short and W. T. Schaller for mineral determinations; to W. D. Johnston, jr., for several photomicrographs; and to G. F. Loughlin, chief of the section, who has read and criticized the manuscript in detail and has not only given suggestions of much value, which have been incorporated in the report, but has saved the writer from many faults of arrangement and expression.

PREVIOUS WORK

Very little has been written on the geology of the Alleghany district. Short notes have from time to time appeared in the Mining and Scientific Press and the Engineering and Mining Journal, and notes on the early history of the district are to be found in some of the early publications on California, particularly those of the State mineralogist. The following are the only publications known to the writer which deal to any extent with the bedrock geology and lode deposits of the region:

Turner, H. W., Further contributions to the geology of the Sierra Nevada: U. S. Geol. Survey Seventeenth Ann. Rept., pt. 1, pp. 521–762, 1896. (Downieville area, pp. 591–625.)

Turner, H. W., U. S. Geol. Survey Geol. Atlas, Downieville folio (No. 37), 1897.

Lindgren, Waldemar, U. S. Geol. Survey, Geol. Atlas, Colfax folio (No. 66) 1900.

Ferguson, H. G., Lode deposits of the Alleghany district, California: U. S. Geol. Survey Bull. 580, pp. 153–182, 1914.

MacBoyle, Errol, Mines and mineral resources of Sierra County, California State Min. Bur., 1920.

Simkins, W. A., The Alleghany district of California: Pacific Mining News of Eng. and Min. Jour., vol. 2, pp. 288–291, 1923.

Ferguson, H. G., and Gannett, R. W., Gold-quartz veins of the Alleghany district, California: Am. Inst. Min. and Met. Eng. Tech. Pub. 211, 1929.

The following publications are devoted chiefly to the study of the Tertiary gravel of the region and contain only incidental reference to the geology of the bedrock formations and lode deposits.

Lindgren, Waldemar, The Tertiary gravels of the Sierra Nevada of California: U. S. Geol. Survey Prof. Paper 73, 1911.

Alling, M. N., Ancient auriferous gravel channels of Sierra County, Calif.: Am. Inst. Min. Eng. Trans., vol. 49, pp. 238–257, 1915.

ROCK FORMATIONS

The older rocks of the Alleghany district (see pl. 1) include metamorphosed sediments and volcanic rocks which are cut by later basic and acidic intrusive rocks. These are exposed principally in the stream valleys. Along the interstream ridges they are capped by Tertiary gravel and andesitic breccia, which at one locality are overlain by basalt of probable Pleistocene age. The convenient nomenclature used in the gold-belt folios has been followed in this report, the former group being called the "Bedrock series" and the latter the "Superjacent series."

As the "Superjacent series" is later than the veins, whose study was the primary purpose of this investigation, the formations belonging to this series have been given only the most cursory treatment.

The rocks are, on the whole, not well exposed, and in many places the thick cover of soil and float and areas of almost impenetrable brush make accurate detailed mapping impossible, at least without spending far more time than would be justified. In a few places where outcrops are especially good (fig. 2) or in areas of especial economic importance (fig. 24 and pl. 49) the surface was mapped in considerable detail, and the complexity is seen to be much greater than is represented on the geologic map as a whole.

In the study of a geologic map of the scale of Plate 1 information as to the degree of accuracy may be important. For the most part the following conventions have been used in representing the geologic boundaries: Where outcrops furnished good control, though not necessarily actual contacts, and the limit of probable error was thought to be not much in excess of 50 feet (0.05 inch on the map) the geologic boundary is shown by a full line. Boundaries which could be followed out with fair assurance but along which paucity of outcrops rendered the exact position uncertain, perhaps within a zone 200 feet in width, are shown by a dashed line. Dotted lines represent geologic boundaries in regions where the outcrops were inadequate to give close control or boundaries that were not traced in detail owing to lack of economic importance, even though a more accurate control might have been obtained had more time been spent.

"BEDROCK SERIES"

Two classes of rocks are included in the "Bedrock series"—(a) highly tilted formations, consisting of metamorphosed sedimentary and volcanic rocks, and (b) deep-seated igneous rocks of various kinds which are intrusive into these.

The sedimentary and volcanic rocks consist of five formations—from east to west the Blue Canyon formation, the Tightner formation, the Kanaka formation, the Relief quartzite, and the Cape Horn slate. The Blue Canyon, Relief, and Cape Horn formations are mapped and described in the Colfax folio; the Tightner and Kanaka formations are new units required by the more detailed nature of the present study. All lie between the base and top of the Calaveras formation as mapped in the Colfax and Downieville folios (pl. 3) and for this reason are regarded as of Carboniferous age. They crop out as belts striking in a general north-northwesterly direction and stand vertically or dip at high angles, generally to the east.

The rocks intrusive into the older formations are mapped in three units: (a) Serpentines, (b) basic intrusive rocks consisting chiefly of gabbro but including also a little diorite and quartz diorite, and (c) granitic rocks.

BLUE CANYON FORMATION

In the Alleghany district the Blue Canyon formation[2] occupies only a narrow strip along the eastern margin, northward from the Middle Fork of the Yuba River at the mouth of Wolf Creek to the ridge north of Kanaka Creek. It is separated by a fault from the formations on the west. As it contains no ore deposits of importance it was not studied in detail.

The greater part of the formation within the area covered by Plate 1 consists of dark fine-grained fissile slate (pl. 2, B), which becomes nearly white on weathering; in places it grades into a fine-grained shaly micaceous schist. Here and there are thin beds of very fine-grained gray quartzite. Near the Mammoth Springs mine there is a small amount of rather fine-grained conglomerate, not found in outcrop but present in the talus fragments. This rock contains unsized and stretched pebbles, an inch or less in length, of slate and a much altered volcanic rock, probably andesite, in a matrix of the fine-grained slate that makes up the bulk of the formation. No limestone beds such as have been described by Lindgren[3] were found in the Blue Canyon formation in the Alleghany area, except for a single lens of impure limestone 5 feet in width and over 60 feet in length at the Independence mine.

The strike of the formation is northerly, at a slight angle to the fault contact that bounds it on the west, and the dip is generally steep to the west, flattening toward the fault. Wherever the bedding could be distinctly seen it was found that on the whole the slaty cleavage closely conforms to the bedding planes in dip and strike, crossing it only on the local folds.

The great width of the area covered by the Blue Canyon formation in the Colfax quadrangle implies an enormous thickness, even if allowance is made for possible duplication by folding. Only a very small part is included in the Alleghany district, but the thickness of the rocks belonging to the Blue Canyon formation within the limits of the area mapped on Plate 1 is well over 1,000 feet.

TIGHTNER FORMATION

The Tightner formation, named from the Tightner mine (now a part of the Sixteen to One mine), where it forms the principal wall rock, adjoins the Blue Canyon formation on the west, although except in the valley of Kanaka Creek basic intrusive rocks separate the two. As the Tightner is here in fault contact with the Blue Canyon formation, the age relations of the two are not absolutely determinable, but as nothing was found to invalidate Lindgren's hypothesis[4] that the formations mapped in the Colfax folio under the general heading "Calaveras formation" form a continuous series from east to west, it is assumed to be younger than the Blue Canyon formation.

The width of the belt occupied by the Tightner formation suggests a thickness of about 7,000 feet. The outcrops are generally poor, and it is quite possible that undetected close folding may exaggerate the apparent thickness. Where individual beds could

[2] For description of extent and lithology see Lindgren, Waldemar, U. S. Geol. Survey Geol. Atlas, Colfax folio (No. 66), p. 2, maps, 1900.

[3] Idem, p. 2, column 1.
[4] Idem, p. 1.

be followed along the strike, however, no evidence of pronounced folding was found, and the two small folds that were discovered do not increase the apparent thickness by more than a few hundred feet at most. Moreover, possibility of error in the opposite direction arises from the fact that the veins crossing the area occupy planes of reverse faulting, dipping northeast at angles of 30° to 60°. The total compression of the section from this cause can hardly exceed 600 or 800 feet.

The Tightner formation was also found to crop out in the valley of the South Fork of the Yuba River 5 miles south of Alleghany and in the valley of the North Fork of the Yuba River 6 miles north of Alleghany. Along the South Fork the distance across the strike from its fault contact with the Blue Canyon formation to the base of the Kanaka formation is about a mile, including the intervening serpentine dike, and along the State highway north of the North Fork the Tightner formation crops out for a distance of about 1½ miles westward from Downieville. It is possible that this formation may be found to be of even more widespread distribution in the California gold belt, for the rocks from the Mother Lode region described by Knopf [5] under the heading "Amphibolite schist associated with the Calaveras formation" seem to be similar to those of the Tightner formation.

The formation consists predominantly of fine-grained greenish schist with varying amounts of quartz. (See pls. 2, A, C; 4, A.) Where least metamorphosed this schist has the appearance of an altered lava or tuff. The bulk of the schist is too highly metamorphosed to permit definite conclusions as to the original rock. Rarely the outcrop shows a faint suggestion of an original pyroclastic structure, but, as a rule, rather massive beds of amphibolite schist prevail, and the difference in aspect of the beds is due to small differences in texture or to greater amounts of quartz or greater resistance to alteration in some parts than others. The chemical composition, so far as indicated by the mineral content, suggests that the rocks may have been originally andesitic. The quartz is in part of later introduction. The major introduction of quartz of the type shown in Plate 4, A, was contemporaneous with the formation of the quartz veins, but there may also have been an earlier introduction of quartz, possibly contemporaneous with the metamorphism that caused the recrystallization of the rock. The hornblende of the schist is in part altered to chlorite, particularly near the veins. Bands of hornblende schist that contain a larger proportion of quartz than the average cross the ridge between the forks of Kanaka Creek. These hold a fairly constant width across the whole region between the lava caps to the north and south, and the amount of quartz present seems about constant in different parts of the same

band. It is therefore thought probable that in these bands the quartz, though now recrystallized, was originally present in the rock and that these bands of more quartzose schist have resulted from the alteration of siliceous igneous rocks. Interbedded with the hornblende schist are small amounts of schist containing both andalusite and hornblende and rare glaucophane schist. There are also in a few places rocks of sedimentary origin, including here and there outcrops of quartzose or chloritic mica schist, more rarely beds of glistening black carbonaceous slate, and in the well-exposed rocks along the Middle Fork of the Yuba River a lens of conglomerate.

In places there are also lenses of coarsely crystalline white limestone. Most of these are known only in mine workings—one in the third level crosscut of the North Fork mine beneath the Tertiary gravel and lava (fig. 27), one in the eastern crosscut of the Yellow Jacket mine (pl. 55), one at the shaft station of the Eldorado mine (pl. 56), and one on the 2,400-foot level of the Sixteen to One mine. As the rock is soluble surface outcrops are scarce. It is reported that limestone was formerly exposed in the bed of the North Fork of Oregon Creek, near the town of Forest, but is now concealed by placer-mining débris. A recent cave at the portal of the main Eldorado tunnel has revealed the presence of limestone elsewhere concealed by soil and surface float. An outcrop less than 50 feet in length was found in the ravine between the Mariposa and Eldorado mines, south of Kanaka Creek. The best outcrops occur along the bank of the Middle Fork of the Yuba River east of the mouth of Jackass Ravine. Here several small lenses were found. (See fig. 2.) As the limestone does not ordinarily form outcrops, it is not certain how far these different occurrences may be continuous beneath the surface, but it is thought likely that they are a series of lenses. Possibly the occurrence in the North Fork crosscut is continuous with that near Forest, and that in the Yellow Jacket crosscut may be continuous with those of the Eldorado mine. Altogether the limestone forms only a very minute proportion of the total mass of the formation, though it is probable that there are undiscovered lenses, because the rock, being soluble, forms outcrops only where erosion has been most rapid. The occurrence in the North Fork mine (fig. 27), for instance, shows solution channels containing waterworn gravel at a depth of at least 200 feet beneath the base of the lava. The only thin section of the limestone examined, from the outcrop on the bank of the Middle Fork of the Yuba, showed a large amount of diopside.

KANAKA FORMATION

AREAL EXTENT

The Kanaka formation, so named from its outcrops in the valley of Kanaka Creek, adjoins the Tightner

---

[5] Knopf, Adolph, The Mother Lode system of California: U. S. Geol. Survey Prof. Paper 157, pp. 11–12, 1929.

formation on the west and is believed to overlie that formation unconformably, though clear exposures of the contact could not be found. The formation is

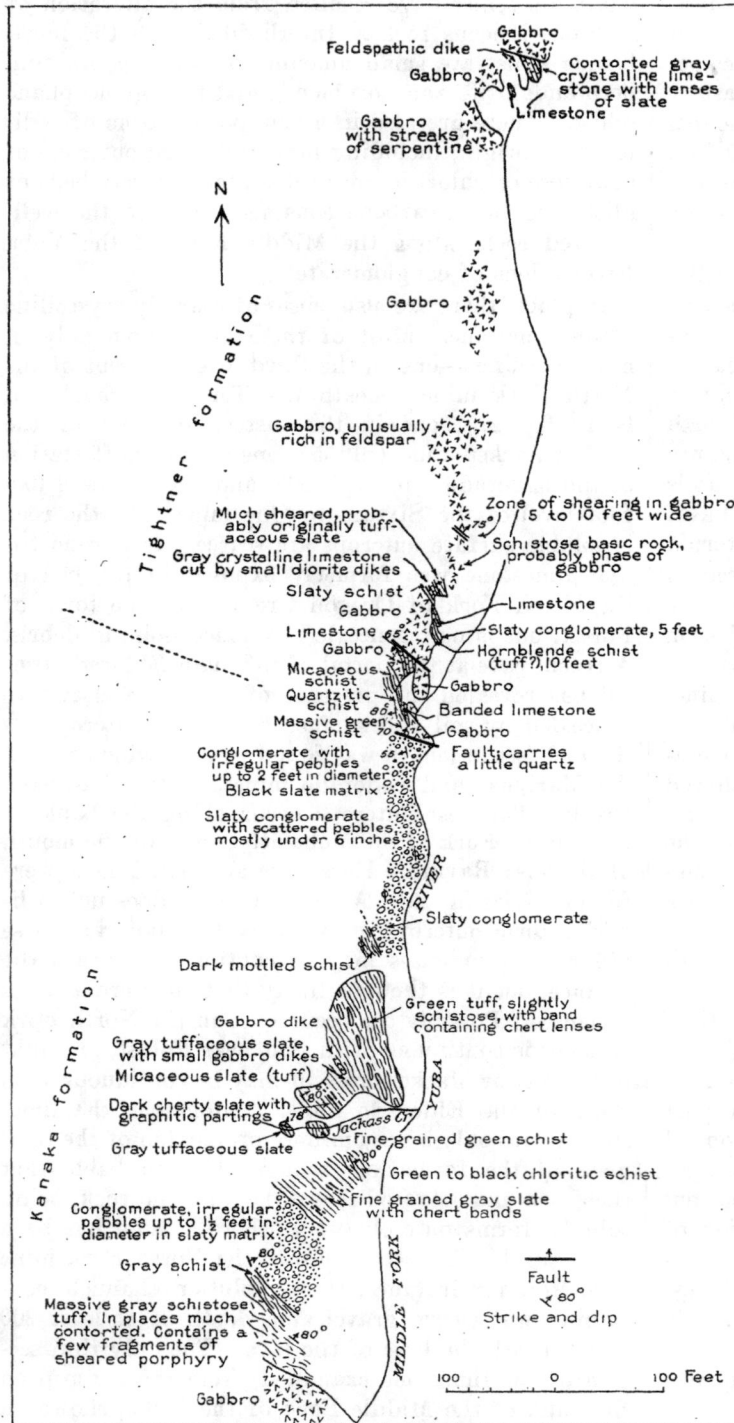

FIGURE 2.—Sketch map of outcrops along the Middle Fork of the Yuba River near Jackass Ravine

also present in the valley of the South Fork of the Yuba River west of the village of Washington and in the valley of the North Fork of the Yuba

along the State highway for a distance of about 2 miles east of Goodyears Bar. On the west the formation is overlain conformably by quartzitic schist of the Relief formation. It is made up of rocks of different lithologic character, including conglomerate, slate, greenstone, and chert. These have been mapped as separate units on Plate 1, but the boundaries between the different members are not everywhere easily distinguishable.

CONGLOMERATE MEMBER

The basal member of the formation in the Alleghany district is conglomeratic and ranges in texture from a black slate with sparsely scattered pebbles to a coarse irregular conglomerate in which abundant pebbles and boulders are surrounded by a matrix of black slate. (See pl. 4, B.) The distribution and thickness of this member vary greatly from place to place. In the valley of the Middle Fork of the Yuba River (fig. 2) a band 100 feet thick is in fault contact with the Tightner formation, and there is a second band of conglomerate 60 feet thick 200 feet south of the first. In the old placer workings on Minnesota Ridge the outcrops are poor and the relations are confused by gabbro dikes, but in places siliceous slate with few pebbles adjoins the Tightner formation and coarse conglomerate crops out a few feet away. The thickness of the conglomerate member here may be as much as 200 feet. On the north side of the lava ridge at Chips Flat the conglomerate is coarse and contains abundant large boulders, and the thickness is probably about 200 feet. In the valley of Kanaka Creek the thickness is greater, but black slate containing scattered pebbles rather than a definite conglomerate forms the greater part of this member. In the old placer workings west of Alleghany there is much conglomerate and relatively little conglomerate slate, but the coarsest conglomerate, which is well exposed in trenches near the transformer house (see pl. 4, B), is not at the base but close to the top of the member. The thrust fault followed by the Sixteen to One vein cuts off a part of the conglomerate and conglomeratic slate. The conglomerate member is well exposed in the valley of Oregon Creek, but the total thickness is not over 200 feet, and the coarse conglomerate is confined to a band about 50 feet in width near the base. Similar conglomerate was found outside the area mapped in detail; in the valley of the South Fork of the Yuba River it occurs next to outcrops of the

Tightner formation, and in the valley of the North Fork of the Yuba it not only adjoins the Tightner formation but occurs in the upper part of the Kanaka formation half a mile east of Goodyear Creek.

No clearly defined contacts of the conglomerate with the underlying Tightner formation were found. In many places, as in the valleys of Oregon and Kanaka Creeks, dikes of gabbro and serpentine separate the two formations. Elsewhere, as in the North Fork crosscut (fig. 27), in the Spiritualist tunnel (pl. 52), and on the bank of the Middle Fork of the Yuba River (fig. 2), the two formations are in contact, but there is evidence of movement along the plane of contact. At one point on Minnesota Ridge a pebbly slate seems to rest directly on the underlying schist, but the exposures are not sufficiently good to preclude the possibility of faulting. It is not considered likely, however, that a fault of any great magnitude separates the two formations, for the relatively thin conglomerate member is in contact with the Tightner formation from the South Fork of the Yuba River to the North Fork, a distance of 14 miles, and the contact makes a sinuous line, whereas the trace of the known fault bounding the Blue Canyon formation on the west is fairly straight. Additional evidence against major faulting is seen in the presence locally in the Tightner formation of small beds of black slate similar to that occurring more abundantly within the Kanaka formation and of a single small lens of conglomerate (fig. 2) with slaty matrix and small pebbles of an altered igneous rock, probably originally of an andesitic nature.

The pebbles of the conglomerate are of all sizes; most of them are 1 or 2 inches in major diameter, but boulders measuring 2 and 3 feet are common, and in the tunnel of the Mugwump mine Mr. Gannett noted one of 14 feet and one of 9 feet. The pebbles of softer rock, such as andesite, are greatly stretched and elongated parallel to the schistosity, but those of harder rocks, such as quartzite, have preserved their original shapes. The pebbles vary greatly in composition, but the most numerous are recognizable under the microscope as originally pyroxene andesite and possibly in part dacite, now much altered, with abundant development of secondary chlorite and hornblende. There are also numerous pebbles of a somewhat altered quartz diorite containing quartz, oligoclase, and partly chloritized biotite. Pebbles of arkose containing grains of quartz and oligoclase are distinguishable from the quartz diorite only by examination of thin sections under the microscope. Quartzite pebbles are less abundant than those of igneous rocks but are more easily recognizable. They are not stretched, but some have been broken and veined and rarely faulted slightly. A thin section of a quartzite pebble showed in plain polarized light distinct outlines of the individual rounded quartz grains,

but with crossed nicols it is seen that there has been complete recrystallization; both the clastic grains and the matrix are changed to a mosaic of minute interlocking quartz grains. Well-rounded pebbles are lacking, and subangular forms predominate. The average size of the quartzite pebbles is less than that of the andesite pebbles, though boulders nearly a foot in diameter were found. In composition they range from a rather pure fine-grained quartzite very similar to the quartzite beds found in the Blue Canyon formation to a rather coarse-grained rock containing, besides quartz, a small amount of oligoclase, thus grading in composition toward the arkose pebbles. In all the pebbles of feldspathic rocks within the conglomerate the feldspars are fresh and unaltered, whereas in the much younger intrusive gabbro and diorite masses identifiable remnants of the original feldspar are rarely seen.

In the bed of Kanaka Creek opposite the portal of the Spiritualist tunnel there is an outcrop of a bed of arkose about 20 feet thick within the conglomerate. This arkose is traceable up the north bank for about 100 feet but was not found south of the creek. In composition it differs from the arkose pebbles in the conglomerate in containing, besides oligoclase, a considerable amount of orthoclase.

No pebbles of the hornblende schist, which forms by far the larger part of the adjoining and presumably underlying Tightner formation, were found within this conglomerate, but a comparison of the rocks of the least metamorphosed parts of the Tightner formation with the most highly altered andesitic pebbles of the conglomerate reveals a similarity, the schist retaining a faint suggestion of its original igneous texture and the altered and squeezed pebbles showing abundant secondary hornblende and chlorite. It is therefore thought likely that the andesitic pebbles of the conglomerate may have been originally derived from the Tightner formation and owe their comparative freedom from metamorphism to being embedded in a relatively plastic and impervious medium. Although none of the specimens from the Tightner formation examined under the microscope give positive confirmation, the quartz diorite and arkose pebbles may also be derived from the Tightner formation and represent unmetamorphosed equivalents of an original quartz diorite intrusion and the arkose resulting from its erosion, now metamorphosed in the Tightner formation to quartzose phases of the hornblende schist; or these pebbles and the arkose may be derived from deep-seated intrusions long antedating the Sierra batholith. The fine-grained quartzite might well have come from the Blue Canyon formation.

It is evident that the conglomerate, together with the conglomeratic slate described above, is not the result of ordinary marine or fluviatile deposition.

Four possible modes of origin are suggested; it may be a fanglomerate, a volcanic breccia similar in origin to the Tertiary breccia that caps the ridges, a tillite, or the result of a mud flow.

Origin as a fanglomerate or fossil talus slope seems to be excluded by the fact that the pebbles rarely touch each other and are separated by varying amounts of black slate. In places the volume of the pebbles exceeds that of the matrix; elsewhere the pebbles are sparsely and irregularly scattered through the black slate. Moreover, a large proportion of the pebbles must have been derived from a distant source.

The abundance of volcanic pebbles suggests that the rock may be of pyroclastic origin. It has some resemblance to the Tertiary andesitic breccia that caps the interstream areas, and such a breccia, an original volcanic mud flow of mainly andesitic pebbles, if metamorphosed to the same degree would certainly resemble the conglomerate of the Kanaka formation. A count of pebbles in one of the outcrops of the conglomerate near the Alleghany transformer house showed 110 of volcanic origin (chiefly andesitic) against 56 of quartzite, slate, and granite, or 66 per cent of volcanic origin. In similar counts of the pebbles in the Tertiary volcanic breccia the andesite pebbles range from 71 to 89 per cent of the whole, the remainder consisting of pebbles of rocks of the " bedrock series " and rhyolite. Therefore the presence of abundant pebbles of nonvolcanic origin can not be regarded as evidence against the origin of the conglomerate as a volcanic breccia, particularly because many of the andesite pebbles included in the count of the Tertiary breccia were doubtless derived, like the rhyolite, from lavas erupted prior to the beginning of the breccia flows. The conglomerate of the Kanaka formation and the Tertiary andesite breccia are also alike in the lack of sizing of pebbles and boulders and the presence of angular boulders, though the pebbles and boulders within the Tertiary formation are, on the whole, better rounded. Also the succeeding members of the Kanaka formation contain rocks of undoubted pyroclastic origin. There is one important difference, however. The matrix of the Tertiary breccia consists of comminuted andesitic material; that of the conglomerate in the Kanaka formation, though containing small fragments of probably detrital quartz, as well as secondary quartz, chlorite, and carbonate, consists largely of very fine-grained opaque black material, such as might have been derived from finely powdered slate either of the Tightner or of the Blue Canyon formation or from mud containing organic matter.

In favor of the possibility that the conglomerate is of glacial origin are the heterogeneity in size and composition of the pebbles and boulders, their angularity, and the fact that the only individuals which could not possibly have been derived from the Tightner formation are the resistant quartzite and granite fragments. A search for striated pebbles was made, but no striations or grooves other than those which might have been caused by movement after consolidation were found. Indeed, the high degree of metamorphism that has resulted in distorting all but the quartzite pebbles renders it unlikely that glacial striae would have been preserved. Moreover, the pebbles tend to break out with a coating of thin slaty matrix, which can not be freed without removing also the outer surface of the pebble, thus rendering it unlikely that striations would be found. The strong resemblance between this conglomerate and a conglomerate in the Mariposa slate (Upper Jurassic) exposed in the railway cut at Colfax station, which was found by Lawson[6] to rest on a glaciated surface, also favors the hypothesis of glacial origin.

Blackwelder[7] has described mud-flow deposits in semiarid climates, not associated with volcanism, and shown how closely these may resemble deposits of glacial origin. The character of the conglomerate is not incompatible with such an origin, and positive evidence in its favor are the occurrence of the conglomerate at different stratigraphic horizons in different areas, suggesting purely local origin, and the presence of a considerable amount of slate containing scattered pebbles. On the other hand, the coarse conglomerate is in places as much as 50 or 100 feet thick without obvious planes of stratification.[8]

LOWER SLATE AND GREENSTONE MEMBER

South of Kanaka Creek and in the valley of Oregon Creek the conglomeratic member of the formation grades toward the west into a series of interbedded dark slates and chloritic greenstones. The greenstones range from chloritized amphibolite schists lithologically indistinguishable from similar rocks in the Tightner formation to altered andesitic and possibly also dacitic tuffs and breccias that retain distinct traces of their original pyroclastic texture. There are also certain beds which are interpreted as flows (and possibly in part intrusive sheets) of much altered andesite and dacite. In places black slates, without noticeable mixture of volcanic material, predominate. Usually, however, there is an alternation between black slate

[6] Lawson, A. C., personal communication.
[7] Blackwelder, Eliot, Mud flow as a geologic agent in semiarid mountains: Geol. Soc. America Bull., vol. 39, pp. 465–484, 1928.
[8] The origin of the conglomerate was a source of friendly discussion throughout the field season. I inclined toward the possibility of glacial origin, while Mr. Gannett favored the hypothesis of origin from a volcanic mud flow and considered that I gave too much weight to the lack of identifiable volcanic material in the matrix and too little to the absence of striated pebbles or glaciated basement. The hypothesis of nonvolcanic mud flow was not taken into consideration in the field.— H. G. F.

A. CRUMPLED HORNBLENDE SCHIST OF TIGHTNER FORMATION BETWEEN MORNING GLORY AND OSCEOLA MINES

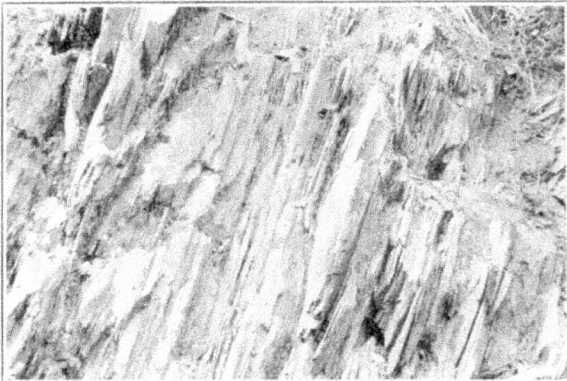

B. SLATE OF THE BLUE CANYON FORMATION NEAR MAMMOTH SPRINGS MINE

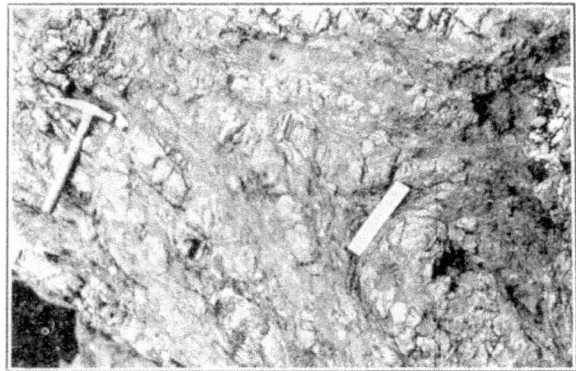

C. CRUMPLED HORNBLENDE SCHIST, SILICEOUS FACIES, OF TIGHTNER FORMATION, VALLEY OF KANAKA CREEK

D. VIEW FROM ALLEGHANY, LOOKING DOWN VALLEY OF KANAKA CREEK

A. HORNBLENDE SCHIST WITH STREAKS OF QUARTZ, TIGHTNER FORMATION, NEAR LOWER OSCEOLA TUNNEL

B. BASAL CONGLOMERATE (TILLITE?) OF KANAKA FORMATION NEAR TRANSFORMER HOUSE, ALLEGHANY

C. CHERT OF KANAKA FORMATION, SOUTH FORK OF YUBA RIVER NEAR WASHINGTON

D. DETAIL SHOWING CONTORTED BANDING IN CHERT, KANAKA FORMATION, SOUTH FORK OF YUBA RIVER NEAR WASHINGTON

A. RELIEF QUARTZITE, FOLD IN QUARTZITIC SLATE, KANAKA CREEK, NEAR RAPPS RAVINE

B. TYPICAL EXPOSURE OF SERPENTINE, NEARLY BARE OF VEGETATION, VALLEY OF KANAKA CREEK

C. EOCENE GRAVEL AT OLD HYDRAULIC WORKINGS SOUTH OF ALLEGHANY

D. FLOWS OF PLEISTOCENE BASALT CAPPING MIOCENE ANDESITIC BRECCIA, BALD MOUNTAIN

Part of town of Forest in foreground.

and greenstone, with the individual bands varying greatly in thickness. Commonly the beds are several feet thick, but in places the alternate bands of green and black are less than an inch wide, giving the rock a peculiar streaky appearance. The green tuffaceous beds commonly contain narrow partings of black slate. In a few places thin lenses of fine-grained conglomerate are present in the slates.

CHERT MEMBER

In the valley of Kanaka Creek the slate and greenstone pinch out north of the Rainbow flume, and north of the creek the conglomerate is bordered on the west by a band consisting dominantly of dense gray contorted chert (pl. 4, C, D), the individual bands averaging a couple of inches in thickness. This member also contains a considerable amount of black slate interbedded with the cherty bands, and in places chert lenses occur in the slate members. Here and there the siliceous beds are slightly coarser in grain, suggesting fine-grained quartzite.

The chert member has a thickness of nearly 300 feet at the old placer diggings south of Chips Flat. It is much thinner on the north side of the valley. In the bedrock exposures at the old placer diggings near the transformer house the chert fades out into greenstone and slate along its strike. Chert does not appear in the section exposed in Oregon Creek, but a narrow band interbedded with slate was seen in the tunnel of the North Fork mine. (See fig. 27.) Only narrow bands of chert were found in the areas underlain by the Kanaka formation in the valley of the Middle Fork of the Yuba River.

Outside the district, in the section exposed in the canyon of the South Fork of the Yuba River, the chert attains a much greater thickness than in the Alleghany district. Only a small amount of chert is present in the Kanaka formation in the valley of the North Fork of the Yuba, but an outcrop about 50 feet thick was found in a cut on the highway 1.7 miles east of Goodyears Bar.

Chert also occurs in other formations corresponding in part to the Calaveras formation. Lindgren [9] notes its presence in the Blue Canyon, Relief, and Clipper Gap formations, and chert with a thickness of about 50 feet was found interbedded with slate belonging to the Cape Horn formation in the valley of the North Fork of the Yuba River 1½ miles west of Goodyears Creek.

UPPER SLATE AND GREENSTONE MEMBER

West of the chert member in the valley of Kanaka Creek comes a succession of slate and greenstone similar to that described above. There is a distinct

[9] Lindgren, Waldemar, U. S. Geol. Survey Geol. Atlas, Colfax folio (No. 66), p. 2, map legend, 1900.

increase in the proportion of slate toward the west, and in some outcrops the resemblance to the slates of the Cape Horn formation is very close. Near the western border beds of quartzitic schist are found, indicating a gradation to the quartzitic Relief formation. The western boundary of the formation is consequently not definite, and the line on the map is arbitrarily drawn where quartzite and quartzitic schist appear to predominate.

SECTIONS

The rocks of the lower part of the Kanaka formation are well exposed in the valley of Oregon Creek a short distance below Forest. Here for a short distance exposures are sufficiently good to permit rough measurement of a section through the lower part of the formation. Eastward from the gabbro contact downstream from the Mugwump mine there is 200 feet of black slate, followed by 200 feet of greenstone schists with small streaks of black slate, which are most abundant at the east end. For about 200 feet eastward from this point there are no outcrops, but the talus consists of green schist and dark slate. Near the Mugwump tunnel dump is a large outcrop showing a thickness of 80 feet of dark slate with small streaks of schistose greenstone and a few lenses of conglomerate, the largest 40 feet long and 1 foot thick. Again for 300 feet eastward along the stream there are no exposures, but beginning at this point outcrops are nearly continuous to the serpentine near Forest, giving the following section:

Generalized section of Kanaka formation near Forest

	Feet
Black slate, conglomeratic in places; few outcrops but talus continuous along the stream	675
Schistose greenstone with streaks of bluish quartz parallel to schistosity	150
Concealed	20
Green schist with thin streaks of dark slate	60
Schistose greenstone	50
Concealed	40
Greenish schist (tuff?) with thin, closely spaced black stripes	50
Greenstone, schistose with streaks of dark schistose slate at top, massive at base	45
Dark schistose slate, more siliceous than usual	10
Greenstone	20
Concealed	20
Greenstone with band of dark conglomeratic slate, 4 feet thick, near the west end	30
Dark slate containing a few much-compressed pebbles	30
Fine-grained greenstone, not markedly schistose	20
Coarse conglomerate with dark slate matrix. Pebbles of siliceous igneous rock, much compressed, with a few of quartzite	30
Dark conglomeratic slate containing abundant white fragments of an altered igneous rock; adjoins serpentine mass on west	100

The total thickness of the Kanaka formation exposed on Oregon Creek is therefore estimated to be about 2,000 feet.

The approximate thickness of the different members of the formation outcropping in the valley of Kanaka Creek is as follows:

Generalized section of Kanaka formation north of Kanaka Creek, east-northeast to the Sixteen to One mill

	Feet
Tuff, slate, etc	700
Chert with varying amount of slate	200
Conglomerate and conglomeratic black slate	350
	1,250

Generalized section of Kanaka formation south of Kanaka Creek

	Feet
Tuff, slate, etc	800
Chert, etc	300
Interbedded tuff and slate	100
Conglomerate and conglomeratic slate	200
	1,400

Only a small part of the formation is exposed along the Middle Fork of the Yuba River (fig. 2), as it is cut off by gabbro a short distance south of Jackass Ravine. The data shown in Figure 2 may be generalized as follows:

Generalized section of Kanaka formation along Middle Fork of Yuba River

	Feet
Gabbro.	
Gray schistose tuff	60
Conglomerate with black slate matrix	60
Chiefly green to gray chlorite schist, in large part probably originally tuff, with bands carrying chert lenses. Grades into slate at top	200
Conglomerate, with black slate matrix, pebbles larger and more abundant near base, in fault contact with Tightner formation	100

The section in the canyon of the South Fork of the Yuba River west of Washington is apparently much thicker, for here rocks of the same character, dipping at high angles and with strike transverse to the course of the river, crop out for a distance of over a mile in the stream bed. In the valley of the North Fork of the Yuba River rocks resembling the Kanaka formation are exposed in cuts along the highway for nearly 2 miles eastward from Goodyears Creek.

RELIEF QUARTZITE

The geologic map of the Colfax quadrangle shows that the Relief formation extends in a narrow belt, nowhere exceeding a mile in width, from a point south of Bear River, near Dutch Flat, 20 miles south of Alleghany, northward to Relief Hill, north of the South Fork of the Yuba River about 5 miles south of the portion of the quadrangle covered by the present report. (See pl. 3.) Lindgren [10] describes the formation as a very fine grained quartzite alternating with streaks of siliceous clay slates, but his map legend reads "Relief formation (chert and quartzite)." The Cape Horn slate adjoins the Relief formation on the west.

In the Alleghany district small areas of sediments which are for the most part quartzitic in composition (pl. 5, *A*), lying between the Kanaka formation and the Cape Horn slate, are considered to represent a northward extension of the Relief formation. These appear to be more highly metamorphosed than the main mass of the Relief quartzite to the south, possibly owing to the granitic dikes, which are particularly abundant in this part of the Alleghany area. The clay slate is here altered to mica schist, and the quartzite has a distinctly schistose appearance, accentuated by partings of muscovite. Rarely, as at the Oriental mine, there are thin beds of dark slate similar to those of the Kanaka formation.

Nothing corresponding to the Relief quartzite was found in a hurried reconnaissance along the North Fork of the Yuba River, but there is a gap nearly a mile in width between the westernmost exposure of the Kanaka formation, half a mile east of Goodyears Creek, and the easternmost outcrop of the Cape Horn formation, west of the creek. Quartzitic slate, either interbedded with or underlying the dark slate of the Cape Horn formation, occurs in contact with the serpentine in the valley of Woodruff Creek, on the road leading to Goodyears Bar south of the North Fork of the Yuba River.

CAPE HORN SLATE

The Colfax geologic map shows the Cape Horn formation as a belt 2 to 5 miles in width extending the whole length of the quadrangle, a few miles east of the western border. In the Alleghany district this formation is found only in the northwestern part, from the Mountain House saddle southward to the Kate Hardy mine. As seen in the valley of Oregon Creek and the road near Mountain House, the prevailing rock is a very dense black slate similar in general appearance to the slates of the Blue Canyon formation but denser and more carbonaceous. In places the slate is much crumpled and lacks the prominent slaty cleavage of the Blue Canyon slate. No quartzitic bands were seen in the Cape Horn formation in the Alleghany district. In places bands of lighter-colored material less than a quarter of an inch in thickness may represent volcanic ash.

In the Alleghany district the Cape Horn slate is separated from the Relief quartzite by an intrusion

---

[10] Lindgren, Waldemar, op. cit. (Colfax folio), p. 2, column 2.

of serpentine, but in the valley of Kanaka Creek, just west of the area mapped, rather poor exposures appear to indicate a narrow zone of gradation between quartzitic schists of the Relief formation and the characteristic dark Cape Horn slate, and a similar gradation may be indicated by the quartzitic slate in the valley of Woodruff Creek, noted above.

SERPENTINE

The Alleghany district extends across the "great serpentine belt," and serpentine occupies a large area in the portion of the district from which the Tertiary rocks have been eroded. Two large masses extend across the entire district along the eastern and western margins, and many smaller dikes cut the older rocks in the area between them. The belts of serpentine, which follow the general northerly trend of the other rock formations, are a prominent feature of the topography, for the serpentine supports only a very sparse vegetation and the broad belts of bare rock stand out in strong contrast to the densely wooded hillsides on both sides. (See pl. 5, B.) The serpentine masses tend to form the "high points" of the lava contacts, and the early gravel does not as a rule rest on serpentine bedrock, implying that under the conditions of subdued topography existing at the time of deposition of the early auriferous gravel the soft yet dense rock was resistant to erosion and tended to form low ridges, which guided the courses of the early streams. This inference, of course, does not hold true for the superposed drainage pattern followed by the streams that deposited the intervolcanic gravel. It has been held that these "high points" are in part due to the pressure exerted by later hydration and consequent expansion of the serpentine or by pressure on the much-sheared rock by the more massive rocks at its sides, but no evidence of distortion of the lava above the serpentine was observed, and, as shown below, there is reason to believe that the process of serpentinization was essentially complete long prior to Tertiary time. Locke,[11] however, ascribes faulting of the Tertiary lava on Tuolumne Mountain, in the Mother Lode region, to renewed movement along an underlying pre-Tertiary fault, resulting from increase of volume by serpentinization.

In the present valleys the serpentine bands tend to form salients bordered by small ravines, which are commonly cut, not in the serpentine but in the bordering rock close to the contact.

The large eastern serpentine belt extends from the Middle Fork of the Yuba River at the mouth of Wolf Creek northward to the head of Buckeye Ravine, bordered on the east by the Blue Canyon slate,

with which it is in fault contact, and on the west by gabbro. The serpentine does not form a single mass farther north, but several smaller masses are intricately mixed with the gabbro on the slope south of Kanaka Creek and cut the schists on the hillside north of the creek. The western serpentine belt borders the western gabbro mass on the west in the valleys of Kanaka and Oregon Creeks and near the northwest corner of the district, near the Brush Creek mine. It is about 3,000 feet wide in the valley of Oregon Creek and narrows to about 1,500 feet at the northern border. Between the two major belts are several smaller masses, the two largest of which, like the major belts, occur at the margins of the gabbro masses. Other smaller masses are present as lenses with long axes roughly parallel to the schistosity of the inclosing rocks of the Tightner and Kanaka formations in the valley of Kanaka Creek. These lenses, as shown by the structure sections (pl. 10), dip to the east, thus cutting the structure of the older formations, but the dip is much steeper than that of the veins.

The relations between gabbro and serpentine are not everywhere clear. In many places dikelike masses of serpentine penetrate the gabbro, but elsewhere the reverse is true, and in some areas, as on the south side of Kanaka Creek, west of the Kinselbach mine, the contacts appear to be gradational. Knopf[12] finds that in the Mother Lode region the serpentine is cut by dikes of hornblendite and gabbro.

The marginal position of the serpentine relative to the gabbro suggests that the basic rocks from which the serpentine was derived formed a basic marginal differentiate of the gabbro mass and that the two rocks are essentially parts of the same intrusion. The relation of the serpentine area to those of the rocks mapped in the folios as amphibolite and to the gabbro intrusions in other parts of the great serpentine belt suggests that this may also hold true over a larger area.

The serpentine is the most distinctive and readily recognized of the rocks of the "Bedrock series." It is dark to light green and owing to its softness and highly sheared condition has a distinctly greasy feel. Commonly it is so completely traversed by shearing planes, along which it is distinctly lighter colored, that it breaks readily into small ellipsoidal fragments 3 inches or less in major diameter, each ellipsoid consisting of darker serpentine relatively free from visible shearing. In places larger lenses, some of them several feet in diameter, of dark unsheared rock suggest remnants of original peridotite, but examination under the microscope shows that these, though they have escaped the more intense shearing, are likewise completely altered to serpentine.

---

[11] Locke, Augustus, Tuolumne Table Mountain: Min. and Sci. Press, vol. 105, p. 85, 1912.

[12] Knopf, Adolph, The Mother Lode system of California: U. S. Geol. Survey Prof. Paper 157, p. 20, 1929.

Little can be inferred as to the original composition of the serpentine. It is probable that rocks composed of pyroxene, of olivine, and of mixtures of the two were originally present. Gabbro grading toward pyroxenite and containing a little olivine was observed. In the thin sections of serpentine specimens from a few localities residual grains of olivine and less commonly of pyroxene were seen, and in a few places the structure of the serpentine suggests derivation from one or the other, but the data were insufficient for any definite conclusion. Bastite plates, which are abundant in places, may indicate derivation from pyroxene.

Here and there magnetite is abundant in the serpentine, and everywhere on the serpentine belts it is present in sufficient amount to cause a considerable deviation of the compass needle. Ilmenite may be present also, and one thin section of serpentine showed minute rutile needles interstitial between the antigorite plates. Segregations of chromite in the serpentine belt crossing Oregon Creek east of the Kate Hardy mine were mined in a small way during the war period. The chromite occurs in lenticular veinlike masses that follow a distinct zone of shearing in the serpentine, but seem to have crystallized prior to serpentinization, as the chromite grains are crystalline against the surrounding serpentine and are veined by the serpentine. The serpentine veinlets cross the chromite, and in the vicinity of the chromite there are abundant small crystals of the chromium-bearing garnet, uvarovite. There are also, in the neighborhood of the larger chromite grains, abundant small specks of what appears to be chromite, but the particles are too small to be determined certainly. If they are chromite they must have crystallized after serpentinization, for they border the antigorite plates.

Further alteration of serpentine under the influence of vein-forming solutions has in places produced talc, but more commonly the serpentine has been replaced by a mixture of ferromagnesian carbonate, probably chiefly ankerite, and mariposite.

That the alteration of the original basic rock to serpentine was a deep-seated process is shown by the fact that the veins do not persist for any great distance in the serpentine, and the character of the mineralization in the serpentine is different from that of the neighboring gabbro and pyroxenites. The hydration to serpentine must therefore in large measure have taken place prior to or contemporaneously with the formation of the quartz veins. It may have been initiated by the water and carbon dioxide given off after the consolidation of the later intrusions of the more silicic rocks.

GABBRO AND DIORITE

A group of dark-colored rocks that range in composition from amphibolitic rocks derived from gabbro to schistose diorite and quartz diorite occupies large areas in the Alleghany district. These rocks were intruded after the older sedimentary and volcanic rocks had received their schistosity, and though in a broad way the intrusions follow the strike of the older formations, exposures of actual contacts show cutting of the older schistosity by the gabbro. (See fig. 3.) The

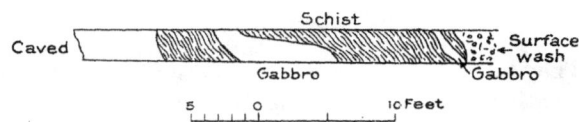

FIGURE 3.—Section along prospect tunnel south of Irelan mine, showing gabbro cutting greenstone schist of Kanaka formation

small bodies of dark-colored diorite and quartz diorite are thought to be merely gradational phases of the gabbro, without sharply marked boundaries.

A large mass of gabbro with an average width of 4,000 feet crosses the western part of the area covered by Plate 1, and there is another large but irregular mass in the valley of the South Fork of the Yuba River, extending northward into the valley of Kanaka Creek. Besides these there are many small dikes which are most abundant in the areas occupied by the Tightner and Kanaka formations.

The gabbro as seen in the hand specimen is typically a dark-green rock of granular texture, in which the usual size of grain is 1 to 2 millimeters. In many parts of the area a rough alinement of alternate streaks relatively rich in hornblende and feldspar gives the gabbro a schistose appearance. This seems to be most common around the margins of the gabbro masses and in the smaller dikes, such as those cutting the schist in the vicinity of Alleghany. Nowhere is a schistosity comparable in degree to that of the schists of the Tightner formation found in the areas occupied by gabbro. The more schistose phases show a banding of thin light and dark streaks; where schistosity is poorly developed the rock has a "pepper-and-salt" appearance.

The principal visible mineral consists of dark-green hornblende in small closely spaced crystals set in a matrix of light-green feldspar, which has a waxy appearance and does not show cleavage faces. The relative proportions of hornblende and feldspar vary widely, but commonly the hornblende exceeds the feldspar in volume. In a few specimens feldspar is present in so small a proportion that the rock should properly be classed as an amphibolite derived from an original pyroxenite. More rarely, as at points on the Foote road between Alleghany and the Oriental mine, feldspar is in marked excess, the rock thus grading to a diorite. The gabbro in many places contains small patches of pyrrhotite.

Thin sections show that little is left of the original constituents of the gabbro. In most specimens the feldspar is completely altered to the characteristic aggregate of alteration products known as saussurite. In the rare specimens that show small remnants of probably original feldspars these are less calcic than would be expected in a rock of this type; andesine was definitely recognized, and very rarely what appears to be primary oligoclase was observed. The remnants of original feldspar are too few to give this observation much significance. It may indicate that the more sodic feldspars, or portions of feldspars, were somewhat less susceptible to saussuritization.

A little quartz that is apparently distinct from the later quartz introduced during the period of vein formation was noted in several of the thin sections of the gabbro and possibly indicates gradation toward the quartz diorite, described below. If the quartz is original, however, it has undergone recrystallization, for it is confined to the feldspathic zones of the schistose types, and the quartz grains are penetrated by needles of hornblende.

The hornblende, which is so prominent a feature of the rock, is chiefly if not entirely a product of metamorphism and was developed at the expense of the original augite, of which there are only small remnants left. The schistosity generally observable in the rock is therefore probably not an original structure but is due to a rough linear segregation of zones containing abundant hornblende and zones in which hornblende and saussurite are present in varying proportions.

Biotite may have been present in small amount as an original mineral in the facies of the gabbro grading toward diorite, but in most specimens that contain recognizable biotite it appears to be later than the hornblende, which has itself developed at the expense of the original augite. The biotite-bearing facies of the gabbro was found only near the granite, suggesting that the biotite was formed as the result of contact metamorphism.

In the more basic facies of the rock the areas of altered feldspar are less prominent, and small remnants of original olivine are found. Specimens that show olivine come from border zones between the more typical gabbro and the serpentine masses, showing that these rocks are, at least in part, transitional in character. In places, particularly where olivine is present, there has been a little serpentinization. This has taken place later than the development of the hornblende, for hornblende is also replaced by serpentine to a very slight extent, though it has, on the whole, been very resistant to serpentinization, whereas augite has yielded readily. The fact that the hornblendized gabbro, even where so poor in alteration products of feldspar as to point to its having been originally close to a pyroxenite, is not affected by serpentinization also indicates that serpentinization was later than the development of the hornblende at the expense of the augite.

Small patches of pyrrhotite and less commonly pyrite in places suggest that these minerals were original constituents of the rock or at least were developed prior to the hornblende. Other opaque accessories include ilmenite or titaniferous magnetite, which gives rise to abundant leucoxene, and probably magnetite.

Further alteration superposed on that already described has resulted in the development of abundant chlorite, chiefly at the expense of the ferromagnesian minerals. Such chloritization seems to be confined to the general vicinity of the veins. Specimens obtained near the veins show still further alteration, the most prominent features of which are the development of a second generation of chlorite at the expense of the hornblende and earlier chlorite and the development of albite in the saussurite areas. Close to the veins there may be also complete or partial replacement by quartz or by carbonate and mica. This type of alteration is considered in more detail in the discussion of the mineralogy of the ore deposits.

QUARTZ DIORITE

Outcrops of a rock closely resembling gabbro but containing visible original quartz were found in a few places in the western gabbro area. Owing to its close general similarity to the gabbro and the fact that the quartz is inconspicuous in the hand specimen, it is possible that there exist still other masses not found in the course of the present investigation.

The known areas of quartz diorite lie within the western gabbro mass, apparently as long, narrow lenses parallel to its general trend, a few degrees west of north. Four such lenses were noted, one crossing the Moores Flat trail near the eastern margin of the gabbro, a second about 1,000 feet to the west, extending from a point south of the Moores Flat trail to Lafayette Ridge, a third north of the ridge near the Kenton mine, possibly the northward extension of the second, and the fourth in the bed of Kanaka Creek near the Spoohn mine. There is no apparent contact-metamorphic effect on the surrounding gabbro, although generally in the vicinity of the quartz diorite the schistosity is more developed than usual.

Examined under the microscope, rocks of this type are seen to be closely allied with the gabbro and diorite described above. The principal difference is the greater abundance of quartz. In the less metamorphosed phases intergrown quartz and andesine indicate that the quartz is original. In other specimens there has apparently been a recrystallization of the quartz, as the quartz grains are penetrated by needles of hornblende which has recrystallized from augite. The relative proportion of light-colored to dark-colored minerals is higher in the quartz diorite than in the gabbro. Biotite may be an original accessory mineral in the quartz diorite. Decreasing quartz content gives a gradation between the quartz diorite and the dioritic phases of the gabbro.

The fact that the localities in which these lenticular bodies of quartz diorite occur are along the strike of the Relief quartzite, the only member of the older rocks containing abundant quartz, suggests the possibility that the quartz content of the quartz diorite may be due to assimilation of these quartz-rich sediments by the gabbro magma.

GRANITE AND APLITE

Although large masses of granitic rocks crop out at short distances from Alleghany (pl. 3), there are only small areas of such rocks within the district. The principal outcrops of granite and aplite lie within a rather narrow zone crossing the area in a northerly direction. The largest masses found are in Wet Ravine near the Oriental shaft (pl. 49) and in the Oriental mine (pls. 50 and 51), where granite forms the footwall of the vein. Other small masses were found in the valley of Kanaka Creek near the Spoohn mine, and farther south, outside the area mapped, along the Middle Fork of the Yuba River. In the Oregon Creek drainage area the small granitic dikes at the Evans prospect and near the head of Sandusky Creek form the northward extension of this zone.

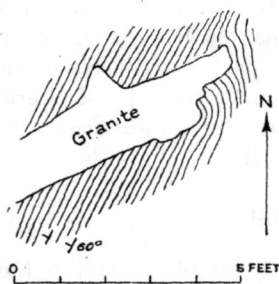

FIGURE 4.—Granite dike cutting schistose gabbro, French Ravine

Outside this zone only a few outcrops of granite were found. A few small dikes, chiefly aplitic, none over 10 feet in width, cut the Tightner formation in the valley of Kanaka Creek, and small outcrops of granite and aplite were found near the Kate Hardy mine.

The granite and aplite are clearly later than the gabbro and have been intruded at a later date than the metamorphism that developed the hornblende in the gabbro and gave the gabbro its schistose structure. (See fig. 4.) Near the contacts of gabbro and granite biotite has been developed in the gabbro at the expense of the hornblende. Similarly the hornblende schists of the Tightner formation contain a little biotite where they are cut by small granitic dikes.

The granite and aplite, even in small outcrops, are easily identified in the field, as they are not schistose and differ from other members of the "Bedrock series" in that fresh specimens are nearly white. The granite is medium to coarse grained, and the only prominent dark-colored mineral is biotite, which is nowhere abundant, the rock being composed essentially of feldspar and quartz, the former distinctly in excess in all slides examined. The feldspars show considerable range in composition; oligoclase, albite, microcline, and orthoclase are all present, and rarely all four varieties can be identified in a single thin section. Albite seems to be the most abundant and oligoclase the rarest, but the proportions vary so greatly that a single thin section is not a fair sample of the rock. The quartz is distinctly less abundant than the feldspar and is intergrown with it. Biotite, usually in large part chloritized, was observed in small amount in most of the thin sections examined, and fairly large plates of muscovite in a few. Other original accessory minerals noted are ilmenite, which is in part altered to rutile, and apatite, zircon, and garnet, the last in part altered to chlorite.

The small aplitic dikes are similar to the granite with the exception of their finer grain. Orthoclase, however, appears to be the more abundant feldspar, and a little muscovite is commonly present; biotite is rare.

Probably allied to the granite are the small irregular veins of quartz and feldspar, which cut the older formations, particularly the gabbro. Study of a thin section of one of these veins showed an aggregate of quartz and calcic albite near oligoclase. Other similar veinlets contain only feldspar.

The granite and aplite show no schistosity, and the only sign of yielding to external pressure is the presence of irregular zones in which the quartz and feldspar have been minutely fractured. Likewise the granite and aplite have not suffered mineralogic changes due to regional metamorphism, such as have so completely changed the character of the gabbro and diorite. All specimens examined, however, showed more or less alteration of the sort considered to have accompanied ore deposition, chiefly the alteration of the feldspars and development of carbonate. These changes are described in connection with the mineralogy of the ore deposits (pp. 38–52).

AGE OF THE "BEDROCK SERIES"

No data were obtained that furnish direct evidence as to the geologic age of the rock formations of the "Bedrock series" in the Alleghany district. Nothing, however, was found which modified the age relations as given by Lindgren in the Colfax folio—namely, that the formations there grouped as the Calaveras formation are of Carboniferous age and that the age sequence begins on the east, the westernmost being the most recent. The detailed mapping of the Alleghany district offers no evidence as to age relations between the Blue Canyon and Tightner formations, for they are in fault contact. There appears to be an unconformity between the Tightner formation and the overlying Kanaka formation, but the Relief quartzite overlies the Kanaka conformably. The Relief and Cape Horn formations are probably conformable but were not found in contact.

The basic intrusive rocks, gabbro and serpentine, are probably essentially contemporaneous but seem to be considerably older than the granitic intrusives. In the Alleghany district the basic rocks and granite cut only rocks of Carboniferous age, but elsewhere in the area covered by the Colfax folio they are intrusive into the Mariposa slate, of Upper Jurassic age, and are older than the Knoxville formation, generally considered of Lower Cretaceous age, though the basal portion is regarded by some geologists as of late Jurassic age.[13] If the granitic intrusions are taken as marking the time boundary between Jurassic and Cretaceous, the basic rocks might be considered to belong within the Upper Jurassic.

"SUPERJACENT SERIES"

As study of the "Bedrock series" was deemed of prime importance in this investigation, the areas of Tertiary gravel shown on the map are restricted to those portions of the gravel-covered ground where the underlying bedrock is concealed. Such ground includes areas in which bedrock was washed clear of gravel and later covered with the débris resulting from hydraulic mining. The authors have attempted but not carried out consistently a separation of the Tertiary gravel into the older gravel, of probable Eocene age, and the later intervolcanic gravel, of Miocene age.

No attempt was made to subdivide the great thickness of volcanic breccia that caps the interstream ridges, although the presence of interbedded gravel in a few places indicates that intervals of stream erosion separated the volcanic eruptions which gave rise to the breccia. Nor was any detailed study given to the remnants of the basalt flows that cap the breccia north of Forest or to the basalt dikes found here and there throughout the area.

The following notes on the formations of the "Superjacent series" are intended rather as an explanation of the units used on the geologic map (pl. 1) than a contribution to the Tertiary and Pleistocene geology of the district. For a more complete description of such geology the reader is referred to the Colfax folio and to Professional Paper 73.

EARLY GRAVEL

The pebbles of the oldest gravel, as would be expected from streams in a region of topographic old age, consist of resistant material, largely vein quartz,

[13] Smith, J. P., The geologic formations of California: California State Min. Bur. Bull. 72, pp. 31–33, 1916.

quartzite, and quartzitic slate. This material tends to be coarse and bouldery at the base but grades upward into fine gravel and sand. At Minnesota and Alleghany (pl. 5, C) there is a layer of "pipe clay" near the top. This probably contains rhyolitic material.

The principal deposits of early gravel in the Alleghany area were laid down by two southward-flowing streams tributary to the main Tertiary river, which in this region flowed westward, south of the present course of the Middle Fork of the Yuba River. (See pl. 43.) One of these streams deposited the gravel at Forest, Alleghany, Chips Flat, and Minnesota. The course of this stream follows very closely the belt of conglomerate at the base of the Kanaka formation. The course of the second stream, which seems to have joined the first above Minnesota, is marked by the gravel deposits near the Mammoth Springs mine and at Balsam Flat. The gravel of the first stream, which crossed the part of the district in which the lode mines have been most productive, carried far more gold and a larger proportion of vein-quartz pebbles and boulders than that of the second, which shows abundant pebbles of quartzite and quartzitic slate derived from the more resistant members of the Blue Canyon formation.

Other small deposits of gravel which were not observed to carry pebbles of the volcanic rocks and are therefore assigned to this formation were noted beneath the lava at Oak Flat and north of Alleghany.

The present valleys of Oregon and Kanaka Creeks and the Middle Fork of the Yuba River cross the valleys of these older streams, and the cross section of the Tertiary valleys thus given shows that the valleys were low, with gently sloping sides. Their courses show a general concordance with the strike of the underlying rocks. Between Forest and Minnesota, a distance of slightly over 3 miles, the difference of altitude of the base of the gravel is 200 feet. As the course is about at right angles to the direction of later tilting this difference probably represents approximately the original grade of this tributary stream.

INTERVOLCANIC GRAVEL

The gravel of intervolcanic age differs from the earlier gravel in being markedly less quartzose. It contains not only a large proportion of the less resistant rocks of the "Bedrock series," which are lacking in the earlier gravel, but also pebbles and boulders of volcanic rocks, both rhyolite and andesite. In the major intervolcanic valley of the Alleghany district clay, containing abundant leaves, overlies the gravel. A small collection of these leaves was submitted to the late F. H. Knowlton, who reported:

The single locality represented is on Lucky Dog Creek 1 mile east of Forest, Calif. The following forms have been noted:

Equisetum sp.
Ilex prunifolia Lesquereux.
Carpinus sp.
Ulmus pseudo-fulva Lesquereux?
Rhus sp.

These are auriferous-gravel species and Miocene in age.

The principal outcrops of intervolcanic gravel are on the sides of Oregon Creek Valley. The major stream of this epoch within the Alleghany district flowed southward on serpentine bedrock east of the Kate Hardy mine. Gravel of intervolcanic age, deposited by a minor stream, crops out in the bed of Sandusky Creek and the valley of Oregon Creek west of the Brown Bear mine. Other outcrops of intervolcanic gravel were noted in Wet Ravine near the Oriental mine and a short distance northwest of Minnesota Flat.

The section of the principal intervolcanic valley given by the valley of Oregon Creek shows an entirely different topographic condition from that prevailing at the time of the deposition of the earlier gravel. The stream is cut on serpentine bedrock, which implies that the drainage is here superposed, because elsewhere along the lava contact serpentine tends to form the "high points," showing that low serpentine ridges were a feature of the older topography. Moreover, the valley is of different character here; instead of the gently sloping sides characteristic of the Eocene valleys, the stream has cut a canyonlike valley 1,000 to 2,000 feet wide, with steep walls 200 to 400 feet or more high. The gravel tells the same story of rejuvenated drainage. It is bouldery and in part not well rounded, and in contrast to the earlier gravel, which is composed entirely of rocks resistant to weathering, all types of bedrock material as well as rhyolite and andesite are represented.

Between the deposition of the early gravel and that of the intervolcanic gravel the old mature landscape must have been partly covered with lava, which in the Alleghany district was removed in the intervolcanic erosion epoch but which forced the streams into new courses, independent of bedrock and presumably determined by initial irregularities in the lava surface.

Later deposits of stream gravel intercalated between flows of the andesitic breccia mark the courses of streams that flowed during the epoch of andesitic eruptions. Such gravel deposits were mapped only in one place, in the lavas above the town of Alleghany.

ANDESITIC BRECCIA

Other than to map its contact with the older formations, little attention was paid to the great mass of andesitic breccia that caps the interstream ridges. The material consists of pebbles and boulders of various rocks, in part angular, in part waterworn, loosely set in a matrix of andesitic material. The pebbles are largely andesite but include also rhyolite as well as representatives of the bedrock formations. The latter are largely waterworn and were presumably plucked from older stream channels. Counts of pebbles in different parts of the breccia show that andesite forms from 71 to 89 per cent of the total. Lindgren [14] regards the andesitic breccias as the result of a succession of mud flows of unconsolidated volcanic material derived from volcanoes near the crest of the range to the east. These flowed westward down the slope of the range, obliterating the older topography.

The andesite breccia is thickest northeast of Forest, where it has been preserved from erosion by the overlying Pleistocene basalt of Bald Mountain. Here the thickness is about 1,000 feet.

BASALT

A succession of flows of dense fine-grained olivine basalt about 200 feet thick forms the summit of Bald Mountain, northeast of Forest. (See pl. 5, *D*.) This is regarded by Lindgren [15] as of Pleistocene age. Here and there throughout the area are small dikes and masses of basalt, which may represent conduits that fed the Pleistocene basaltic volcanoes. The largest encountered cuts the andesitic breccia about three-quarters of a mile west-northwest of the Irelan mine and shows well-developed columnar structure normal to the contact.

The most interesting occurrence of basalt is that in the Plumbago mine. (See pl. 58.) On the three highest levels the vein is cut by an intrusion of basalt which enters from the footwall side of the B drift at an altitude of 4,075 feet and follows the vein fissure at least as far as the No. 1 tunnel, 400 feet higher. The mine map shows that the basalt occupies the site of the vein for horizontal distances ranging from 75 to 150 feet on each level. This portion of the mine has not recently been worked and the dike could be seen only on the No. 2 and No. 3 levels at altitudes of about 4,280 and 4,180 feet, 200 to 300 feet below the base of the andesite.

The basalt cuts the vein sharply and has not displaced it by wedging it aside but rather occupies the site of the vein. On the No. 2 level about half the space from which the vein is removed is occupied by basalt, which contains numerous fragments of quartz and gabbro and a few andesitic pebbles that must have been derived from the breccia above. The remainder of the drift space consists of bedded material, resting against the basalt, but without evidence of baking by

---

[14] Lindgren, Waldemar, op. cit. (Colfax folio), p. 6.
[15] Idem, pp. 6–7.

the basalt. This material is largely clay and fine sand but also includes pebbly layers in which fragments of scoriaceous basalt, andesite, quartz, granite, and schist were noted. The bedding is generally about horizontal but in places shows irregular dips of as much as 20°, presumably due to slumping.

In the No. 3 tunnel the proportion of basalt in the drift is larger, but otherwise conditions are the same. The main mass of the basalt contains numerous foreign inclusions, but at the contact with the quartz a little dike of dark scoriaceous basalt without foreign material, 1 inch to 3 inches in width, ramifies irregularly through the vein quartz for a distance of at least 5 feet. Another similar small dike extends along the wall of the vein for an unknown distance.

The open cavity shown on the map could not be reached. According to the description given, it was irregular in outline, bordered by vein and country rock on the far side and extending an unknown distance down in the direction of the dip of the vein. It apparently does not extend to the point where the basalt was cut in the B level.

The surface exposures are poor, but boulders of basalt were found over only a few feet of the surface, implying that the dimension of the intrusion transverse to the vein was not greater than the drift length as exposed underground.

The peculiar features of the intrusion are best explained by supposing that the area occupied by basalt and mud was first the site of a steam explosion. Material from the vein and adjoining country rock was blown out toward the surface, which at that time must have been from 800 to 1,300 feet above No. 3 level and possibly even higher. Material consisting of fragments both from the vein and country rock and from the overlying andesitic breccia fell back into the vent. Next came the intrusion of the basalt, which worked its way upward through the loosely packed rubble. Part of this rubble was displaced upward and part included as fragments in the lava. The rising lava, however, never filled more than a part of the neck. After outflow ceased there was a certain amount of shrinkage of the basalt, due to its consolidation, and the space thus provided, combined with the loose, incoherent condition of the rubble within the neck, permitted the access of surface waters in sufficient amount to effect the bedding of the material against the lava. The open cavity on the No. 3 level may have been formed at the time of the first explosive eruption, or it may be due to withdrawal of the basalt that once filled it.

PLEISTOCENE AND RECENT GRAVEL

The geologic history of the district after the outflow of the basalt is to be read principally in the carving out of its present topography by erosion rather than in the presence of deposits capable of being shown on the geologic map. Here and there deposits of gravel at varying altitudes mark stages in the present stream erosion. These deposits have been mined for their gold content and are shown on the geologic map only where the accumulation of the bouldery waste from such mining obscures the relations of members of the "Bedrock series." In places, as in the broad valley north of Forest, there is a deep accumulation of soil containing many boulders derived from the andesitic breccia and boulders of the breccia itself. This is reworked material and not a result of weathering of the breccia in place, but in the course of mapping no separation from the breccia was attempted.

The gravel shown on the map along the courses of the present streams is not, for the most part, gravel resulting from normal stream erosion but waste from the hydraulic mining of the Tertiary gravel, which the present streams have not yet been able to carry away.

Mine dumps in places obscure considerable areas of bedrock, but they have been shown on the geologic map only where their presence makes doubtful the interpretation of the boundaries of the underlying bedrock formations.

GEOLOGIC STRUCTURE AND HISTORY

The major episodes of the geologic history are to be deduced in part from the constituents of the rocks themselves, in part from their position, which is the net result of the forces to which they have been subjected, and finally from the topographic forms which the region now presents. The character of the different rock formations has been described in the preceding section. In this section an attempt will be made to give a historical interpretation to the lithologic character of the rocks, to sketch the reasons for their present position and attitude, and to interpret the present topographic features, as illustrating the last stage in the geologic history. It is believed that the record and interpretation in chronologic order of such scattered fragments of the history of the rocks as are furnished by the rocks themselves—by their composition, their present attitude, and the evidences of metamorphism they have undergone—will give the reader a better idea of the geologic history of the region than he would obtain from a separate presentation of lithologic and structural data.

STRUCTURE OF THE "BEDROCK SERIES"

The slates and sandstones of the Blue Canyon formation, which are the oldest rocks of the district, indicate deposition in shallow water. From the great thickness of this formation, as mapped in the Colfax folio, it is evident that the condition which they record prevailed throughout a long time. Fossils found in other parts of the formation in the Colfax quadrangle [16] indicate that it is a marine deposit of Paleozoic age. The higher Clipper Gap formation contains fossils of lower Carboniferous (Mississippian) age, so that the beginning of the geologic history as recorded in the rocks of the Alleghany district may date only from the Carboniferous or may be older.

As the Blue Canyon formation is in fault contact with the Tightner formation, which adjoins it on the west, there is a gap in the record of unknown extent. The metamorphosed rocks of the Tightner formation are interpreted as being dominantly of volcanic origin, giving a hint of crustal disturbances in the region at this early period. Granitic rocks may have been intruded at this time, though not within the district itself. On the other hand, within the Tightner formation are lenses of limestone and slaty schist which suggest that marine deposition continued in intervals between volcanic activity and that the lavas and accompanying pyroclastic materials either were poured out on land that was close to sea level or may have resulted from submarine eruptions.

Certainly a change in the topography of the region ushered in the period in which the conglomerate at the base of the Kanaka formation was deposited. If the conglomerate is of glacial origin, there must have been glacial activity here within or prior to the Mississippian epoch. Whatever the origin of the conglomerate, its nature seems to exclude a marine origin and to imply prior uplift. The source of the arkose and granite pebbles found in this conglomerate is unknown. Their presence suggests that during the accumulation of the lavas of the Tightner formation, or earlier, there had been intrusions of deep-seated igneous rocks. This in turn requires a considerable period of erosion to bring these rocks to the surface before the deposition of the conglomerate of the Kanaka formation. The upper members of the Kanaka formation give a record of periods of volcanism alternating with sedimentation. The chert may have resulted from precipitation from silica-rich waters associated with the lavas.

As a whole, the lithology of the Tightner and Kanaka formations suggests major earth movements during this early period, the effects of which are now largely masked by later distortion and metamorphism.

The rocks from the base of the Relief quartzite through the portion of the Cape Horn slate included in this area seem to record uninterrupted sedimentation, dominantly siliceous during the time in which the rocks now forming the quartzites and quartz-mica schists of the Relief formation were laid down and dominantly argillaceous during the period of deposition of the Cape Horn slate.

The general strike of the sedimentary and pyroclastic formations is from north to north-northwest, and the dip is commonly either vertical or steep to the east, indicating slight overturn if the assumed age sequence is correct. Steep westerly dips are present in a few places. Here and there unusually favorable exposures reveal the presence of minor folds (pl. 5, A), and doubtless many which escaped observation are also present. The degree of metamorphism in the older formations naturally varies with the type of rock; it is most intense in the volcanic and pyroclastic rocks of the Tightner and Kanaka formations and least in the conglomerates with slaty matrix and in the siliceous sediments, such as the quartzites of the Relief formation and the chert member of the Kanaka formation. The schistosity throughout these formations is generally accordant with the strike and dip.

It is not definitely known how long a gap intervened between the deposition of the youngest member of the Calaveras formation and its folding and metamorphism. Certainly there was a period of major folding prior to the deposition of the Sailor Canyon formation, which crops out in the higher mountains about 12 miles west of the Alleghany district. This rests unconformably on the Blue Canyon formation, and Lindgren [17] thought that it probably contains beds ranging from Upper Triassic to Lower Jurassic in age. The later faunal studies of J. P. Smith, however, have led to its assignment to the Upper Triassic. A comparison of the formations of Calaveras age with the overlying Mariposa slate of Upper Jurassic age, as exposed in the vicinity of Colfax, about 24 miles south of Alleghany, shows a far greater degree of metamorphism of these older rocks than of the Mariposa formation. Knopf [18] also finds that in the Mother Lode region "the Calaveras formation was subjected to a metamorphic action that was not experienced by the Mariposa rocks." It is believed, therefore, that folding of considerable magnitude intervened between the deposition of the formations found at Alleghany and that of the Sailor Canyon and Mariposa formations.

On the basis of comparison of the degree of metamorphism and distortion of the rocks of the Calaveras formation with those of the Mariposa slate, it is thought probable that so far as movement of the earth's crust alone, as distinguished from igneous intrusion, is concerned, the folding during this period may have been of greater magnitude than that which took place at the end of the Jurassic. This folding, which must have occurred near the end of the Car-

16 Lindgren, Waldemar, op. cit. (Colfax folio), p. 2.

17 Idem, p. 3.
18 Knopf, Adolph, op. cit., p. 11.

boniferous or in the early part of the Triassic period, may have been contemporaneous with the Appalachian revolution.

As the southward extensions of the gabbro and serpentine of Alleghany cut the Mariposa slate, their intrusion could not have dated from this early period of folding. In the conglomerates of the Mariposa formation, however, there are pebbles of igneous rocks. The source of these pebbles is unknown, but their presence suggests that some igneous activity, though certainly minor in amount as compared to the later intrusions, accompanied this early folding and metamorphism.

The Mariposa formation contains irregular conglomerates for which torrential and glacial origin has been suggested.[19] At Colfax a conglomerate of the Mariposa formation rests on a glaciated surface.[20] Knopf[21] has also found a probable tillite in the Mariposa formation in the Mother Lode region. The presence of such conglomerates indicates a land area, presumably an upland, and suggests unstable crustal conditions premonitory of the great revolution soon to follow.

The Sailor Canyon and Mariposa formations are themselves highly folded, and this folding took place before the great intrusions of igneous rocks reached their present positions. The strike, but not the dip, of the Mariposa formation is on the whole accordant with that of the older sediments, suggesting that the later diastrophism followed the same general lines as the earlier. The geologic maps of the Bidwell Bar, Downieville, Nevada City, and Colfax folios indicate that the post-Calaveras sediments occur in two parallel belts, each striking a few degrees west of north, about equidistant east and west from Alleghany. It is perhaps a fair inference that these belts represent the major synclines of the post-Jurassic folding and that the wide strip of older rocks between these belts represents a major anticline.

Faulting may have accompanied the folding of both periods, but the evidence is obscured by the later faulting, subsequent to the intrusions of the basic and silicic rocks.

The next event in the history of the region was the intrusion of the basic rocks, the gabbro and the rocks now altered to serpentine. These are roughly, though not in detail, accordant with the strike of the older rocks. The belt of basic intrusions of the Alleghany district is a small section of a great zone of similar intrusions extending for many miles, through the Placerville, Colfax, and Downieville quadrangles. Lindgren[22] refers to this as " the great serpentine belt."

In both the Alleghany district and the Mother Lode region this belt of serpentine intrusions is bordered on the east by a fault, and it seems likely that a major fault existed prior to the intrusion of the serpentine and determined its position.[23]

After the intrusion of the basic rocks there must have been an interval with continued earth movements before the intrusion of the great masses of granitic rocks, for the schistosity of the gabbro in the Alleghany district was developed prior to the granite intrusion.

The main granitic batholith of the Sierra Nevada lies to the east of the Alleghany district and subordinate satellitic batholiths to the west. (See pl. 3.) If, as suggested above, the absence of Jurassic rocks from the belt inclosing the Alleghany district indicates the position of a major anticline, the great intrusions of granitic rocks, which cut the folds, took place along the flanks of this anticline and within the synclinal areas rather than beneath its crest.

If the intrusion of the granite and allied rock is taken as marking the end of the Jurassic period, it is logical to consider that the folding of the Mariposa slate and the intrusion of the basic rocks took place within Upper Jurassic time.

FAULTING

The last major event in the pre-Tertiary history of the district was the production of a complicated fault pattern and the introduction of quartz veins along the planes of the faults. It is probable that faulting also took place during and after earlier diastrophic periods, but the evidence is concealed by renewed movement after the last great intrusions.

Only one of the faults, that which borders the Blue Canyon formation in the eastern part of the Alleghany district, is a major structural feature. This has an average strike of about N. 20° W. and a dip of 80° E. to 90°. What appears to be the same fault forms the western boundary of the Blue Canyon formation in the valley of the South Fork of the Yuba River. In the valley of the North Fork of the Yuba the Tightner formation crops out west of Downieville and the Blue Canyon formation to the east, and the nearly straight southerly course of the North Fork of the North Fork suggests that this fault may be traceable for some distance farther north. As the same formations were not found on opposite sides of this fault its throw is not determinable. The presence of the supposedly older formation on the hanging-wall side suggests that the vertical component of movement was in the reverse direction. Although some movement along the fault has taken place since the intrusion of the serpentine, the fact that for at least 10

[19] Moody, C. L., The breccias of the Mariposa formation in the vicinity of Colfax, Calif.: California Univ. Dept. Geology Bull., vol. 10, pp. 383–420, 1917.
[20] Lawson, A. C., personal communication.
[21] Knopf, Adolph, op. cit., p. 14.
[22] Lindgren, Waldemar, op. cit. (Colfax folio), p. 3.

[23] Benson, W. N., The tectonic conditions accompanying the intrusion of basic and ultrabasic igneous rocks: Nat. Acad. Sci. Mem., vol. 19, pp. 75–76, 1927.

miles it forms the eastern border of the great serpentine belt suggests the possibility that it may have been in existence prior to the intrusion of the basic rocks and may have determined the position of the basic intrusions.

Knopf[24] has shown that a major fault borders the Mother Lode region on the east in a manner similar to the eastern fault at Alleghany. Definite evidence of such a fault was found by him at Plymouth and Jackson, in the northern part of the Mother Lode region, and at the Eagle Shawmut mine farther south. Plymouth is 60 miles south of the South Fork of the Yuba River, and no such fault has been recognized in the intervening territory, but the agreement in strike and the persistence of the elongate intrusions of serpentine along a line connecting the two are suggestive. (See pl. 6.) A hasty examination along the railroad east of Towle showed nothing incompatible with the presence of such a fault. It is possible, therefore, that a major reverse fault with easterly dip extends along practically the entire California gold belt and forms the eastern limit of the productive portion of this belt.

The fissure system followed by the Brush Creek and other veins, which extends at least from the North Fork of the Yuba River at Goodyears Creek to the South Fork of the Yuba River, a distance of about 15 miles, forms a symmetrical break along the western border of the serpentine belt. There is no evidence of notable faulting along this system, but inasmuch as the Relief and Cape Horn formations are not in contact in the Alleghany district, it is possible that the western serpentine mass was intruded along a fault.

An alternative explanation of the structure of the district is therefore possible. The basal conglomerate or tillite of the Kanaka formation resembles, except for its greater degree of metamorphism, the tillite of the Mariposa formation as exposed in the cut near the Colfax railroad station. No fossils were found in the Kanaka formation, and thus its age is determinable only by its relations to the underlying and overlying beds. It is conceivable, therefore, that the Kanaka and Relief formations as mapped in the Alleghany district are equivalent to a part of the Mariposa slate and that the serpentine in the western part of the district was intruded along a major fault separating these formations from the Cape Horn formation. This explanation is, however, based on nothing more substantial than the possible identity of the two supposed tillites, and it involves difficulties when applied to a larger area. The geologic map of the Colfax quadrangle shows the Relief and Cape Horn formations in contact for a distance of 12 miles, southward from Relief Hill, without any intervening serpentine;

therefore on the foregoing assumption either a fault contact between these formations has escaped observation or the present authors are in error in assigning the quartzitic sediments east of the Cape Horn formation in the Alleghany district to the Relief quartzite.

The other faults of the district are of minor importance as structural elements but are economically important because they are the sites of the principal veins of the district. A description of the vein system is given on pages 28–32, and the speculation as to the origin of the fissure system is included with other material bearing on the origin of the ore deposits (pp. 70–72), but in order to give a complete summary of the geologic history the salient structural features will be here briefly outlined.

There are three types of fissures followed by the veins. One group has in general northwesterly strikes and easterly dips that are commonly less than 50°; along these fissures there has been reverse movement, and the maximum displacement where determinable is about 900 feet in the direction of the dip, but the average displacement is much smaller. On the hanging-wall side of the principal fissure are fissures which join it but along which the displacement appears to have been a few feet in the normal direction. The fissures of the third group strike a few degrees west of north and have steep westerly dips, generally over 60°. These are reverse faults and displace the eastward dipping reverse faults, but the displacement is probably smaller; where measurable it is less than 250 feet.

As all these faults carry quartz, movement on both sets of fissures must have preceded mineralization. The position of the major fissures relative to the fault along the eastern edge of the district suggests that these fractures may be auxiliary to the larger fault.

Movement continued during and after the deposition of the quartz, but such movement was probably small in amount. Possibly this may have been in part caused by increase in volume during serpentinization.

TERTIARY HISTORY

During Cretaceous time and most of Eocene time the district was undergoing erosion. It is estimated (see pp. 61–70) that the amount of cover removed in the Alleghany district during these periods was at least 10,000 feet and possibly much more.

Near the end of the Eocene, in the stage represented by the earliest auriferous gravel, the land was reduced to a condition approaching topographic old age. Prior to this, however, probably several erosion cycles had run at least partial courses. A condition approaching peneplanation may have existed at the end of the Cretaceous period, and possibly an intermediate erosion

---

[24] Knopf, Adolph, op. cit., p. 46.

stage, as far advanced as that during which the auriferous gravel was laid down, was developed early in Eocene time.

Near the end of the Eocene epoch began the volcanic eruptions, which continued intermittently until the Pleistocene. The only trace of the earliest rhyolitic eruptions in the Alleghany district is in the stream-laid material forming the " pipe clay " above the auriferous gravel at Alleghany and Minnesota, yet the area must have been at one time covered with lava or tuff, for the canyon of the intervolcanic stream of Miocene age indicates a superposed stream.

The Miocene intervolcanic gravel affords evidence of an interlude in the volcanic activity. This was followed by the great outflows of andesitic breccia, which not only filled the valleys but probably spread out over the whole area of the Alleghany district. Although in places these breccias are as much as 1,000 feet thick, the original thickness must have been even greater, for they are easily eroded and were exposed to erosion for a considerable interval prior to the last volcanic activity, which is represented by the basalt of Bald Mountain.

The only flows of effusive basalt in the Alleghany area are those that now form Bald Mountain, but the small dikes of basalt found here and there suggest that a group of basaltic volcanoes may have once been active in the district. The basalt intrusion at the Plumbago mine marks the neck of such a volcano.

The major Miocene streams which yielded the intervolcanic gravel had, like the earlier rivers of the auriferous-gravel epoch, general westerly courses determined by the westward tilt of the Sierra Nevada. Local irregularities of a lava-covered surface, however, may have determined the courses of the tributary valleys, such as those which cross the Alleghany district, for, although these valleys are cut in the rocks of the " Bedrock series," they are independent of the rock structure. Similarly there must have been a tendency for the Pleistocene basalt to flow down the pre-existing valleys. If so, the position of the basalt on Bald Mountain at altitudes between 5,500 and 5,600 feet may represent a minimum level for the position of the top of the andesite breccia at about the beginning of the Pleistocene epoch. Lindgren [25] observes that erosion of the andesite was well advanced before the basalt eruptions. To what extent this altitude is a measure of Pleistocene uplift is uncertain, but if, as seems probable,[26] a postmature topography had been developed at about the end of the Pliocene epoch, the early Pleistocene uplift at Alleghany must have been of the magnitude of at least 2,000 or 3,000 feet, large in itself but smaller for this region than for the crest

of the Sierra Nevada at this latitude, as the uplift involved a tilt of the block toward the west. Also the uplift is less in this portion of the Sierra as a whole than farther to the south.

DEVELOPMENT OF THE PRESENT TOPOGRAPHY

The story told by the hard-rock formations of the district ends with the basalt flows of Bald Mountain and the small dikes and plugs of Lafayette Ridge and the Plumbago mine. The more recent history is to be read by inferences drawn from the forms of the surface features.

The limit to which the valley glaciers from the high Sierra extended is far to the east of the Alleghany district, so that glaciation does not enter directly into the physiographic history of the district, but the much greater volume of water flowing down the Middle Fork of the Yuba River during periods of the melting of the glaciers may account in part for the greater deepening of its canyon relative to Kanaka and Oregon Creeks, which do not reach the glaciated region.

A study of the physiographic history of the Alleghany district, to be of value, should include observations over a much larger area than that studied in this investigation, and it should also include a more detailed study of the later features in the district itself than the authors were able to give. A few observations made in connection with the bedrock study, however, may be worth noting. The present drainage, which is inherited from a period when the entire surface was covered with lava, is naturally independent of the rock structure and owes its general southwesterly course to the tilt of the great Sierra fault block. Locally the courses of the streams were no doubt influenced and in places greatly changed by the basalt eruptions. There is a suggestion in the course of Brush Creek and the continuation of the same course in the valleys of Woodruff and Goodyear Creeks, in the Downieville quadrangle, that early in the physiographic history of the district there was an important north-south line of drainage, now captured by the westward-flowing streams. The southward-flowing Brush and Goodyears Creeks have relatively broad valleys and gentle grades, in contrast with the steep grade of the northward-flowing Woodruff Creek.

Throughout the district there is a distinct tendency toward gentler slopes above an altitude of about 4,000 feet. This is well marked in the valleys of Oregon Creek and the North Fork of Kanaka Creek but is only faintly seen in the valleys of Kanaka Creek and the Middle Fork of the Yuba River. The greater maturity of topographic expression is more pronounced where the streams cut the easily eroded andesitic breccia but is also apparent below the lava in the valleys of Brush Creek and the North Fork of Kanaka Creek and on the higher slopes of the valley

[25] Lindgren, Waldemar, op. cit. (Colfax folio), p. 6.
[26] Matthes, F. E., Geologic history of the Yosemite Valley: U. S. Geol. Survey Prof. Paper 160, p. 29, 1930.

of Kanaka Creek west of Wet Ravine. In several of the valleys stretches of fairly gentle gradient are separated by short stretches of steeper grades. As this steepening is independent of the rock structure it must indicate a relatively recent stimulation of erosion. Wet Ravine shows a notable steepening of grade between the Oriental tunnel and a point a short distance above the Foote road, with its upper reaches again at a gentle grade, and the North Fork of Kanaka Creek displays a similar steepening near the Osceola mine. At about the same altitude on Kanaka Creek there is a stretch of steeper grade between the Golden King mine and the forks west of the Kinselbach mine. The new erosion has not yet entered the portion of Oregon Creek lying within the district studied, but the Colfax topographic map indicates a stretch of steep grade at a point about 4 miles west of the mouth of Brush Creek. On the Middle Fork of the Yuba River the change in grade is far to the east of the Alleghany district. The Downieville topographic map indicates a marked steepening in grade near Milton, at about the western limit of glaciation.

It is thought that this stimulation of erosion must be due more to renewal of uplift of the great Sierra fault block than to the increased volume of water furnished to the streams at the end of the glacial epoch, for Oregon and Kanaka Creeks not only did not reach the glaciated region but with their small drainage areas could hardly have received any very great increment of water from this source.

A minor feature of the topography not well brought out even in so detailed a topographic map as that of the Alleghany district is the fairly constant presence of a distinct bench, from a few feet to 200 or 300 feet in width, along the base of the andesitic breccia. (See pl. 7, A.) This bench is followed by roads in many parts of the district—for instance, parts of the main road between Forest and Mountain House and the branch that swings up the valley of Little Sandusky Creek, the road along the west side of Wet Ravine, and parts of the road following the North Fork of Kanaka Creek, north of Alleghany. This feature is believed to be due to the greater ease of erosion of the soft andesitic breccia than of the underlying rock, even though this is in places deeply weathered, so that the breccia retreats faster than the streams and rills can establish the same grade in it as in the rocks of the bedrock formations.

Part 2. ORE DEPOSITS

HISTORY AND PRODUCTION

Gold is obtained in the Alleghany district from both placer and lode deposits. In the early days of gold mining the placers were most productive. First the gravel of the present streams and low-level benches was washed. Then the Tertiary gravel of the upland ridges was followed by drifting beneath the lava cap and later hydraulicked wherever the situation was favorable. In recent years the production from placer mining has decreased to almost nothing, and there seems little prospect of any great increase from the placers of the Alleghany district, even if there were to be a resumption of hydraulic mining. Lode mining, on the other hand, although there have been several periods of stagnation, has increased in importance, and there seems good reason to expect a large production from the district for many years to come, though owing to the fact that at Alleghany the gold occurs in small shoots of exceedingly rich ore, the output may be expected to vary greatly from year to year.

The stream gravel worked at first was undoubtedly very rich, as it derived gold both from the lodes and from the Tertiary gravel. No record exists of this early work. It was undoubtedly extensive, as the old "dry walls," made by the miners in removing boulders from the stream beds, still exist along the banks of even the smallest streams.

It was not long, however, before the mining of the Tertiary gravel deposits was begun. Those at Minnesota, near the present Irelan mine, were worked as early as 1852, and by the next year 400 miners were at work there.[27] At about the same time, or slightly later, the deposits of gravel at Chips Flat, Balsam Flat, Alleghany, and Forest were worked. Certainly all were being mined by 1854.[28]

The available water in Kanaka and Oregon Creeks at the required altitude was not sufficient for effective hydraulic mining, and it was not until a short time before the enforced stoppage of hydraulic mining in 1884 that ditches from the Yuba River had been constructed and large-scale hydraulic operations begun.

The gravel deposits available for hydraulic mining were less extensive though probably much richer than those of the main Tertiary channel south of the present Middle Fork of the Yuba River; consequently the greatest production was won from drift mines which followed the auriferous channels beneath the lava cap. According to MacBoyle[29] drift mining at the Uncle Sam mine, 2½ miles north of Alleghany, began in 1850. Tunnels followed not only the major channels between Minnesota and Chips Flat, between Alleghany and Forest, and north of Forest, but various minor tributary channels and the channels of intervolcanic age. Apparently the crest of production was reached at Alleghany about 1861, though the drift mining in the area north of Forest was extremely profitable at a later date. In 1858 there were 18 tunnel companies in profitable operation at Alleghany, but by 1868 only 25 men were at work.[30] The magnitude of operations is shown by the fact that there were formerly underground connections by means of these drift tunnels between the towns of Alleghany and Forest and between Minnesota and Chips Flat. According to Browne,[31] drift mines at Alleghany had produced $400,000 to 1867; the Live Yankee mine, near Forest, between 1854 and 1863 produced $713,777, of which $336,459 was paid in dividends; and the Highland and Masonic mine, working a stretch of the channel between Forest and Alleghany, had produced over $300,000 prior to 1867.

The development of the more deeply buried and less accessible gravel deposits north of Forest was somewhat later. In 1874 Forest was said to be "one of the liveliest and most prosperous mineral sections of California."[32] The area of gravel available for drift mining in the vicinity of Forest was much greater than that at Alleghany, and work continued to be active there for a much longer time and has not entirely ceased even to-day. The most famous of the drift mines, the Bald Mountain, began operations in 1872 with an original expenditure for opening the mine of $20,000, and from that time to 1887 it produced $3,100,000, of which $1,300,000 was paid out in dividends.[33] Most of this production was made prior to 1881.[34] Other drift mines in this region produced notable amounts, but none approached the record of the Bald Mountain. The greatest period of drift mining had passed by 1883, and in 1888, the first year for which production figures are available, the output

---

[27] Browne, J. R., Mineral resources of the States and Territories west of the Rocky Mountains for 1867, p. 139, 1868.

[28] Trask, J. B., Report on the geology of the coast mountains and part of the Sierra Nevada, p. 64, Sacramento, 1854.

[29] MacBoyle, Errol, Mines and mineral resources of Sierra County. p. 60, California State Min. Bur., 1920.

[30] Browne, J. R., op. cit., p. 139.

[31] Idem, pp. 139–140.

[32] Min. and Sci. Press, vol. 19, p. 130, Aug. 29, 1874.

[33] MacBoyle, Errol, op. cit., p. 31.

[34] Min. and Sci. Press, vol. 40, p. 209, Apr. 2, 1881. Burchard, H. C., Report of the Director of the Mint on production of precious metals for 1882, p. 120, 1883.

from this source was only a fraction of its former size. The writer believes that $10,000,000 is a fair estimate for gold won from drift mining in the entire district.

The amounts of gold recovered by hydraulic mining and from the gravel of the present streams are more difficult to estimate, as there are not even fragmentary data such as are available for the drift-mining operations. Presumably, as the area hydraulicked was small, the production was much less, possibly between $2,000,000 and $4,000,000. The production from the old surface placers of Oregon Creek, Kanaka Creek, and the smaller ravines of the area is unknown, but it

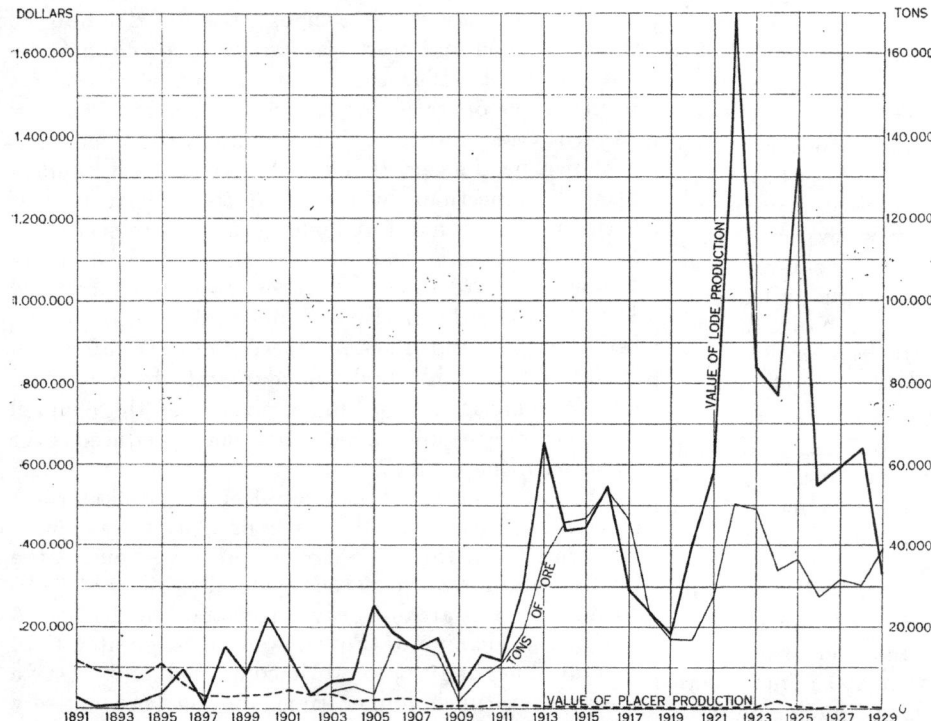

FIGURE 5.—Gold produced from lode and placer mines of the Alleghany district, 1891–1927, and ore mined, 1903–1927

may have been as great as that of the drift mines, and the gravel of the Middle Fork of the Yuba River at the southern edge of the district must have yielded a large amount of gold. Even at the present time in periods of low water " snipers " pan the material from the crevices on bedrock.

Lode mining seems to have begun in 1853 at the German Bar mine, on the south bank of the Middle Fork of the Yuba River,[35] and the Irelan mine is said to have been located in the same year.[36] Veins carrying arsenopyrite rich in free gold were discovered in the valley of Oregon Creek prior to 1854.[37] A

stamp mill was in operation at the Rainbow mine in 1858.[38] The Brush Creek, Docile, Twenty-one, Independence, Oriental, Plumbago, and Gold Canyon mines were all worked before 1870. No data as to the early lode production are available. The output was probably not large, yet it must have been of fair size to divert the miners even to a small extent from the rich placer deposits. The first definite figure for production is $95,000 for the Brush Creek mine in the year ending June 1, 1870.[39] By 1875 the Brush Creek mine is said to have produced $1,000,000 to a depth of 520 feet; and from 1872 to 1875 the Plumbago mine produced $100,000, all from hand mortars.[40] The lodes worked in this period, with the exception of the Rainbow, were those that were exposed at the present surface. According to local tradition, which seems reasonable, the drift miners, familiar with the tendency of the placer gold to work its way into crevices of the weathered bedrock, regarded the gold of the outcrop of the veins as of the same origin and took out only the upper 2 feet or so. It is probable, moreover, that few quartz veins were discovered in drifts actually following the buried channels of the Eocene auriferous gravel, because, owing to the subdued topography prevailing at the time of deposition of the gravel, the quartz outcrops would tend to form low ridges and would rarely be cut by the channels of the Eocene streams. This would less likely be true of the intervolcanic gravel, for it was deposited by superposed streams and hence was independent of the bedrock structure.

More favorable opportunities for discoveries beneath the lava were offered by tunnels that crosscut the upper part of the bedrock in order to reach gravel channels. The first recorded discovery of a vein beneath the lava cap was that of the Rainbow mine in 1858.[41] This vein was cut in the gravel tunnel 2,000 feet from the portal, and an incline was sunk on the vein. Some rich ore was obtained, but there was trouble with water and the work was soon abandoned. Apparently

[35] Trask, J. B., Report on the geology of northern and southern California, p. 63, Sacramento, 1856.
[36] Eng. and Min. Jour., vol. 87, p. 1301, 1909.
[37] Trask, J. B., Report on the geology of the coast mountains and part of the Sierra Nevada, p. 63, Sacramento, 1854.

[38] Browne, J. R., op. cit., p. 138.
[39] Raymond, R. W., Mineral resources of the States and Territories west of the Rocky Mountains for 1870, p. 45, 1871.
[40] Min. and Sci. Press, vol. 31, p. 378, 1875.
[41] Browne, J. R., op. cit., p. 147.

A. NORTH SIDE OF VALLEY OF KANAKA CREEK LOOKING TOWARD SIXTEEN TO ONE MILL, SHOWING SHOULDER BELOW LAVA

B. OUTCROP OF VEIN ON FOOTWALL SIDE OF SIXTEEN TO ONE VEIN

In road cut near mine office. Drag of the slates on hanging-wall side indicates reverse movement.

C. JUNCTION OF SIXTEEN TO ONE VEIN AND VEIN ALONG STEEP FAULT, 1,800-FOOT LEVEL

Shows contemporaneity of quartz along both fissure systems.

some further work was done later, for Burchard [42] in 1883 refers to the mine as having been worked for many years. In the later part of 1881 high-grade ore was discovered not far below the lava, and the shoot yielded $350,000 to 1884.[43] It is said that $60,000 was taken out in a single day and $100,000 in a single month.[44] A single slab of gold-bearing quartz calculated to contain $20,468 in gold was exhibited in San Francisco in 1882.[45]

The next discovery of ore beneath the lava seems to have been that of the North Fork mine in 1875.[46] The direction of a bedrock tunnel had to be changed owing to swelling ground encountered in a belt of serpentine, and as a consequence a vein was struck which was explored for a distance of 800 feet and a rich shoot was discovered. The gold occurred in association with arsenopyrite, pyrite, and some galena. The oxidized outcrop was so decomposed that the material taken out could be sluiced, but apparently the oxidation did not extend to any appreciable depth. The harder ore was so rich that two men working with hand mortars pounded out $8,000 in gold in a week, and on one occasion over $20,000 is said to have been recovered in a single day's work.[47] Some sinking was done on the ledge in 1877, but the exploration of the vein was soon abandoned. The amount taken from the rich shoot is variously given as $100,000 and $40,000.[48] Of late years there have been many attempts to recover veins cut in old gravel tunnels and reported to have contained high-grade ore. The single outstanding success was the rediscovery of the Tightner vein in 1907 by H. L. Johnson. Other veins discovered in tunnels run for gravel on which there has been more or less exploration in recent years are those of the Red Star, Dead River, North Fork, South Fork, and Mugwump mines. The discoveries of rich ore in the Tightner and Sixteen to One caused renewed activity throughout the camp, old mines were reopened, and since 1912 production has continued on a greatly increased scale.

In spite of the richness of the ore shoots mined, lode mining has proceeded in rather desultory fashion, and several mines have been worked for short periods and then closed down, to be later reopened when an important "strike" gave new impetus to mining. This seems to have been in large part due to the nature of the deposits. A shoot of enormously rich high-grade ore would occasionally be encountered, and in the period of flush production that followed it proved impossible for the owners, for the most part men of small means, to refrain from "cashing in" a large part of the proceeds. Consequently after the shoot was exhausted the mine was left with a surplus inadequate for the amount of unproductive exploration necessary to discover the next shoot. Therefore, except for a few mines which have operated fairly consistently, the history of the district has been one of marked ups and downs. The production recorded, beginning in 1891, shows an output to 1930 of more than $13,000,000. For the years prior to 1891 the only data are scattered references in technical publications, principally the early issues of the Mining and Scientific Press, the reports of the Director of the Mint, and official publications of the State of California. These publications have been used in compiling notes on the history of the individual mines and, combined with such local information as appears to be reliable, indicate that the total lode production of the district exceeds $20,000,000.

The following table, compiled from the records of the San Francisco office of the Bureau of Mines, through the courtesy of J. M. Hill, V. C. Heikes, and Miss Helen M. Gaylord, shows the production of the district as far as records are available. The same data are presented graphically in Figure 5.

Gold produced in the Alleghany district

[No records for 1889 and 1890]

Year	Lode mines		Placer mines
	Ore (tons)	Value	
1888			$114,808
1891		$25,570	117,074
1892		1,916	92,532
1893		5,914	81,726
1894		17,500	75,018
1895		36,238	110,748
1896		98,526	60,604
1897		8,530	30,535
1898		154,837	28,800
1899		81,920	31,471
1900		227,066	33,713
1901		132,111	46,795
1902		31,200	31,427
1903	4,282	68,135	35,844
1904	5,520	72,348	18,841
1905	3,508	255,455	21,693
1906	16,442	182,873	24,762
1907	15,130	148,184	22,258
1908	13,626	174,583	3,590
1909	1,523	41,792	4,165
1910	7,587	137,898	3,596
1911	10,975	118,700	11,011
1912	18,442	295,509	8,557
1913	37,030	652,424	4,747
1914	45,903	436,466	2,464
1915	46,731	443,511	181
1916	54,082	551,391	1,751
1917	46,801	293,254	830
1918	23,271	240,039	1,084
1919	17,119	185,646	1,167
1920	16,992	400,555	0
1921	27,100	578,070	1,393
1922	50,231	1,710,521	1,587

[42] Burchard, H. C., Report of the Director of the Mint on production of precious metals for 1882, p. 121, 1883.

[43] Idem for 1884, p. 153, 1885.

[44] Idem for 1883, p. 216, 1884.

[45] Hanks, H. G., Second report of the State mineralogist, p. 149, Sacramento, 1882.

[46] Min. and Sci. Press, vol. 31, p. 377, 1875.

[47] Min. and Sci. Press, vol. 34, p. 319, May 19, 1877.

[48] Idem, p. 309; vol. 35, p. 86, Aug. 11, 1877. Information received from residents of the district indicates that the shoot produced about $100,000 but that only $40,000 was sent to the mint; the remainder was used locally as bullion in payment of debts of the company.

Gold produced in the Alleghany district—Continued

| Year | Lode mines | | Placer mines |
	Ore (tons)	Value	
1923	49, 262	$839, 225	$1, 946
1924	33, 848	770, 766	18, 311
1925	36, 845	1, 344, 178	3, 741
1926	27, 572	548, 424	0
1927	31, 614	595, 862	2, 278
1928	30, 240	634, 341	3, 940
1929	38, 919	327, 553	2, 534
1930	44, 717	507, 189	1, 022
		13, 376, 220	1, 058, 544

The total for lode production is certainly too low by an unknown amount, possibly of the magnitude of 5 or 10 per cent, for a district producing ore of this type is always cursed with the "high grader," and in the early years, before methods of collecting statistics had reached their present degree of accuracy, there were undoubtedly many small mines from which production reports were not received.

The ratio of value of gold produced ($12,554,892) to tonnage of quartz mined from 1903 (755,312 tons) is about $16.60 to the ton, but this figure is merely the average figure for the total amount of nearly barren quartz mined in the course of exploration and the production derived from small shoots of high grade. It does indicate, however, that on the average, in spite of the smallness of the ore shoots, the returns from lode mining have been higher than in most other California districts. The graph (fig. 5) shows a distinct lag of the curve for tonnage of quartz mined behind that of production. This is believed to reflect the spasmodic increase in exploration that follows discovery of high-grade ore in any one mine.

THE VEINS

SCOPE OF DESCRIPTIONS

It is proposed in this section to describe the principal features of the veins in the Alleghany district, their position, their relation to the different kinds of country rock, and their principal characteristics. As far as possible the material here presented will be descriptive, and the deductions that may be drawn as to the origin both of the fissure system and of the veins that follow the fissures will be reserved for the general discussion of the origin of the deposits (pp. 70–72).

As the region is one of considerable relief and most of the veins have low dips, the trace of the outcrops as plotted on the geologic map does not give a true idea of the distribution and general relations of the veins; therefore they are shown on Plate 8 projected to a horizontal datum plane. The plane of the 3,700-foot contour has been selected for the reason that the mine workings near this altitude are the most extensive, and hence less error is involved in the projection, for the dips of the veins as well as the strikes vary greatly from place to place. In only two mines, the Gold Canyon and German Bar, are the principal workings below an altitude of 3,700 feet.

Where the data are derived from mapped drifts differing less than 100 feet in altitude from the 3,700-foot plane the projected veins are shown by full lines, and only these should be considered to be more than approximately correct in position. Veins developed by drifts but at altitudes differing more than 100 feet from the datum plane are indicated by dashed lines. The distance involved in any particular projection and the consequent probable degree of accuracy can be ascertained by comparison with Plate 9, on which the approximate altitudes of the principal mine workings are shown. Veins in which the data for projection are derived from outcrops and crosscuts are shown by dotted lines, and their positions on the projection should be considered as only approximate, particularly where a difference in altitude exceeding 200 feet is involved. To emphasize the distribution of the veins belonging to the different systems those with easterly dips (generally less than 45° except for the veins following the fault along the eastern boundary of the district) are shown in red and those with westerly dips (exceeding 60°) in blue.

In the section giving mine descriptions maps of the mines are reproduced, and in the following pages occasional reference is made to these maps. In order to give a comprehensive view of the extent of development in the district the workings of the different mines, in so far as they were accessible and maps of them were available, are shown on Plate 9. The workings have necessarily been generalized to the scale of the map, and for the most part only drifts are shown. All raises and stopes have been omitted, and crosscuts are shown only where necessary to bring out the relations of the workings to the surface. In many of the older mines much work was done which is now inaccessible and for which no maps exist. The workings of the old drift placer mines, as far as maps were available, have been included, more to preserve information that is in danger of being permanently lost than for any bearing it may have on the geology of the veins. Where apparently reliable data as to the location of mining claims could be obtained these have been shown on the map, but no attempt has been made to make this feature complete. Moreover, it was not everywhere possible to obtain a close check between the position of a claim relative to the section corners as shown by patent surveys and the known position of the claim stakes with reference to the topography. Hence the position of the claims as shown on the map should be regarded rather as a guide to their shape and location than as an authoritative statement of their exact position and extent.

DISTRIBUTION

There are two principal series of veins—those with northwesterly to westerly strike and northeasterly to northerly dips of 25° to 40°, and those with northerly and northwesterly strike and westerly dips, commonly close to vertical and everywhere exceeding 60°. A fairly well defined alternation of belts in which one or the other type of vein predominates is evident from Plate 8.

The fault that bounds the district on the east is accompanied by fissures containing quartz veins, and the fault plane itself carries quartz at the Independence mine. At the Kinselbach and Mammoth Springs mines the veins occur within a few feet of the fault and parallel to it in strike and dip. This fault, if present north of the district, in the valley of the North Fork of the Yuba River, does not appear to be accompanied by quartz veins, but the Red Paint mine, near Washington, in the valley of the South Fork of the Yuba (see Colfax folio), is probably along the same fault. The possible continuity of this fault with the fault along the east side of the Mother Lode region has already been suggested.

In the valley of Kanaka Creek, just west of this fault, there is a group of minor veins that do not appear to be continuous over long distances. Exploration has been less thorough here, however, than in other parts of the district. Of these the Docile, Cedar, and Eastern Cross belong to the group with northwesterly strike and gentle northeasterly dips, but the other veins have northerly strikes and steep westerly dips, except the Belmont, which is a vertical vein with nearly westerly strike. The Cedar vein is faulted by a fissure belonging to the steeply dipping group, but the other intersections have not been explored. Possible northward extension of this group is indicated by the westward-dipping veins cut in the South Fork tunnel.

West of this zone of steeply dipping veins lies a belt containing a group of veins with northwesterly strikes and gentle northeasterly dips which have furnished the major part of the production of the Alleghany district. In the valley of the Middle Fork of the Yuba River the veins of this group include the Plumbago, Arcade, Irelan, Oriflamme, Gold Canyon, and German Bar. The apex of the Rainbow vein is beneath the lava of Lafayette Ridge. The Sixteen to One, Twenty-one, Morning Glory, Ophir, Mariposa, Osceola, Colorado, Bullion, and Eldorado crop out in the valley of Kanaka Creek. The structure sections on Plate 10 show the geologic relations of this group between the Sixteen to One and Eldorado veins and across the Plumbago vein. Northward under the lava cap there are other veins with similar strike and dip,

including the Red Star, South Fork, and Amethyst, and an eastward-dipping vein cut in the North Fork crosscut.

As seen on the projection (pl. 8) the veins of this group tend to fork southward, and the belt appears to be wider at the south than in the northern part. The country beneath the lava north of Kanaka Creek, however, has not yet been sufficiently explored to make certain of this inference.

The Plumbago vein, on the east side of this belt, follows a fault whose maximum displacement may be of the magnitude of 400 feet or more, as shown by the presence of the Tightner formation on the footwall in the two upper levels. (See sec. J–J', pl. 10, and pl. 58.)

The southern part of the Eldorado vein is roughly accordant in strike and dip with the Plumbago, whose northernmost known extension is 2,300 feet southeast of the southernmost Eldorado workings. The two veins probably follow the same fissure system, but it is doubtful whether they form parts of a single vein, for, although the southernmost workings on the Eldorado show a strong vein, the northernmost development on the Plumbago indicates a feathering out of that vein toward the north. The displacement on the Eldorado vein (secs. H–H' and I–I', pl. 10) is in the reverse direction, but the amount is much less than on the Plumbago, probably not over 100 feet, and decreases northward. The sharp bend in the Eldorado vein (pls. 8, 55, and 56) coincides with the crossing of one of the small serpentine bands and may be due to the deflection of the original fissure by the serpentine. The direction of later movement, as shown by the striae on the walls (pl. 56), seems to have been normal to the average strike of the vein rather than directly up the dip in the region of the bend.

North of the Eldorado mine the same fissure continues as the Colorado vein. It has been developed in the Yellow Jacket mine but is not as well defined there as farther south.

The Sixteen to One vein (secs. A–A' to G–G', pl. 10) is about 2,000 feet west of the Eldorado and has been explored for a distance of nearly a mile along its strike in the Tightner, Sixteen to One, and Twenty-one mines, now included in the Sixteen to One, and the Rainbow Extension mine, south of Kanaka Creek.

The nature of the movement along the fissure is definitely shown by the thrusting of the Tightner formation over the Kanaka formation and by the displacement of the serpentines. The maximum displacement may be as great as 900 feet (sec. E–E', pl. 10), but this appears to decrease rather abruptly to the south, to perhaps 300 feet along the line of section G–G', and more gradually to the north, to about 500 feet along the line of section A–A'. Data are not adequate for determination of the structure in its southern part, but the relation of the vein to the serpentine in the Rainbow Extension workings implies a decreasing amount of reverse movement.

In the Twenty-one mine and to the south there are two parallel veins, the western known as the Twenty-one vein. A vein on the footwall side of the Sixteen to One vein has been exposed in the road cut near the mine office (pl. 7, *B*) but has not been developed. The distortion of the schistose slate along the wall indicates that the movement along this fissure was also in the reverse direction.

The Rainbow veins beneath the lava south of Kanaka Creek follow the same strike and dip as the Sixteen to One vein but are terminated on the north by the vertical Clinton vein. The projection of the Oriflamme, Irelan, and Arcade veins relative to the Rainbow suggests branching of the fissures toward the south. The two Irelan veins seem to be branches from a single trunk, and the projection indicates the possibility that the fissures which these veins occupy, though not the veins themselves, are continuous with those occupied by the Gold Canyon and German Bar veins. There has been reverse faulting along both the Rainbow and Irelan veins and probably along the German Bar, but the data are not sufficient for determining the amount of throw. At its south end the Gold Canyon vein frays out into the serpentine and apparently displaces the contact slightly, also in the reverse direction. No data are available for a determination of the amount or direction of faulting of the other veins of this group.

Between the Sixteen to One and the Eldorado are other veins with northwesterly strikes and easterly dips. These include the Osceola, Minnie D, Colorado, Bullion, Morning Glory, Eclipse, Ophir, and Mariposa. These appear to follow fissures that branch off from the Sixteen to One fissure on the hanging-wall side, but only the Ophir has been followed to the point of junction. In the neighborhood of the junction of the Ophir and Sixteen to One veins, so far as it has been explored, the weaker Ophir vein swings sharply to the strike and dip of the stronger vein, continues parallel for a short distance, and finally merges into the major vein. (See figs. 8 and 31.) The Osceola and Ophir veins show by the displacement of the serpentine dikes which they cross (secs. A–A' and G–G', pl. 10) that they follow normal faults of small displacement, and it is probable that the same is true of the others of this group.

Besides the eastward-dipping veins there are several minor veins with steep westerly dip between the Sixteen to One and Eldorado. These include a steeply dipping fissure which cuts the Colorado vein in the Colorado Extension workings (sec. G–G', pl. 10), the Lucky Larry vein (sec. B–B'), the vein of the Extension of the Minnie D (secs.

D–D' and E–E'), steeply dipping fissures that fault the Morning Glory vein (secs. C–C', D–D', and E–E'), and those which displace the Sixteen to One vein in the lower levels of that mine. The quartz-bearing faults that cut the Sixteen to One vein in the upper levels have a similar northerly strike and steep westerly dip. South of Kanaka Creek is a small fault with westerly dip in the Rainbow Extension workings. The steeply dipping Clinton vein limits the northwestward extension of the Rainbow vein.

Wherever intersections with eastward-dipping veins and fissures have been explored, as in the Sixteen to One, Morning Glory, and Colorado Extension mines, the westward-dipping fissures are found to be the later, though the quartz filling along both sets of fissures is contemporaneous. The Lucky Larry vein, however, does not appear to displace the Osceola and Sixteen to One veins, but it may be that the displacement of the Morning Glory vein at the tunnel level (sec. C–C') is along the fissure followed by that vein. The displacement along the westward-dipping fissures is measurable only for those which fault the Sixteen to One vein. For the group which crosses the vein in the upper levels the aggregate displacement reaches a maximum of 135 feet (secs. B–B' and C–C') but decreases to about 50 feet to the north and south (secs. A–A' and G–G'). The lower group of westward-dipping faults between the 1,800 and 2,100 foot levels displaces the vein more than 150 feet.

West of the Sixteen to One vein there is a belt about 4,000 feet wide in which the fissures dip steeply to the west. These include the North Fork, Diadem, Mugwump, and Dead River veins, north of Kanaka Creek; the Sherman, Spoohn, Happy Jack, and Wonder, which crop out in the Kanaka Creek Valley; and several small veins in the valley of the Middle Fork of the Yuba River. No data are available for determining the amount of displacement along these fissures.

To the west of this belt of fissures with westerly dip there is a group of veins with gentle dips to the east and north, including the Kenton, Oriental, and Alta, in the valley of Kanaka Creek. These, however, do not show either the persistence or the regularity of the eastward-dipping fissures of the group to the east. The Oriental fissure is a reverse fault with a displacement along the dip of about 200 feet, shown by the offset of the contact of granite and gabbro (pl. 51), but data are lacking for measure of the movement on the others. In the same belt as the Oriental and Kenton, in the valley of Oregon Creek, are the fissures followed by the Tomboy, Eureka, and Evans veins, which have southeasterly and northeasterly dips.

The westernmost vein system, followed by the Brush Creek and Kate Hardy veins, is well defined and shows a greater continuity of quartz on the outcrop than the veins of the central part of the district. The system is traceable well beyond the limits of the map. The adit of the Brush Creek mine is a short distance north of the area mapped, in the valley of Woodruff Creek, and it is said that the vein can be followed northward along the serpentine contact at least as far as Goodyears Bar, on the North Fork of the Yuba River. To the south the same system appears in the valley of Kanaka Creek, where it has been developed in the Gold Dollar and Kanaka mines, southwest of the area mapped. The veins follow closely the western contact of the great serpentine mass but for the most part are within the slates of the Cape Horn formation

places the vein-forming solutions entering along a gently dipping fissure faulted by one of the steeply dipping fissures continued along the steep fissure without entering the faulted segment of the flat fissure. Such a hypothesis would explain the abrupt ending of the Rainbow vein against the Clinton.

RELATION TO COUNTRY ROCK

The veins cross all the pre-Tertiary rocks west of the eastern fault except the larger serpentine masses. The portion of the drainage basin of Kanaka Creek occupied by the Tightner formation contains a greater

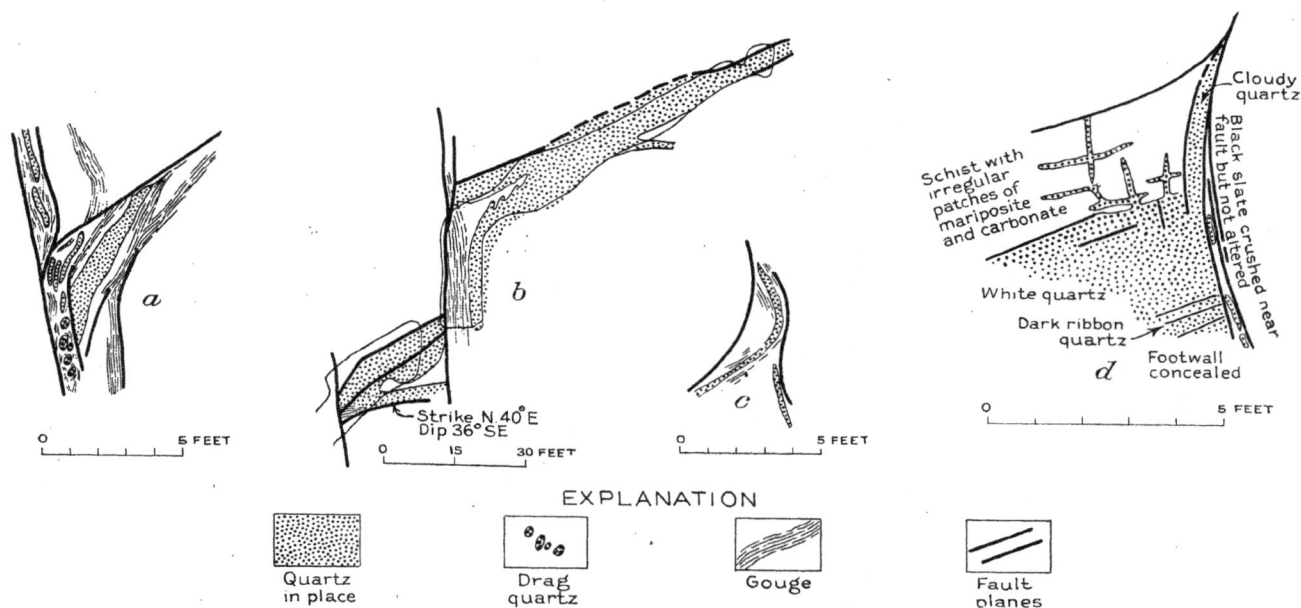

FIGURE 6.—Details of junction of main vein and fault veins, showing contemporaneity of quartz filling, Sixteen to One mine: a, Second sublevel above 250-foot level; b, raise above 250-foot level; c, third sublevel above 250-foot level; d, raise above 300-foot level

rather than directly on the contact. Although the drag of the slates indicates reverse movement, there is no evidence that the veins of this system follow any major fault, but, as suggested on page 22 it may be that the serpentine was intruded along an earlier fault. Later faults without quartz, apparently normal in displacement, cut the Kate Hardy vein. (See pl. 47.)

As will be seen from the foregoing summary there is a rough system of alternate belts containing gently dipping and steeply dipping veins. Wherever the intersections of the fissures have been exposed the gently dipping fissures are faulted by those of steep dip. The maximum known displacement in the lower part of the Sixteen to One vein, 150 feet, is much smaller than that along the earlier gently dipping fissures, for which a maximum displacement of about 900 feet is estimated, but the displacement along most of the steep fissures is unknown.

The quartz that follows the fissures is contemporaneous in both fissure systems. (See pls. 7, C, and 11, B, and fig. 6.) It is possible, therefore, that in

number of veins per unit of area than that occupied by the other formations, but other productive veins, such as the Plumbago, Gold Canyon, and Oriental, lie within the gabbro, and the Kanaka formation forms the walls of the Rainbow vein. The Irelan vein crosses both gabbro and the Kanaka formation. The Brush Creek and Kate Hardy veins lie within the slates of the Cape Horn formation.

The serpentine is the only rock that has exercised a direct effect on the form of the veins. No veins penetrate the larger serpentine masses, but two major veins, the Oriental and Gold Canyon, abut against the serpentine. Close to the contact there is in each vein a marked change in dip and strike, and for a short distance the vein follows and apparently faults the contact of serpentine and gabbro. (See pl. 50; figs. 7 and 46.) Although fissuring along the trend of the vein in places continues within the serpentine for a considerable distance, the vein itself frays out within a few feet into small quartz stringers bordered by serpentine which is replaced by a mixture of carbonate and mariposite. In the Oriental vein there is also a marked feathering of the vein within the gabbro close to the serpentine contact, as shown in Figure 7 and Plate 50. Other veins, including the Plumbago, the Kenton, and several of the westward-dipping veins,

dipping veins without loss of continuity of the veins, but even these small serpentine masses exert a marked influence on the veins. Wherever a well-defined quartz vein is in contact with the serpentine there is evidence of postmineral faulting between the quartz and serpentine; that is, the quartz adjoining the serpentine owes its present position to movement along the plane of the vein. Where the quartz is thus faulted against the serpentine of these small masses the serpentine may show less intense hydrothermal alteration than is found where quartz veins abut against the larger serpentine masses. On the other hand, where the vein is in frozen contact with one of these small serpentine masses the quartz is commonly small in amount or entirely lacking, its place being taken by the mixture of carbonate and mariposite characteristic of alteration of the serpentine close to the vein.

The Sixteen to One vein shows in places a distinct change in dip and strike where it crosses the smaller serpentine dikes, as if the serpentine had caused some deflection of the course of the original fissure, but the position of the serpentine can not account for all the variations. The change in both strike and dip of the Eldorado vein (pls. 55 and 56; sec. I–I', pl. 10) may also be the result of deflection of the fissure by the serpentine. The Osceola and Ophir veins, which branch out from the hanging wall of the Sixteen to One, cross the small serpentine bands with only slight displacement.

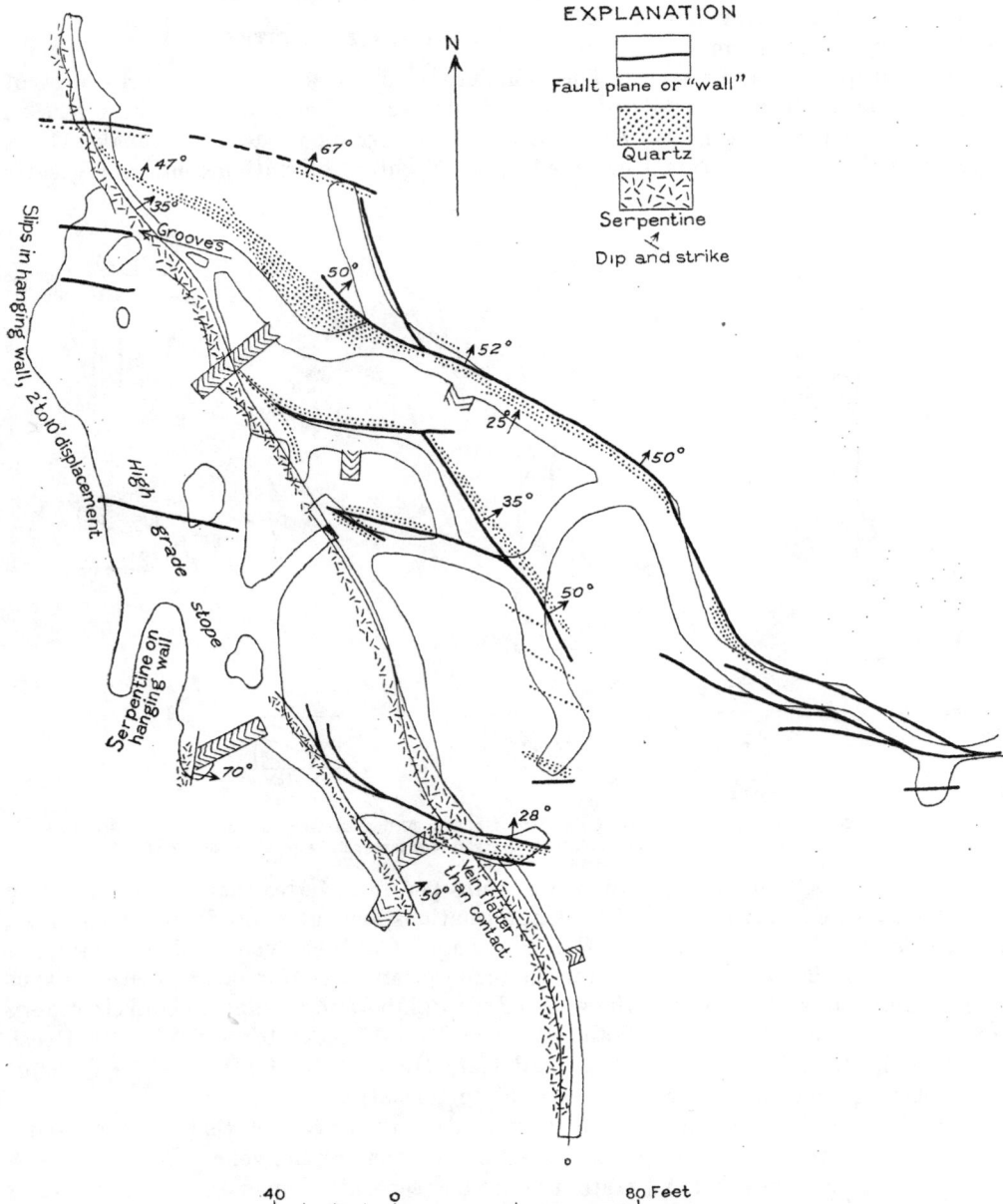

FIGURE 7.—Map of western part of adit level, Oriental mine, showing feathering of vein near serpentine contact

appear to pinch out before reaching the serpentine contact. The Kenton vein (fig. 28) is parallel to the serpentine contact for some distance within the gabbro, and there is a minor vein along the contact of gabbro and serpentine in the Gold Canyon mine (fig. 46).

The smaller serpentine masses in the valley of Kanaka Creek are crossed and displaced by the eastward-

SIZE AND STRUCTURE

In the following paragraphs a short description is given of the major characteristics of the veins. The description of the textural features of both vein and altered country rock will be reserved for the section on mineralogy (pp. 38–52).

The veins of the district are well defined, and the different vein systems cross the entire area covered by the detailed map. The individual veins are less persistent, but a few have been developed over considerable distances. The Sixteen to One vein, for instance, has been developed almost continuously from the southern workings of the Rainbow Extension to the northern part of the Tightner mine, a distance of nearly a mile. Other eastward-dipping veins show probable continuity over comparable distances, though the development on them is less extensive.

The veins likewise vary greatly in thickness. Although in any given level there are parts where little or no quartz is present, corresponding parts of drifts above or below will show quartz, and in most mines the quartz is continuous through the whole developed portion of the vein, though not necessarily so on any single level. On the other hand, there are places, particularly where the vein bows upward or shows a sharp change in strike, where the quartz may attain a considerable thickness. Several veins show such swellings in which the quartz is more than 30 feet thick. Probably 5 or 6 feet is an average thickness for the productive veins, but the variation is so great that any such figure is of little significance. As would be expected in veins in fissures along which there has been reverse movement, there is usually a greater thickness of quartz in portions where the dips are flatter than the average. This relation, however, is by no means universal. Except for the Kate Hardy vein, which is one of the strongest veins of the district, the quartz is less persistent in the steeply dipping veins than in those with gentle dip, and the average thickness of quartz is also much less.

As erosion in the valleys of Kanaka Creek and the Middle Fork of the Yuba River has been rapid, quartz outcrops are not generally conspicuous. In a few veins where the position of the erosion surface coincides with a swelling in the quartz, as in the Osceola vein, there may be a bold outcrop. In the valley of Oregon Creek, which is not yet influenced by the revived erosion, the topography is more mature, and there has been greater opportunity for differential erosion; consequently the quartz veins tend to form more prominent outcrops.

Most of the major veins are somewhat irregular in strike and dip, and in many places the changes are very sharp. In a few places such bends can be correlated with the passage of the vein from one rock to another, particularly where the vein crosses small belts of serpentine or frays out into the larger serpentine masses. The junction of the Ophir and Sixteen to One

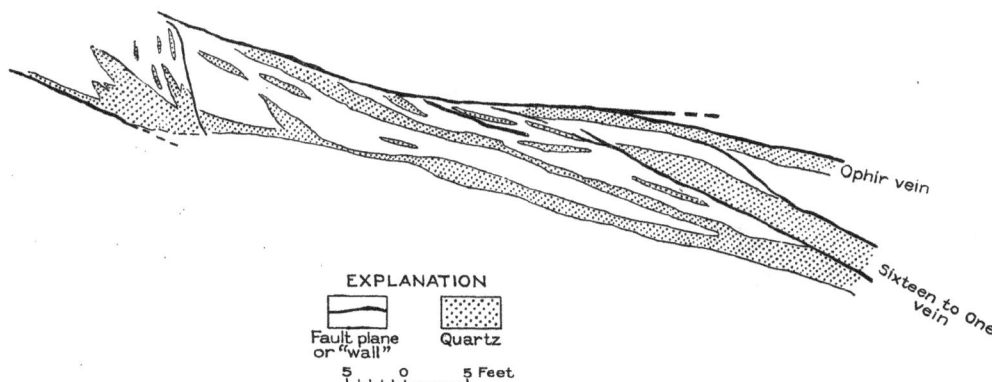

EXPLANATION

Fault plane or "wall"   Quartz

5   0   5 Feet

FIGURE 8.—Section showing detail of junction of Sixteen to One and Ophir veins, Sixteen to One mine

veins (figs. 31 and 32) is the only junction of two productive eastward-dipping veins that has been explored. Here the weaker Ophir vein near the point of junction bends toward parallelism with the major vein and continues for some distance as the hanging-wall strand of the Sixteen to One vein. (See fig. 8.) Apparently the same is true of the junctions of the Sixteen to One with the Eclipse and Morning Glory veins, but these have not been fully explored. Minor branches from the major veins also tend to split off at acute angles. To some extent such junctions control the position of bends in the main vein, as at the junction of the Sixteen to One with the Ophir and Morning Glory. This, of course, does not imply that the junctions caused the bends in the major vein but that bends in the major fissure favored development of subordinate fissures. The faulting of the eastward-dipping veins by those with steep westerly dip does not perceptibly affect either the strike or the dip. The causes of the changes in the course of the veins are as a rule not evident. The steeply dipping veins are distinctly more regular than those with gentle dips.

The Alleghany veins, particularly those with relatively flat dip, rarely show unbroken quartz from wall to wall but are split into two or more "strands" by intervening material. (See pl. 11, A, C.) Such separation of the quartz strands may be original, due to original deposition of the quartz along parallel and closely spaced planes or to successive reopening of the fissure or may be the result of later movement. (See pl. 11, *D*.) The intervening septa may consist of bands of altered country rock several feet thick cut by "crossovers" of quartz. (See pl. 12, *B*.) In places the intervening dark band thins out and two quartz strands coalesce. Where the separation of the quartz strands is due to later movement the septum may consist only of a little gouge along the fault plane, or it may contain later minerals deposited along the fault plane and may cross the vein.

crosscuts commonly show an increasing number of irregular quartz stringers in the country rock toward the vein. (See pl. 13, *A*.) In places (pl. 13, *B*, *C*) there has been prior crushing in the wall rock, and the quartz in the wall rock is in stringers parallel to the major vein; more commonly, however, the stringers in the country rock, both adjoining the vein and at a distance from it (pl. 13, *D*), indicate that at the time of introduction of the quartz the wall rocks were sufficiently massive and unsheared to be capable of breaking irregularly across their cleavage.

Movement during and after the period in which the quartz was deposited is shown not only along the walls of the vein but in the vein itself and also in the adjoining country rock. The boundary between the vein and wall rock is commonly sharply marked by a post-mineral shear and in places also by a breccia of quartz fragments cemented by quartz or carbonates (fig. 9 and pl. 14, *A*, *B*), showing still earlier movement before the completion of mineralization. The later fault planes are locally known as the "walls" of the veins (pl. 15, *A*), and the plane farthest from the vein is commonly considered the "true wall." These "walls" follow the vein for considerable distances and in general tend to wrap around the vein quartz (pls. 15, *B*; 16, *A*), but in many places their strike is at a slight angle to the vein. Under favorable conditions for observation it is possible to trace a given "wall" from the country rock to the vein, where it is deflected along the vein for some distance, then into the vein, where it forms the boundary between two strands of quartz (figs. 10 and 11; pl. 16, *B*), and then to the opposite side, where it is again deflected along the vein, finally passing out into the country rock. As there has been movement along these planes, in general nearly parallel to their dips, one segment of the vein has moved with respect to the other. This has caused thinning of the quartz in some places and piling up of strand upon strand in others, but the great thickenings of the vein, as in the 1,000-foot level of the Tightner mine and the No. 9 level of the Plumbago, seem to be due chiefly to original deposition of a greater thickness of quartz.

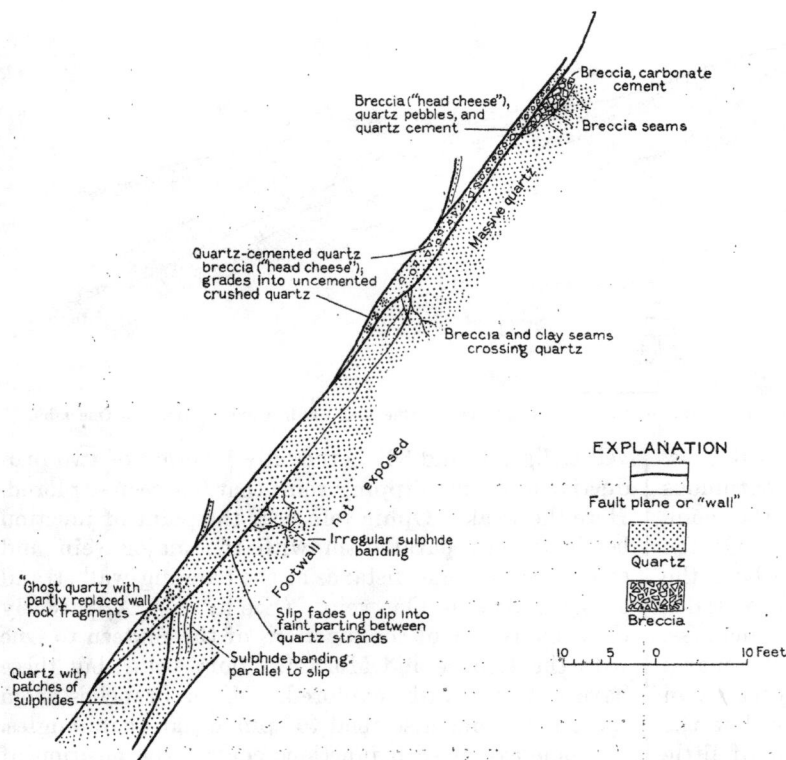

FIGURE 9.—Section along vein in Eldorado mine, on raise above adit level

As later movement within the vein has most readily taken place along the softer country rock between the quartz strands, gradations between the two kinds of septa are common. At one place the quartz may be frozen to the altered country rock that intervenes between the two strands, whereas a short distance away the country rock may be sheared to a gouge as the result of movement between the strands of quartz.

Later movement approximately parallel to the walls of the veins has largely obliterated the original contact of vein and country rock, but where the original wall is preserved there are commonly found a large number of quartz stringers ramifying irregularly through the country rock for a distance of several feet from the original wall. (See pl. 12, *A*, *C*.) Similarly

Of course the shearing is rarely as simple as outlined in the preceding paragraphs. Splits in the fault planes and irregular wedges of quartz and crushed country rock are frequently encountered. In places spear points of the more solid quartz have been driven into the country rock. (See pl. 17, *A*, *B*.) Also the

A. VEIN WITH SEPTA OF ALTERED COUNTRY ROCK SEPARATING QUARTZ STRANDS CUT BY LATER TRANSVERSE VEINLETS, NORTH FORK MINE, THIRD LEVEL

B. JUNCTION OF MAIN VEIN AND FAULT VEIN, SIXTEEN TO ONE MINE, STOPE BELOW 300-FOOT LEVEL

Fault plane, nearly vertical, is behind center of stull.   Stull in upper right-hand corner is in stope on upper segment of main vein.   Hanging-wall slip on lower segment bends to parallelism with fault, and quartz of lower segment extends up along the fault.

C. ALTERED BUT UNSHEARED COUNTRY ROCK SEPARATING QUARTZ STRANDS, RAINBOW MINE, MAIN VEIN, ADIT LEVEL

D. BANDING IN VEIN DUE TO SHEARING, WHICH HAS TAKEN PLACE CHIEFLY ALONG SEPTA OF COUNTRY ROCK, SIXTEEN TO ONE MINE

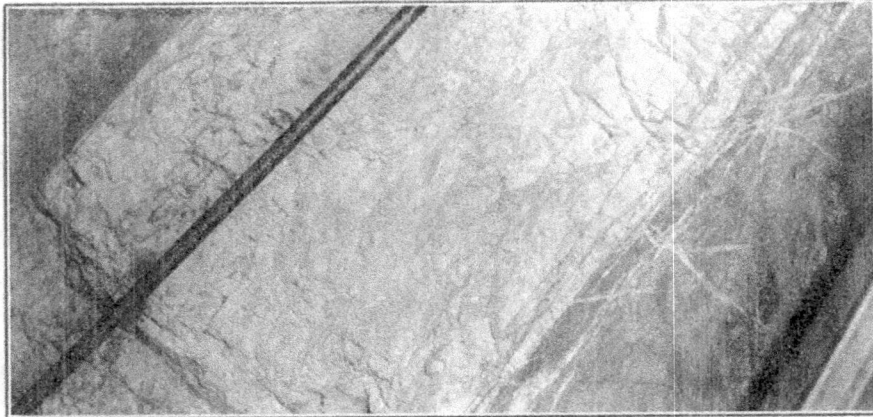

A. QUARTZ STRINGERS IN FOOTWALL OF VEIN, ELDORADO MINE, NEAR BOTTOM OF SHAFT

Stringers are parallel and at angles to the vein but cross the vertical cleavage of the country rock.   Hanging wall marked by later shear.

B. "CROSSOVER" STRINGERS BETWEEN HANGING WALL AND FOOTWALL, KATE HARDY MINE, LOWEST LEVEL

Walls are nearly vertical.

C. VEIN WITH SHARP FOOTWALL DUE TO FAULTING AND FROZEN HANGING WALL WITH QUARTZ STRINGERS, PLUMBAGO MINE, No. 4 LEVEL

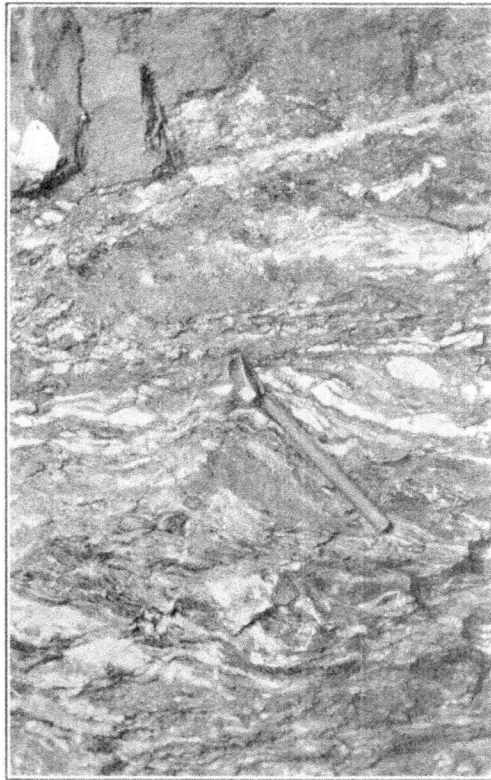

B. LENSES OF QUARTZ IN CRUSHED SLATE ALONG WALL OF CLINTON VEIN, RAIN-
BOW MINE

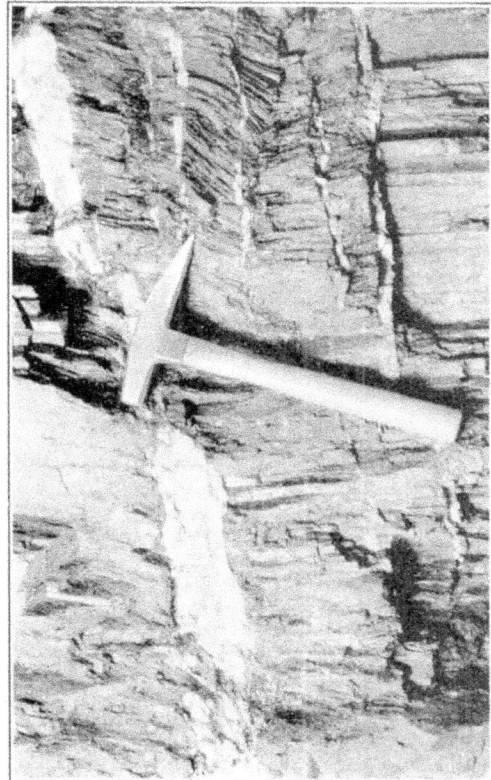

D. QUARTZ STRINGERS CUTTING CAPE HORN SLATE, BRUSH CREEK MINE, WINZE
IN HANGING WALL

A. TRANSVERSE QUARTZ STRINGERS IN CARBONATIZED SLATE CLOSE TO MAIN
VEIN, RAINBOW MINE, ADIT LEVEL

Light-colored streaks and patches in the wall rock are chiefly carbonate.

C. LENSES OF QUARTZ AND CARBONATE IN CRUSHED SLATE, KATE HARDY MINE,
ADIT LEVEL, CROSSCUT INTO HANGING WALL NEAR SHAFT

Later movement is shown by fault planes at point of the pick and along the hanging wall.

A. BRECCIA CONTAINING QUARTZ FRAGMENTS, CEMENTED BY FINE-GRAINED QUARTZ, ELDORADO MINE, ADIT LEVEL

Breccia lies along hanging wall of vein but extends down into the vein along small fissures.

B. "HEADCHEESE" BRECCIA, ELDORADO MINE

Coarsely crystalline white quartz cemented by fine-grained quartz.   Natural size.

A. VEIN WITH WELL-MARKED WALLS, PLUMBAGO MINE, No. 4 LEVEL

Country rock outside of walls is massive and unsheared. Material between the walls consists of sheared and altered rock with lenses of quartz. Left-hand mass of quartz is crushed and broken by later faulting. That on right is roughly banded, following an older fault.

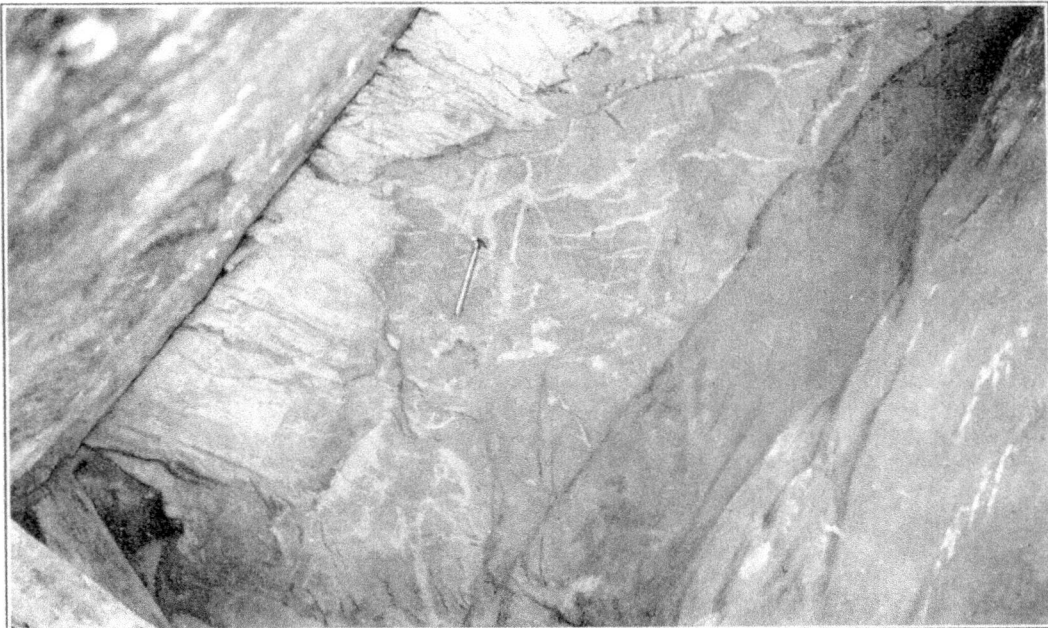

B. WALLS OF VEIN DUE TO POSTMINERAL MOVEMENT, PLUMBAGO MINE, No. 4 LEVEL

The quartz is cut by the hanging wall but on the footwall side grades into stringers, which are cut off by the footwall fault.

A. FOOTWALL OF VEIN MARKED BY CURVING POSTMINERAL FAULT, PLUMBAGO MINE, RAISE ABOVE No. 4 LEVEL

The postmineral movement probably followed an earlier break, for the banding in the quartz (at the right) is nearly parallel to the fault.

B. FAULT PLANE IN QUARTZ PASSING FROM FOOTWALL TO HANGING WALL, SIXTEEN TO ONE MINE, 800-FOOT LEVEL

Vein is here in the unusual condition that neither wall is formed by a fault and stringers extend out into both foot and hanging walls. The straight boundary along a small part of the footwall (to the right of the center) may be due to a fault prior to the introduction of the quartz, since it does not cut the quartz.

A. QUARTZ DISPLACED BY MINOR FAULT WHICH DOES NOT CUT THE FOOTWALL, ELDORADO MINE, UPPER TUNNEL ON BULLION VEIN

B. WEDGE OF QUARTZ IN COUNTRY ROCK, BOUNDED BY FAULTS, ORIENTAL MINE, SHAFT BELOW LEVEL 10

Major hanging-wall fault (on the right) cuts off the minor steep fault.

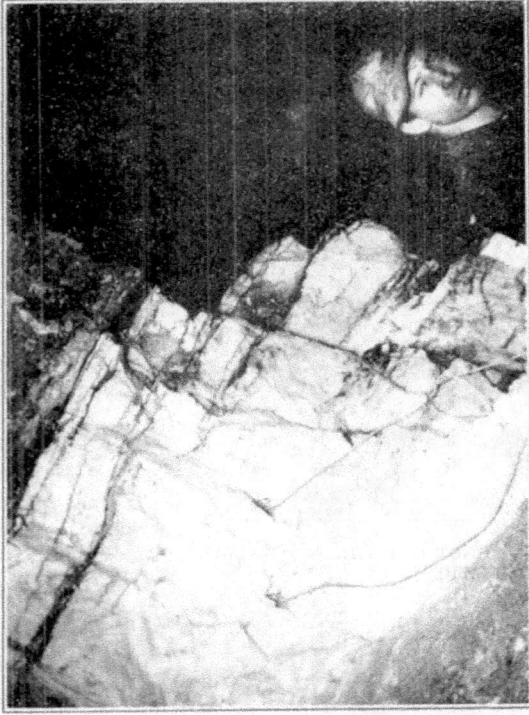

B. RIBBON QUARTZ DUE TO SHEAR PRIOR TO LATEST MOVEMENT BUT PARALLEL TO WALLS OF VEIN, SIXTEEN TO ONE MINE, HIGH-GRADE STOPE IN OLD TIGHTNER WORKINGS

C. RIBBON QUARTZ CUT BY TRANSVERSE SHEETING, KATE HARDY MINE, ADIT LEVEL

A. SPECIMEN OF RIBBON QUARTZ, GERMAN BAR MINE

About 4/5 natural size.

original septa of altered country rock within the vein, being more easily sheared than the main body of the quartz, commonly form planes along which there has been movement. The fact that movement has occurred at many different times also greatly complicates the relations of the different fault veins. It is common to find earlier shears, which may be cemented by

from the bend. On the Bullion vein there was found a younger set of slickensides superposed in the same plane on older grooves showing a variance in direction of as much as 30°.

Wherever the direction of this later movement could be determined it was found to be reverse, but only here and there could definite evidence be obtained. It

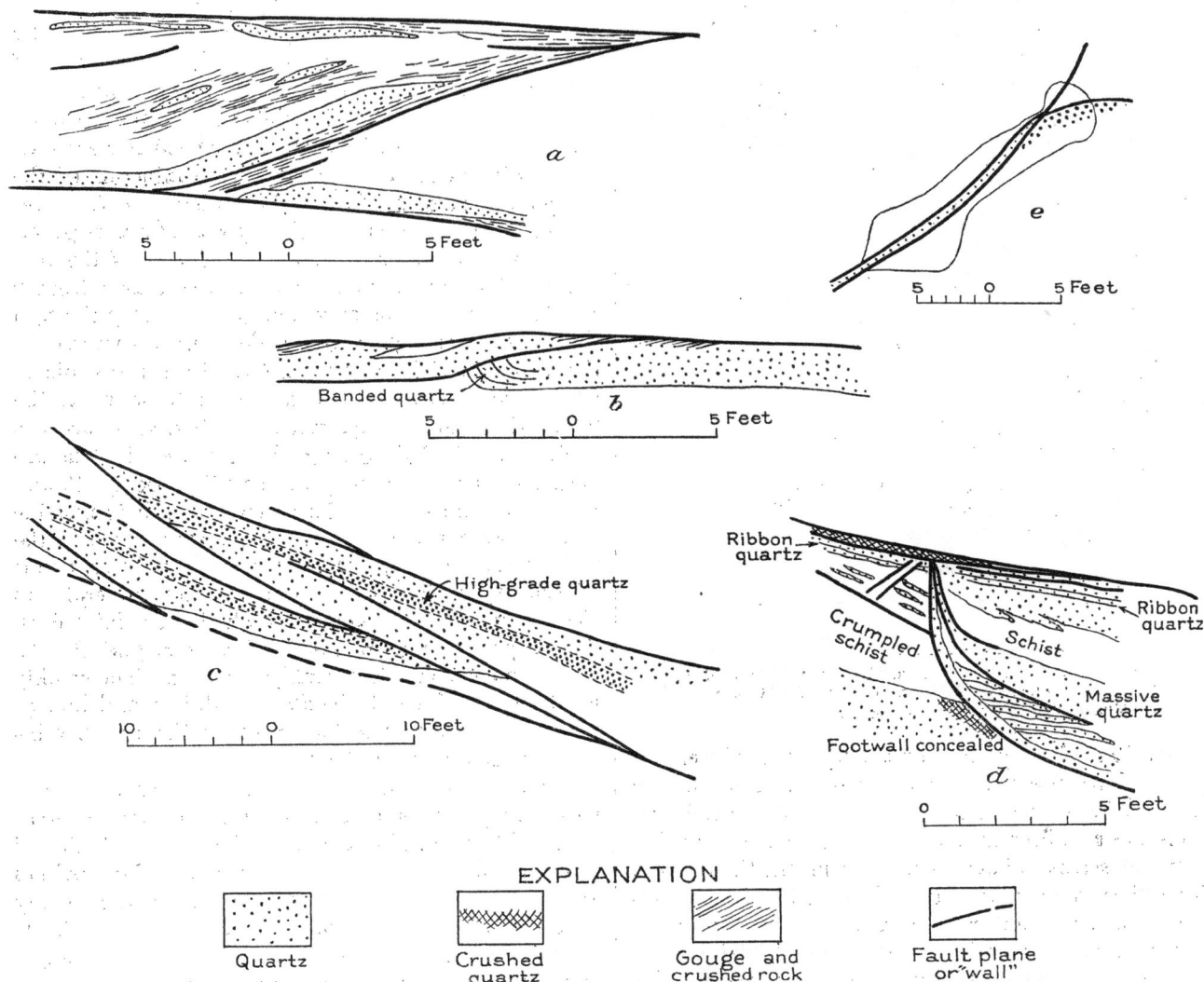

EXPLANATION

Quartz   Crushed quartz   Gouge and crushed rock   Fault plane or "wall"

FIGURE 10.—Sketches showing thrusting in veins: a, b, Looking up dip, Kenton mine, lower tunnel; c, displacement of strand carrying high-grade ore, Sixteen to One mine; d, quartz introduced along early fault, Sixteen to One mine, 300-foot level; e, quartz between fault planes, Extension of Minnie D mine, raise above upper tunnel

quartz or carbonate, cut by one or more later faults, and there is in places evidence (fig. 10, d, e) that a shear which itself cuts the earlier quartz has controlled deposition of later quartz.

The grooves and slickensides resulting from the latest movements are generally in the direction of the dip and accordant with the throw of the fault. In the region of the bend of the Eldorado vein (pl. 56) the grooves tend to be parallel to the average dip of the vein rather than to the diverging dips outward

is possible, therefore, that on some of these walls the movement may have been normal. Even in veins such as the Ophir, Osceola, and Morning Glory, which follow normal faults, at least a part of the postmineral movement in the productive part of the vein was in the reverse direction but very small in amount. It seems, therefore, that the original normal displacement along these veins may have been later somewhat reduced by postmineral reverse movement, particularly in the vicinity of the major veins.

In spite of the fact that the movement along the veins was complex and long continued, it is believed that the total displacement resulting from intermineral and postmineral movement was not large and that most of the movement shown on Plate 10 took place prior to the introduction of the quartz. Mr. Gannett's notes indicate the possibility of a single postmineral displacement of as much as 40 feet in the Sixteen to One mine, but none of the faults shown in

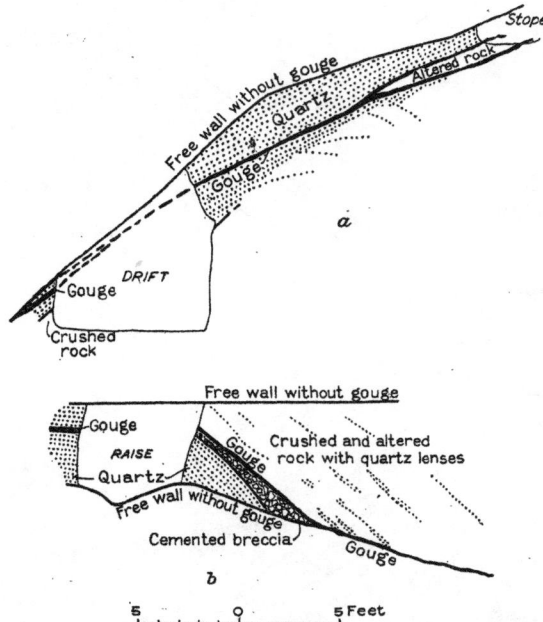

FIGURE 11.—Sections across vein, showing vein crossed by later fault, Plumbago mine, raise above No. 4 level: *a*, Looking along strike; *b*, looking up dip

Figure 10 indicate more than a few feet displacement. The best evidence against major movement is found in a study of the intersections of the flat and steep veins. There has also been movement along the steeper veins, though it seems to have been less complex than in and along those with flatter dip. Therefore if the intermineral and postmineral movements were of any magnitude a complex mosaic of fault blocks should be present in the region of the intersection of the veins of the two systems. The junction of the Rainbow and Clinton veins (fig. 12) does not indicate such a complex of any extent, for the postmineral thrusts along the Rainbow vein are cut off by the Clinton vein, though the quartz of the Clinton vein is contemporaneous with that of the Rainbow vein and older than the latest movement along the Rainbow.

At the junction of the Sixteen to One vein and the veins that follow the faults (pls. 7, *C*; 11, *B*) the postquartz shear parallel to the major vein bends to parallelism with that of the steep fault vein. Similar structure is shown in the sketches of the intersection

in the upper levels given in Figure 6. In the section shown in Figure 13, however, movement along nearly vertical faults seems to have been later than any movement in the plane of the vein, but there are also nearly vertical shears which are cut off abruptly by the "wall" following the vein. Near the drain level this "wall" itself bends toward the vertical fault.

The "ribbon quartz" (pl. 18, *A*) that is so characteristic a feature of the Alleghany veins is due chiefly to later fracture which has permitted the deposition of later minerals along the closely spaced planes. In places the position of the ribbons is determined by septa of country rock along which movement has taken place, with later replacement by the minerals of the carbonate and sulphide stage; elsewhere the fractures forming the "ribbons" are closely spaced planes parallel either to a major shear along the wall of the vein (pl. 18, *B*) or to a shear crossing the quartz (figs. 9 and 10), or the bands may ramify irregularly through the quartz; rarely there is sheeting transverse to the earlier banding (pl. 18, *C*). In certain places where the pressure has been more intense, as at the point of a wedge of quartz, the ribbon structure has itself been folded, giving the crenulated structure locally known as "crinkly banding." (See fig. 19 and pls. 19, *B*, and 42, *C*.) In a few places it is possible to trace gradations from banded quartz in which the intervening dark septa are clearly original thin bands of country rock, now altered to later minerals, to characteristic ribbon quartz in which only faint dark lines separate narrow bands of quartz, and this in turn to quartz showing crinkly banding. The crinkly banding is believed to be due to folding, and its significance in relation to the origin of the quartz veins is discussed on pages 77–80.

Inclusions of wall rock are common. In places one strand may consist entirely of white quartz, whereas its neighbor contains abundant inclusions of altered country rock, some showing sharp angular outlines and others almost completely replaced and visible only as faint discolorations of the quartz. (See pl. 19, *A*, *B*.) In most of the inclusions a linear arrangement and the preservation of some trace of the original schistose texture indicate that the inclusions owe their position in the vein to the deposition of quartz along parallel, closely spaced openings and were isolated by later transverse fissuring and replacement. (See pl. 20, *C*.) Others do not show continuity of structure but are sufficiently close together to suggest the possibility of mutual support (pl. 20, *D*), particularly if there has also been some replacement. Such an arrangement might be due either to the filling of the interspaces of a rubble-filled cavity or to replacement of the finer-grained material in a fault zone containing rolled fragments. For a few isolated inclusions, how-

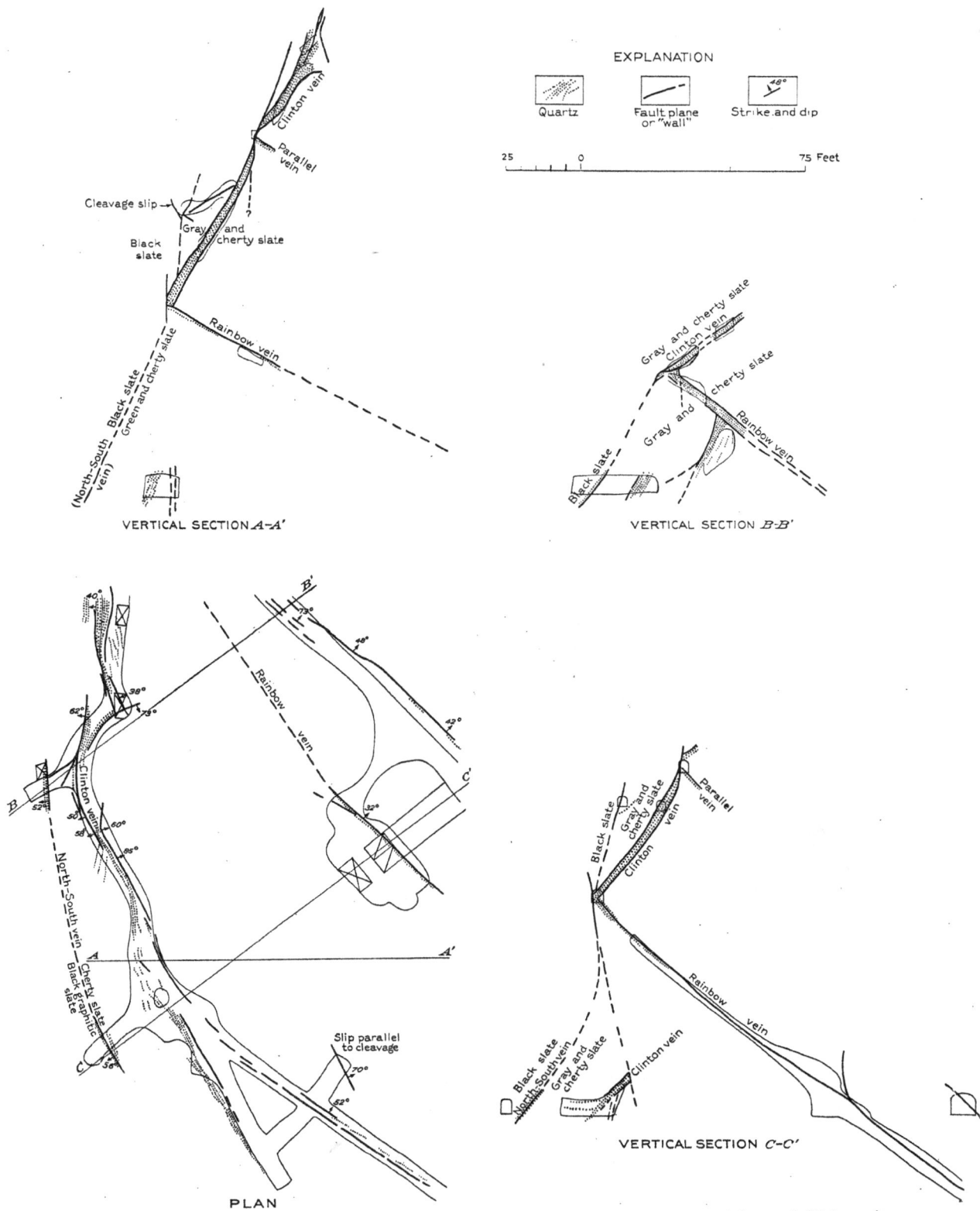

FIGURE 12.—Plan and sections of a part of the Rainbow mine, showing relation of Rainbow and Clinton veins

ever, neither of these explanations is applicable. These are small and are commonly replaced completely by minerals of the carbonate stage. In places there is a growth of quartz crystals pointing outward from the inclusion.

lowed by a more detailed description of the minerals and textures characteristic of each stage.

Chlorite stage.—The earliest stage of mineralization, perhaps regional rather than directly connected with the ore deposits, yet apparently more intense

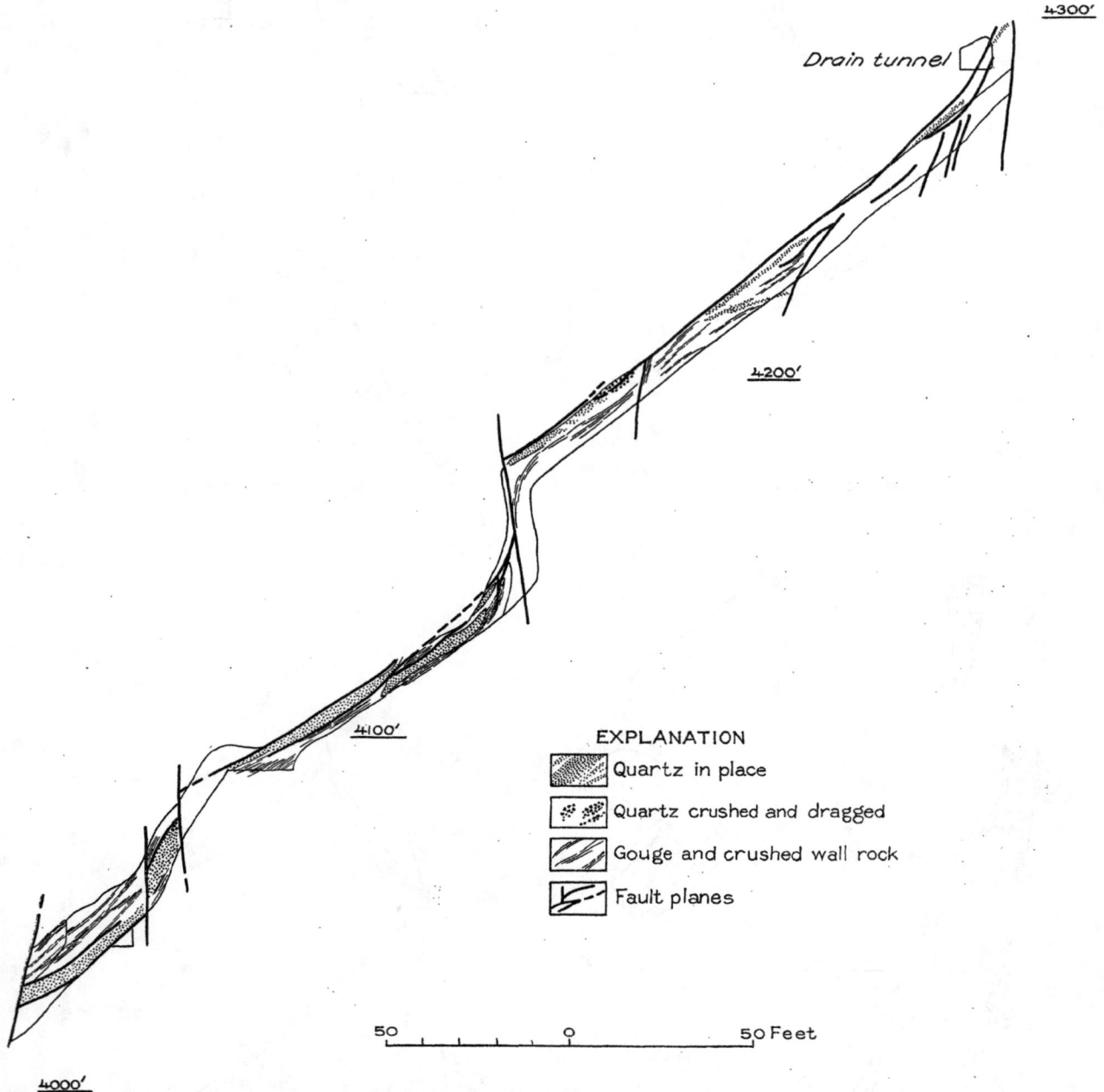

FIGURE 13.—Section of Sixteen to One vein below drain tunnel, showing faulting along and across vein

MINERALOGY

SUMMARY

It has been possible, largely on the basis of study of thin sections under the microscope, to identify several stages in the development of the mineralization of the veins. The summary statement given below is fol-

in the neighborhood of the veins, consisted in the chloritization of the hornblende of the schists and altered gabbro and the development in the plagioclase of the granite and granodiorite of very minute crystals apparently consisting of epidote and a micaceous mineral. Probably also the alteration of at least a part of the peridotite and pyroxenite to serpentine belongs to this stage.

Quartz stage.—Following this, how closely can not be determined, came the main period of quartz filling of the veins. Contemporaneous with the quartz and crystallizing slightly earlier than the quartz were the sulphides, pyrite and arsenopyrite. Early in this stage there was replacement of favorable wall rocks by calcic albite and oligoclase and minor crystallization of albite within the veins. Barite, which is only rarely found, seems to belong to this stage of mineralization and occurs both as a replacement mineral of the wall rock and as a vein mineral contemporary with the final crystallization of the quartz.

Carbonate stage.—The next series of events was complex, and could it be worked out in detail would presumably be found to have included several distinct phases of mineralization, the earlier phase overlapping somewhat on that summarized in the last paragraph. In this stage of mineralization replacement of wall rock was very extensive, and the earlier quartz was also in part replaced. The principal features were the replacement of large amounts of wall rock by carbonate (chiefly ankerite), sericite, mariposite, and minor amounts of chlorite, arsenopyrite, and pyrite. The sporadic development of talc probably belongs to this stage. A variety of minerals, including carbonate, sericite, mariposite, carbon, several sulphide minerals, and free gold, replace the earlier quartz along fissures. The quartz is also cut by little veinlets of silica, usually in the form of quartz but containing also opal and chalcedony; in places all three are observable in the same veinlet. The earlier quartz has locally been brecciated and recemented by the later quartz.

Final stage.—The last minerals, which are of no importance commercially and small in total amount, consist of calcite, very minutely crystalline pyrite, and possibly marcasite. These minerals occur as veinlets in both the altered country rock and the quartz veins and as coatings on the drusy crystals of the older quartz. A supergene origin for these minerals is possible.

CHLORITE STAGE

The preliminary stage of mineralization was essentially one of rearrangement, and little or no new material other than water was introduced. It is assumed that before the introduction of the vein-forming solutions hot waters rising along the fissures that were later to become the sites of veins caused recrystallization of the more susceptible minerals of the wall rocks and altered the peridotite, at least in part, to serpentine.

The widespread alteration of the gabbro to a mixture of "saussurite" and hornblende has been noted in a previous section. In specimens collected near the veins, however, there is seen a further alteration of the hornblende—in places to chlorite which roughly preserves the outline of the hornblende crystals and in places to mixtures of opaque nonmetallic material and chlorite. (See pls. 20, *A*; 34, *D*.) Patches of chlorite in the granite may represent the alteration of original biotite.

The already saussuritized feldspars of the gabbros do not appear to have undergone further alteration at this stage, but where the wall rock is granite, as at the Oriental mine, there were developed in the original oligoclase feldspar very minute crystals of what appears to be epidote together with a micaceous mineral. (See pl. 31, *A*, *B*.) The other minerals characteristic of this stage of mineralization do not involve a change in the chemical composition of the host rock other than the addition of water. It is, therefore, thought likely, inasmuch as the orthoclase and microcline are not similarly altered, that the mica of this stage may be paragonite, the soda mica, rather than sericite.

The relation of the fissures to the serpentine, as described on pages 31–32, indicates that serpentinization preceded or accompanied fissuring. It is possible that the serpentinization of the pyroxenite and peridotite followed closely upon the intrusion of these rocks and occurred prior to the intrusion of the granite and the formation of the veins. The recrystallization of the gabbro, moreover, preceded the formation of the fissures, and the agents which effected it may have also caused or initiated the serpentinization. But the alteration of the wall rocks described in the preceding paragraphs indicates that, after the intrusion of granite, hot waters with carbon dioxide, which could have effected the serpentinization, were available. It is thought likely that although the serpentinization may have been initiated prior to the formation of the fissures—and indeed the relation of the fissures to the serpentine suggests that this was the case—it was continued by the mineralizing solutions. The serpentinization of these basic rocks is not even now complete, for isolated grains of pyroxene and olivine still remain. It is possible, therefore, that the process continued throughout the period of mineralization and even that supergene waters may have played a minor part. But it is evident that serpentinization must have been well advanced before the formation of the quartz. Had the original peridotite remained unaltered up to the stage of vein filling, there is no reason why it as well as the gabbro should not have been cut by the quartz veins. If serpentinization started prior to the introduction of quartz and continued during the crystallization of the quartz it is to be supposed that the increase in volume tended to close fissures already present in the peridotite and prevent the formation of new ones. Certainly, at least near the veins, serpentine was already developed prior to the introduction of minerals of the carbonate group, for it has been replaced by these minerals.

Serpentinization, or at least a readjustment of the serpentine minerals, was later than the development of chromite, for chromite grains are idiomorphic against the antigorite and are cut by little veinlets of antigorite. (See pl. 20, B.) These veinlets contain minute crystals of the chromium-bearing garnet, uvarovite, which presumably derived its chromium content from the chromite.

QUARTZ STAGE

ARSENOPYRITE AND PYRITE

The only sulphide minerals belonging to the quartz stage of mineralization are arsenopyrite and pyrite.

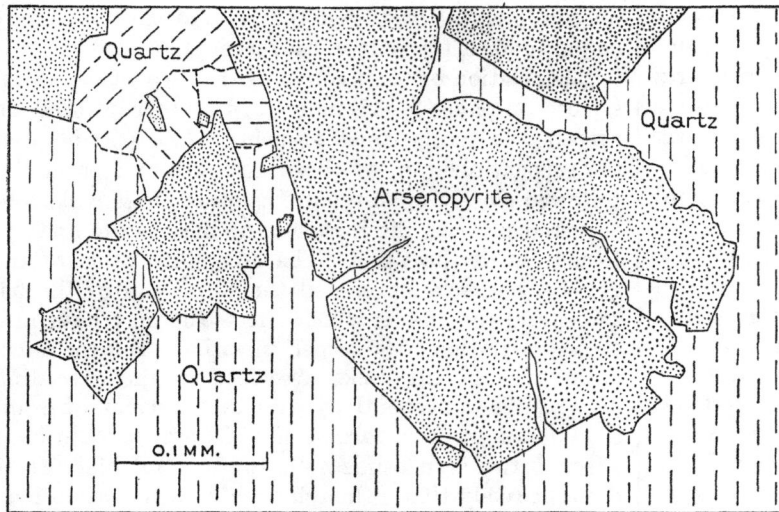

FIGURE 14.—Arsenopyrite veined by quartz

Both are distinguishable from the same minerals of the succeeding stage by their coarse crystallization and by the fact that their crystallization preceded that of the quartz which surrounds them, so that these early sulphides show crystal faces against the quartz and are also veined by the same quartz which forms the body of the vein. (See fig. 14 and pls. 21; 22, A, B; 46, D, E, F.)

Arsenopyrite is the more abundant and of greater economic importance, as it has in many places formed the nucleus for the deposition of the free gold of the next stage. This early arsenopyrite is commonly found in clusters of radiating prismatic crystals as much as several centimeters in length, either growing out from the wall rock or separated from the wall by a narrow strand of quartz. (See pls. 21, B; 46, F.) Within the vein arsenopyrite is found as roughly equidimensional crystalline aggregates, in places as much as 5 centi-

meters in diameter, completely surrounded by quartz. One side of such a crystal aggregate may show an irregular though sharp boundary as if it were a fragment broken off from a larger mass, or there may be scattered angular fragments within the quartz, as if a large crystal had been broken and dragged. (See pl. 46, E.) Other groups of arsenopyrite crystals in the veins suggest by their shapes a possible replacement of wall-rock inclusions or growth outward from an inclusion. The coarse-grained arsenopyrite has been found in all the veins of the district from which any considerable production has been made but seems to be most abundant in veins such as the Oriental, Gold Canyon, and Plumbago, which cut gabbro and serpentine. It is particularly noticeable that wherever a vein is close to serpentine there is likely to be abundant coarsely crystalline arsenopyrite.

Pyrite belonging to this stage of mineralization occurs here and there throughout the veins but is less common than the arsenopyrite. Although in places later sulphides and gold have been observed against and replacing this earlier pyrite, it is not commonly associated with the high-grade ore and seems to be more abundant in the less productive veins. It occurs sporadically throughout the quartz in crystalline masses or groups of contiguous pyritohedrons, with faces a centimeter or less across, and, like the arsenopyrite, is crystalline against and veined by the quartz. (See pl. 22, A, B.) It is unlike arsenopyrite in that its occurrence seems to be entirely independent of the kind of wall rock.

QUARTZ

GENERAL FEATURES

Quartz forms by far the greater part of the filling of the veins and has replaced an unknown amount of wall rock. In the hand specimen the vein quartz is milky white and of a massive appearance. Individual crystals or grains can not be distinguished except where drusy crystals project into vugs. Vugs are uncommon but are occasionally found. They have a maximum diameter of over a foot and seem to be most abundant in the larger swellings in the veins. The size of the drusy crystals tends to increase with the size of the vug. The largest seen showed prisms about 3 inches in greatest length. Such crystals commonly show a central portion which has a milky appearance due to the closely spaced vacuoles and an outer coating of clear quartz.

Under the microscope the average vein quartz is seen to be made up of anhedral interlocking grains of quartz which vary greatly in size. In most thin sections grains more than 10 millimeters in diameter are rare, but most of the section is composed of grains whose diameter exceeds 5 millimeters. In nearly every slide examined there are areas of very fine grain in which individual grains as small as 0.01 millimeter can be seen. Most of the quartz grains show irregular outlines, but rarely a single isolated grain is bounded by crystal faces (pls. 22, *C, D, E*; 28, *C*), and more commonly the crystal outline is incomplete (pl. 22, *D*). In a few areas where later folding is indicated by the crinkly banding the grains may show a common direction of elongation. (See pl. 44, *D*.)

In places small rosettes of outward-pointing quartz crystals (pl. 23, *A, B, C*) are found in the vein surrounded by quartz with the usual allotriomorphic texture. These appear to have grown outward from a nucleus of included wall rock now altered to sericite. Similar quartz crystals normal to the faces of the large crystals of arsenopyrite were also noted. (See pl. 23, *D*.)

Columnar texture normal to the walls (pls. 24, *A*; 28, *E*) was observed only in and along the edges of the small stringers which cut the slate in the vicinity of the larger veins. Rarely small veinlets a fraction of an inch in width show transverse ("cross fiber") structure. (See pl. 24, *B, C*.) The irregular grain of the quartz is due in part to mutual interference at the time of final crystallization and in part to later fracturing of the quartz grains.

Strain twinning (pls. 24, *D*; 29, *D*; 46, *B*) is common and seems to have caused recrystallization and actual movement in the quartz, as the edges of strain-twinned crystals are in places delicately serrate, with the saw teeth accordant with the twin lamellae. Strain-twinned quartz seems to be sporadic in distribution and not closely associated with those portions of the veins where there has been the greatest degree of postmineral movement.

Gradations can be observed between quartz grains in which strain has caused cloudy extinction without noticeable inner boundaries through those in which the areas of slightly different extinction position have definite boundaries to those in which rotation between different parts of the broken grain has given sharply different extinction positions. Grains whose boundaries are the result of fracture are with difficulty distinguished from original grains with irregular boundaries, but in places the fracturing of an original larger grain may be inferred by the preservation of the symmetrical arrangement of vacuole inclusions which was present in the larger grain. (See pl. 24, *E, F*.) Grains formed by the fracturing of larger crystals may be indicated by clear lines free from vacuoles along the boundaries of the grains, as shown in Plate 25, *C. D*.

Very commonly zones of microbrecciation, the result of later crushing, are developed in the quartz. (See pls. 25, *A, B*; 38, *A, B*; 44, *A, B*.) As a rule these zones can be distinguished from those in which the change in size of grain is due to variation of crystallization by the fact that the fracturing of the quartz results in clearing it of the minute cavities which give it a cloudy appearance, and also by the fact that the grains in an area of microbrecciation nowhere show crystal faces. As far as could be determined, microbrecciation did not involve movement other than that necessary to cause fracture and slight rotation of the quartz grains. No evidence was found of faulting of a preexisting structure along a zone of microbrecciation, and such zones commonly end abruptly against the "ribbons" of later minerals. (See pl. 44, *A, B*.) The presence of zones of microbrecciation gives to the quartz in the hand specimen the slightly different luster which distinguishes the "live quartz" of the miners.

VACUOLES

A thin section of vein quartz seen under the low magnification of the microscope has a cloudy appearance. Under the high magnification this cloudiness is resolved into a multitude of little bubblelike cavities. These vacuoles, though most abundant in the quartz, are present in other minerals as well. Both the quartz and the albite of the dikelike lenses in the schist contain abundant very minute vacuoles and they were found sparingly in the original feldspar and quartz of the granite. Among the vein minerals albite in places contains them in abundance (pls. 26; 27, *A*) but elsewhere is clear. They were noted in barite, but the rough appearance of this mineral in thin section makes their identification difficult. The minerals later than the quartz are free from vacuoles, with the exception of very rare minute cavities within the carbonate, although it is certain that a part of the vacuoles in the vein quartz were formed after the introduction of these later minerals. The absence of vacuoles from the carbonate and mica might be due to the well-developed cleavage of these minerals, but the later quartz is also entirely clear.

Although a part of the vacuoles are certainly of later origin than the quartz, it is convenient to depart from the strictly chronologic order of presentation and describe both classes in sequence to the description of the vein quartz itself.

The vacuoles range in diameter from the limit of vision, less than 0.001 millimeter to a maximum of about 0.020 millimeter. (See pl. 27, *B, C*.) The larger ones appear to be nearly filled with liquid and ordinarily contain a bubble. The smaller ones may have no visible bubble and may be filled with either liquid or gas, though under the maximum magnification it is possible to see a bubble even in some of these. The largest cavities, such as those shown on Plate 27, B, C, are found only in quartz crystals of greater than average size. The smaller cavities are generally irregular in shape; the larger ones are in part distinct negative crystals (pl. 27, C) but commonly show peculiar tails or roots (pl. 27, B). The bubbles do not occupy more than a small fraction of the space filled with liquid. Where a definite negative crystal contains both liquid and bubble it is possible to calculate roughly that the volume occupied by the gas is not over 5 per cent of the total cavity. For those of irregular shape with rootlike projections the gas volume is not calculable but appears to be somewhat smaller.

Three modes of distribution were noted. The cavities may be arranged in more or less definite zones or bands within the crystal (pls. 24, E; 27, D; 28, A–C), scattered at random through the quartz (pls. 22, D; 23, C), or in fairly sharp lines that may cross the crystal boundaries (pls. 28, E; 29, A, B). Those of the first group may have been formed contemporaneously with the present crystallization of the quartz; the last mentioned are of later origin and in many places are clearly dependent upon later-formed minerals. (See pl. 30, A.) The random cavities have been formed chiefly at the time of the crystallization of the quartz, but in places the apparent random distribution may be due to the angle at which the plane of the section cuts closely spaced linear cavities.

Among the earlier-formed vacuoles there is no sharp distinction between the random and the zonally arranged cavities. In places (pls. 22, D, E; 27, D; 28, A, B, C) the core of a crystal has a cloudy appearance due to closely spaced vacuoles without apparent arrangement, surrounded by more or less distinct zones of alternate relatively clear and cloudy quartz. This change, though distinct, is not as sharp as that exhibited where the quartz has been rendered clear by later fracturing. In such zoned crystals the core contains abundant vacuoles, and a clear core surrounded by zones of the bubbly quartz was nowhere observed.

The vacuoles showing linear arrangement commonly occur as sharp lines in places crossing the boundaries between adjacent grains. (See pls. 28, D, E; 29.) Many of these are sharply defined and where examined under the highest magnification are seen to consist of closely spaced very minute vacuoles, generally about 0.001 millimeter or less in diameter. The larger ones commonly have their greatest elongation at right angles to the line in which they lie. (See pl. 29, B.) Examination of sections of greater than normal thickness indicates that although most of these lines are

traces of planes in which the vacuoles are irregularly spaced, a number show, as the lens is raised or lowered, a succession of similar lines, as if these lines of vacuoles were independent rays, more or less parallel, or as if, though lying within a plane, they tended to form at the intersections of planes. Such rays or planes are clearly later than the quartz. Not only do they cross from one grain to another, but they cross areas of random or zonally distributed vacuoles (pl. 29, A) and in places can be seen to spread out from minerals later than the quartz, such as later sulphides (pl. 30, A) and carbonate (pl. 33, C). The rays may be parallel across several crystals and conform to the elongation of the later minerals, such as the carbonate shown in Plate 28, D, E, or without apparent orientation, as in Plate 29, B.

Other raylike arrangements of the vacuoles may be of earlier origin. Some rays appear to be dependent on early formed sulphides which are veined by the quartz (pl. 22, B), and in places the lines are confined to a single grain or core of a grain. In some crystals the linear arrangement seems to be related to crystal structure (pl. 24, E), and in places the directions of the rays vary in different crystals.

As explained above, microbrecciation tends to wipe out these cavities and clear the quartz. This process must involve freeing the gas and liquid contained in the smaller broken fragments and closing up the cavities by actual movement of the quartz. The size of the clear fragments (as much as 0.15 millimeter in the specimen shown in pl. 25, B) in the zones of microbrecciation is, however, greater than that of original crystals that contain vacuoles. (See pl. 23, C.) The grains in the microbrecciated area must therefore either have fissures which do not involve rotation and are therefore not visible, or there must have been later cementation and recrystallization. Small threadlike lines of clear quartz without sharp boundaries occur in association with the jamesonite needles that replace the quartz. (See pl. 40, C.) This is taken to indicate that concomitant with the development of the jamesonite in the quartz there were formed channels which permitted the release of the liquid and gas of these minute cavities, followed by healing of the quartz.

Although it is certain that the last-formed vacuoles were developed in the quartz after microbrecciation, as rays and planes of vacuoles are dependent on later-formed minerals, these later lines of vacuoles do not cross areas of microbrecciation, presumably owing to the many channels of escape offered by the minutely fractured quartz.

The relations to strain twinning are more complex. Plate 29, C, D, shows a grain of strain-twinned quartz bordering a grain in which the relief of strain has caused a cloudy extinction and broken appearance,

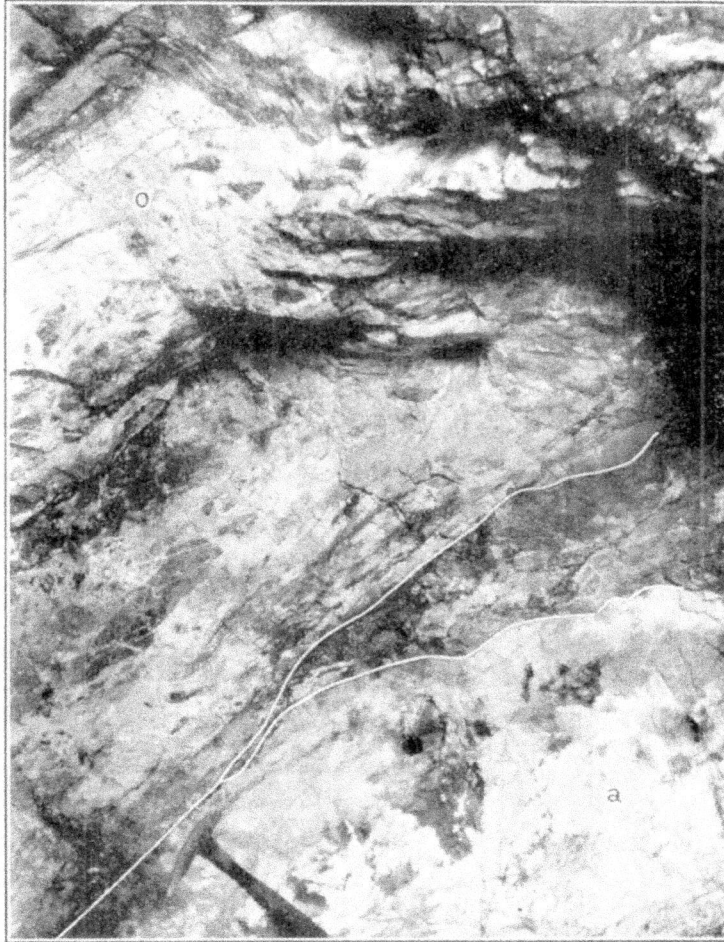

A. JUNCTION OF ALTA VEIN (a) AND ORIENTAL VEIN (o), ORIENTAL MINE,
LEVEL 8

Alta vein is here massive quartz (dark patches are due to powder stains).   Oriental vein near
contact is "ghost quartz" containing both completely replaced fragments which show only
faint outlines and fragments altered to carbonate.   This grades upward into massive
quartz containing wall-rock fragments, possibly replacement residuals.   Wedge of altered
gabbro on right, between the veins.

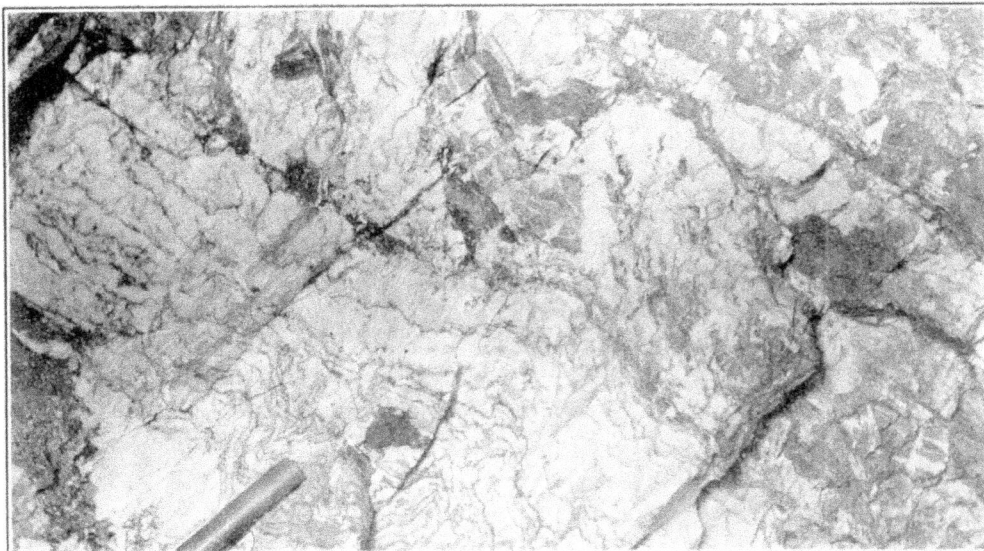

B. VEIN WITH LOWER STRAND SHOWING CRINKLY BANDING AND STRAND ABOVE CONTAINING FRAG-
MENTS OF ALTERED WALL ROCK, ELDORADO MINE, ADIT LEVEL

Along hanging wall in right-hand upper corner is breccia containing quartz fragments in a matrix of carbonate and mariposite.

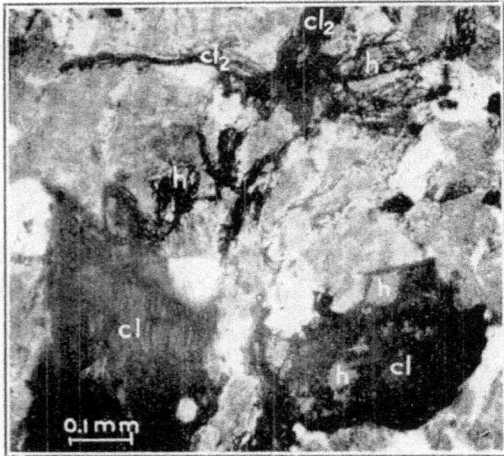

A. DEVELOPMENT OF CHLORITE AND ALBITE IN SAUSSURITIZED GABBRO, ORIENTAL GROUP, PROSPECT TUNNEL WEST OF ALTA TUNNEL

Altered gabbro consisting essentially of saussuritized feldspar (gray) and hornblende (h). Hornblende has been almost completely altered to chlorite (cl). Albite (white) in part replaces the saussuritized feldspar. With the albite are minute grains of apatite (?) not clearly shown in the photograph. Later chlorite (cl₂) cuts both albite and older feldspar. Parallel nicols.

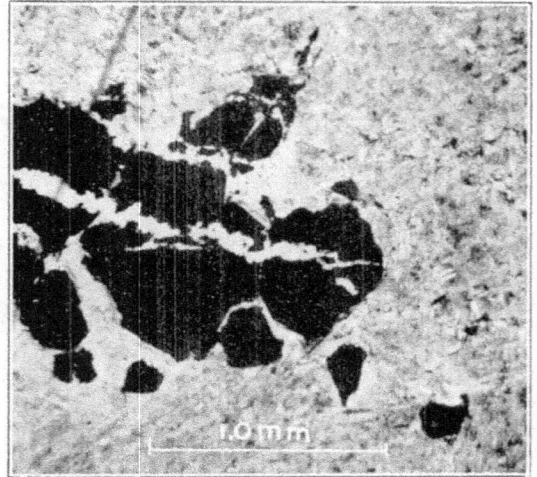

B. GRAIN OF CHROMITE IN SERPENTINE, VEINED BY ANTIGORITE, DORRIS PROSPECT

Crossed nicols.

C. VEIN QUARTZ SHOWING INCLUSIONS ASSOCIATED WITH SEPTA OF COUNTRY ROCK AND PARTIAL REPLACEMENT OF ROCK BY QUARTZ

About 11/13 natural size.

D. VEIN QUARTZ WITH ABUNDANT INCLUSIONS SHOWING PARTIAL REPLACEMENT BY QUARTZ

The residuals consist of a mixture of carbonate and mariposite.   About 5/7 natural size.

A. COARSELY CRYSTALLINE EARLY ARSENOPYRITE CRYSTALLINE AGAINST QUARTZ AND VEINED BY QUARTZ, WITH GOLD (WHITE) REPLACING ARSENOPYRITE, SIXTEEN TO ONE MINE, TIGHTNER WORKINGS

Natural size.

B. ARSENOPYRITE BROKEN AND VEINED BY QUARTZ, WITH GOLD (WHITE) GENERALLY IN CONTACT WITH THE ARSENOPYRITE BUT LARGELY REPLACING QUARTZ, SIXTEEN TO ONE MINE

See Plate 41, *A*, for details.   Natural size.

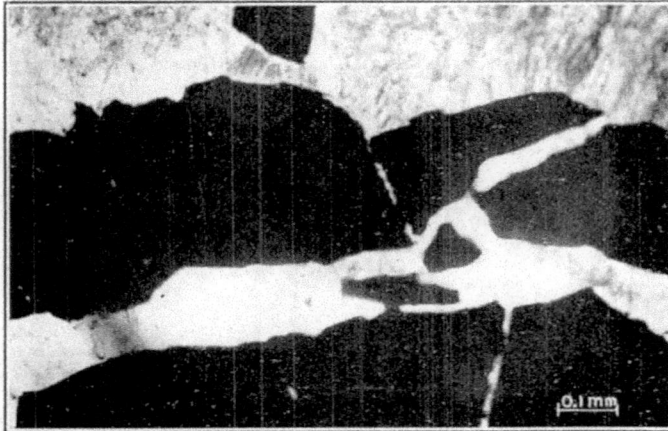

A. PYRITE VEINED BY QUARTZ, ORIENTAL MINE, ALTA SHAFT DUMP

Part of same slide as *B*.   Parallel nicols.

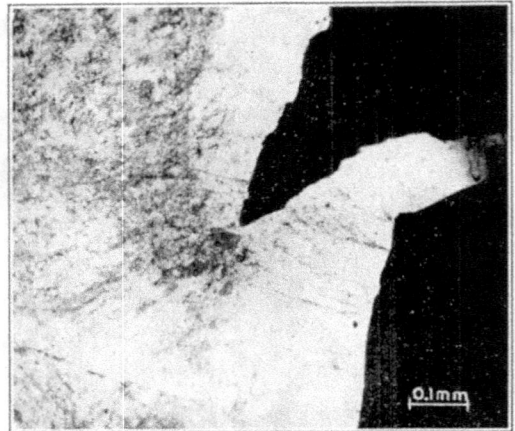

B. PYRITE VEINED BY QUARTZ, ORIENTAL MINE, ALTA SHAFT DUMP

Part of same slide as *A*.   Shows clear zone (not continuous) around border which is in optical continuity with large quartz grain and without sharp boundary.   A few rays of vacuoles extend from the pyrite across the clear quartz.   Parallel nicols.

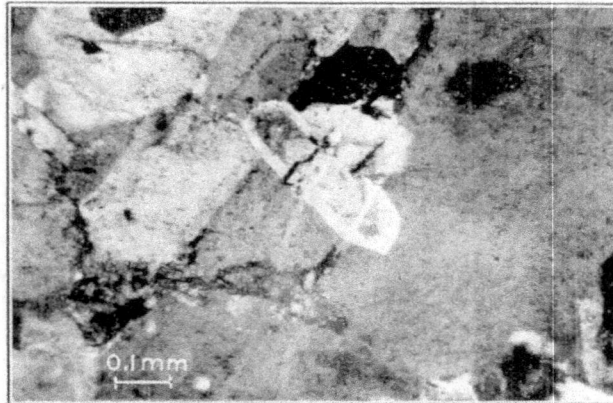

C. QUARTZ GRAIN SHOWING CRYSTAL FACES SURROUNDED BY QUARTZ WITH ALLOTRIOMORPHIC TEXTURE

Crossed nicols.   See also Plate 27, *D*.

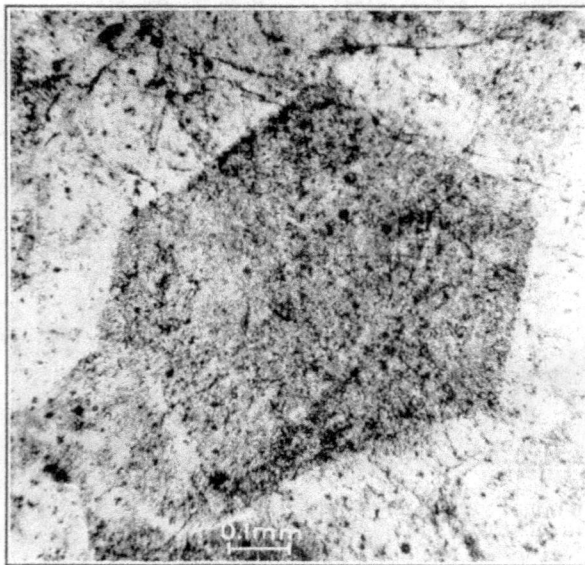

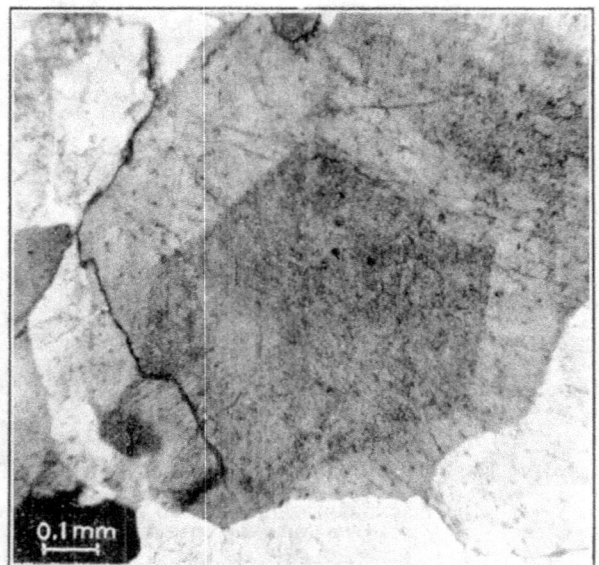

D, E. QUARTZ CRYSTAL WITH CORE CONTAINING ABUNDANT VACUOLES, SURROUNDED BY GROWTH OF RELATIVELY CLEAR QUARTZ, SIXTEEN TO ONE MINE, 800-FOOT LEVEL

D, Parallel nicols; *E*, crossed nicols.   See also Plate 28, *C*.

A. VEIN QUARTZ SHOWING ROSETTE OF OUTWARD-POINT-
ING QUARTZ CRYSTALS, SIXTEEN TO ONE MINE, 800-
FOOT LEVEL

Variation in size of grain is believed to be due to crystallization around
nucleus of wall-rock fragment now represented by a little sericite.
Crossed nicols.

B. DETAIL OF *A,* SHOWING TENDENCY OF
SMALLER QUARTZ CRYSTALS TO POINT OUT-
WARD FROM CENTRAL AREA

Crossed nicols.

C. DETAIL OF *B,* SHOWING QUARTZ CRYSTALS POINTING
OUTWARD FROM AREA CONTAINING SERICITE SHREDS

Vacuole inclusions occur in both large and small quartz crystals,
though more abundant in large crystals.   Crossed nicols.

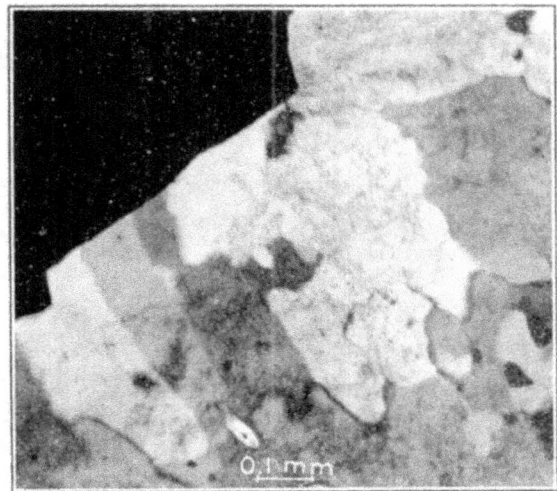

D. GROWTH OF QUARTZ CRYSTALS OUTWARD FROM
SURFACE OF ARSENOPYRITE CRYSTAL, SIXTEEN TO
ONE MINE, 800-FOOT LEVEL

Crossed nicols.

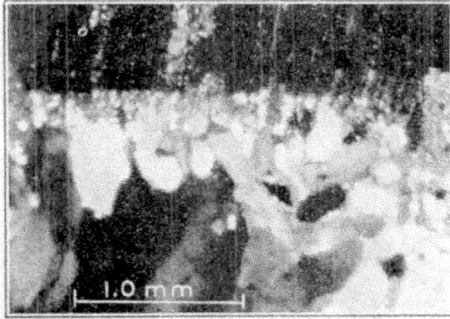

A. QUARTZ CRYSTALS ELONGATE NORMAL TO
WALL OF QUARTZ VEIN CUTTING SLATE,
BRUSH CREEK SHAFT

Fine grained along wall, becoming coarser outward into
vein. Crossed nicols.

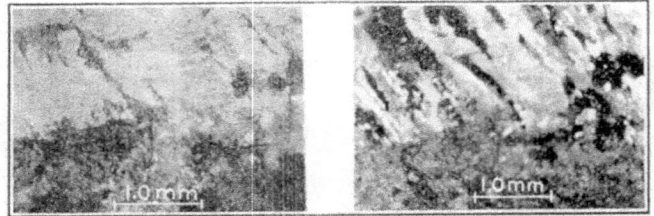

B, C. QUARTZ VEINLET WITH CROSS-FIBER STRUCTURE, RAINBOW
MINE, WINZE ON CLINTON VEIN

Contact with wall rock is blurred by development of later carbonate, which also
penetrates the quartz. *B*, Parallel nicols; *C*, crossed nicols.

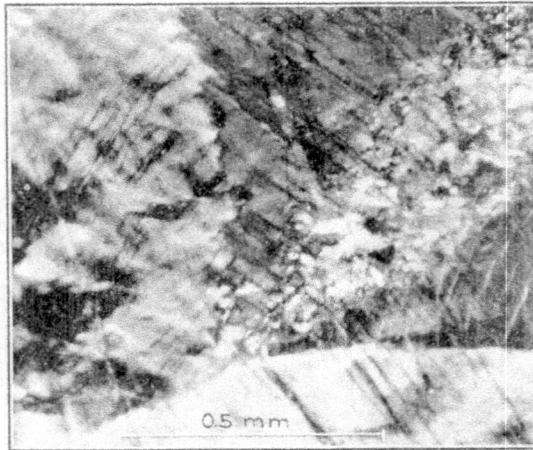

D. VEIN QUARTZ SHOWING STRAIN TWINNING AND
MICROBRECCIATION, RAINBOW MINE, CLINTON VEIN,
LOWER DRIFT

The serrate contact between the left and right grains may indicate
movement resulting from development of the strain twinning.
The straight boundary between the upper and lower grains is
believed to be due to fracture, as it is crossed by one set of strain
lamellae with only slight difference of orientation. But a second
set of lamellae with different orientation has developed in the
upper grain. The area of microbrecciation in the upper right-
hand grain was presumably later than the development of strain
twinning, as the twin lamellae on opposite sides have slightly
different orientation. Crossed nicols.

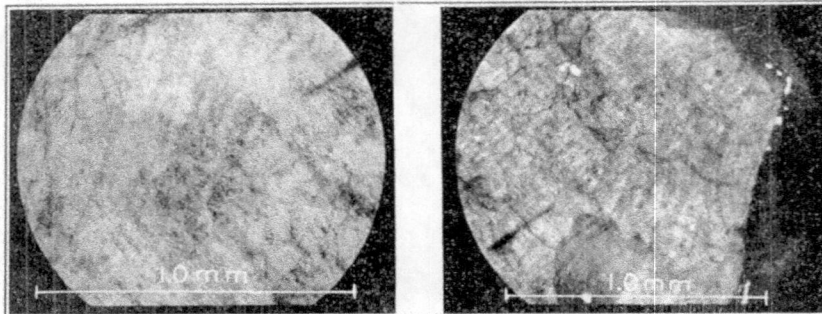

E, F. QUARTZ CRYSTAL SHOWING VACUOLES IN ZONES PARALLEL TO PRISM, BELMONT
MINE

Boundaries of small crystal shown in *F* are believed to be due to later fracture, as the zones of inclusions
are common to both. *E*, Parallel nicols; *F*, crossed nicols.

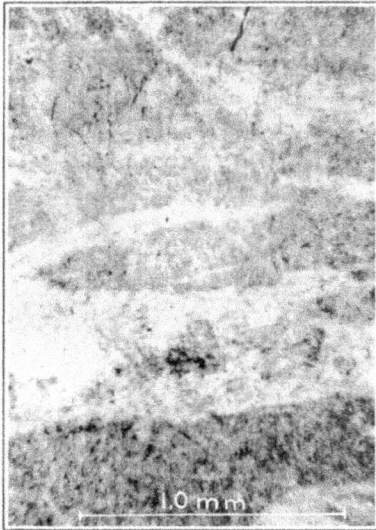

A, B. CLOUDY QUARTZ CLEARED IN ZONE OF MICROBRECCIATION, PLUMBAGO MINE

Boundaries between clear and cloudy areas are less sharp than those produced by later veining by clear quartz. *A,* Parallel nicols; *B,* crossed nicols.

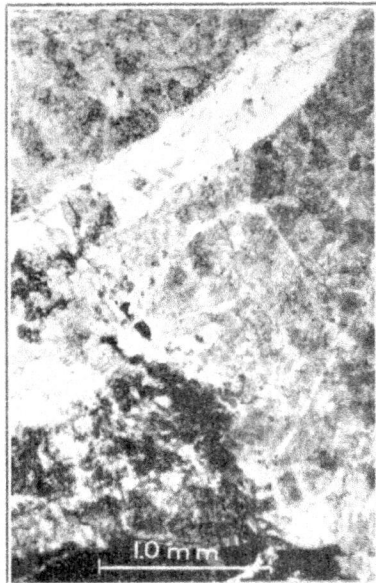

C, D. CHANGE IN GRAIN CLOSE TO WALL OF VEIN, PLUMBAGO MINE, LEVEL 5

Probably due to later fracturing, as grain boundaries are irregular, elongation is parallel to the wall instead of at right angles, and there is a tendency to development of streaks free from vacuoles around the edges of the grains. Veinlet of later clear quartz crossing slide presumably formed prior to the fracturing. Near the wall both quartz and wall rock are replaced by carbonate, which is contemporaneous with the later quartz. *C,* Parallel nicols; *D,* crossed nicols.

A, B. ALBITE, DARK FROM ABUNDANCE OF VACUOLE INCLUSIONS, INTERGROWN WITH QUARTZ, MINNIE D PROSPECT

A, Parallel nicols; *B,* crossed nicols.

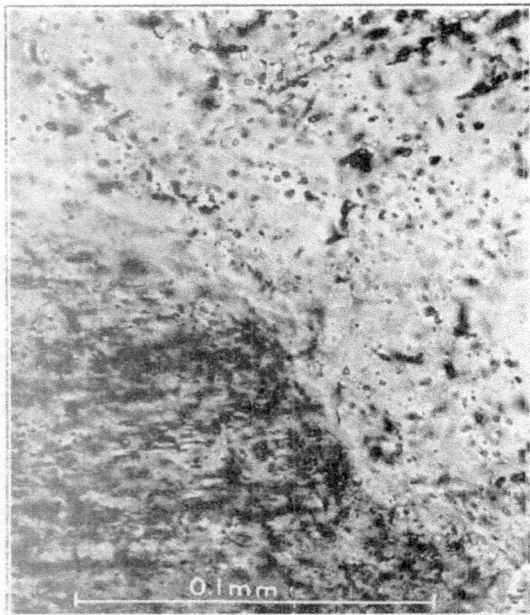

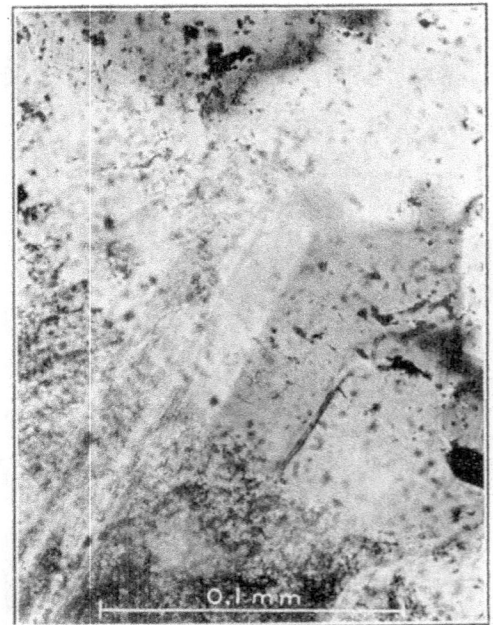

C. DETAIL OF *A,* SHOWING CONTACT OF QUARTZ AND FELDSPAR

Vacuoles in feldspar are closely spaced and follow cleavage planes; those in quartz have a rather indefinite linear arrangement. Parallel nicols.

D. ALBITE WITH ZONING DUE TO VACUOLES, SIXTEEN TO ONE MINE

Inner zone of albite contains abundant vacuole inclusions. Inclusions in outer zone in random distribution and about equally abundant with those in the neighboring quartz grains. Dark grain of rutile on right crosses border of quartz and albite. Crossed nicols.

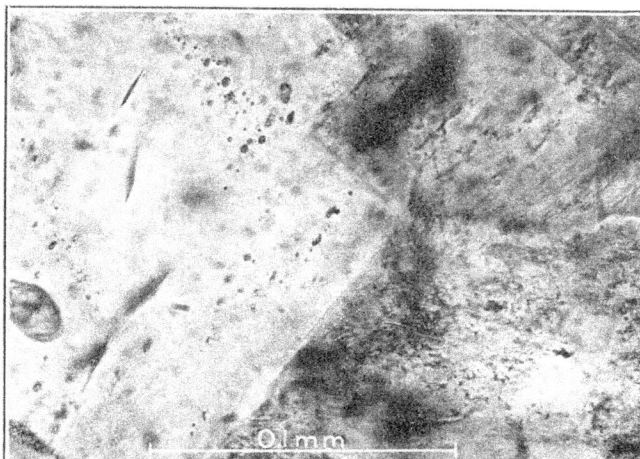

A. INTERGROWN QUARTZ (LIGHT) AND ALBITE (DARK), GOLDEN KING MINE DUMP

Abundant minute vacuoles in albite, in linear arrangement; vacuoles less abundant in quartz and less regularly spaced but larger and showing rough zonal arrangement. Nearly parallel nicols.

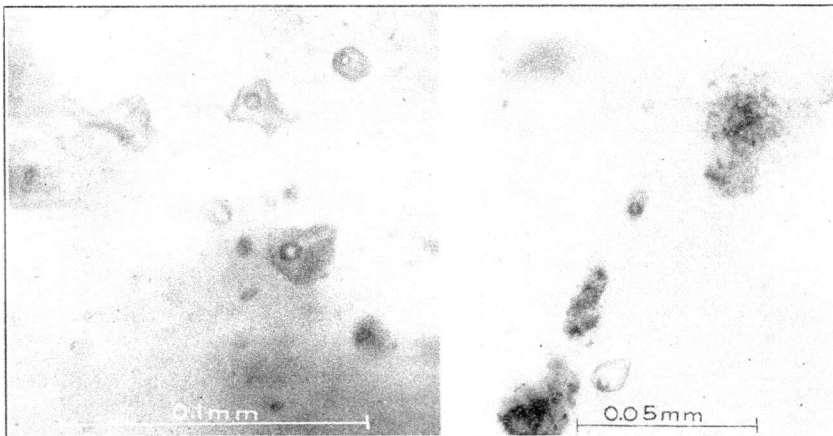

B. VACUOLES WITH BUBBLES IN QUARTZ, ELDORADO MINE, SOUTH RAISE

Crossed nicols.

C. CLUSTERS OF MINUTE INCLUSIONS AND NEGATIVE CRYSTALS WITH BUBBLES IN QUARTZ, ELDORADO MINE, SOUTH RAISE

Parallel nicols.

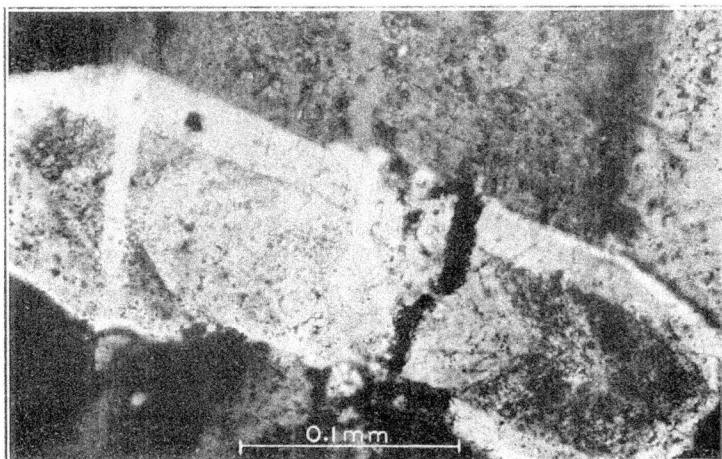

D. CRYSTAL OF OLDER QUARTZ SHOWING ZONAL STRUCTURE DUE TO ARRANGEMENT OF VACUOLES, PLUMBAGO MINE, LEVEL 10

Detail of Plate 22, *C.* Neighboring crystals contain abundant vacuoles in random distribution. Later veinlets of clear quartz and veinlet containing carbonate and arsenopyrite. Crossed nicols.

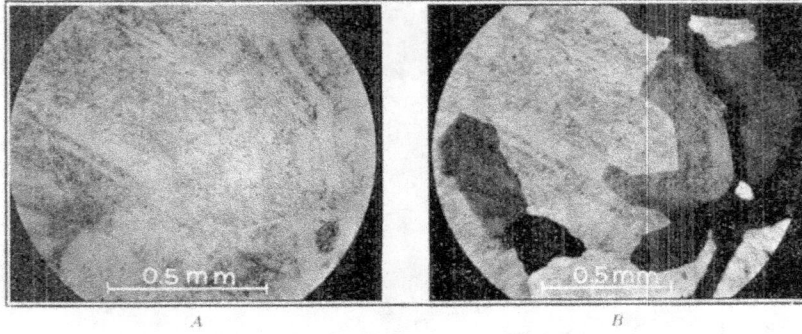

A, B. COMPLEX ZONING IN QUARTZ DUE TO VACUOLE INCLUSIONS, KENTON MINE,
STOPE SOUTH OF ADIT

Outer zone is clear in prismatic portion where optically continuous with core but contains abundant
vacuoles in portion around pyramid, which has a different orientation.   A, Parallel nicols; B, crossed
nicols.

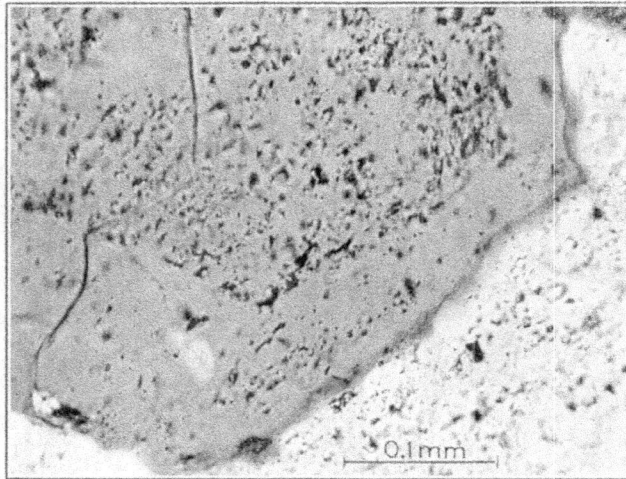

C. QUARTZ GRAIN SHOWING CRYSTAL FACES, WITH INNER CORE
CONTAINING ABUNDANT INCLUSIONS, SIXTEEN TO ONE MINE,
800-FOOT LEVEL

Outer core of relatively clear quartz.   Same specimen as shown in Plate 22, D, E.
Crossed nicols.

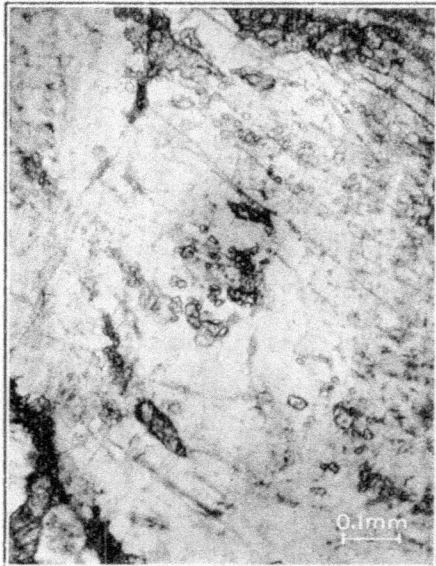

D, E. QUARTZ CRYSTALS ELONGATE NORMAL TO WALL OF VEIN, CROSSED BY LINES OF VACUOLES,
RAINBOW MINE, LOWER LEVEL, WINZE

Shows irregular replacement by carbonate. Lines of vacuoles are roughly accordant with the carbonate masses and most
numerous next to carbonate. D, Parallel nicols; E, crossed nicols. E shows lines of vacuoles crossing crystal boundaries.

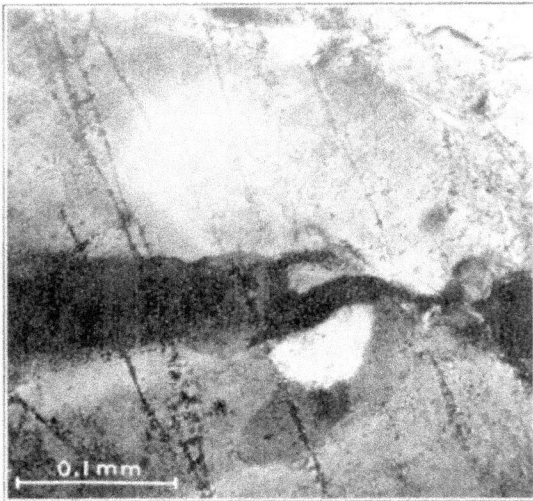

A. LINES OF VACUOLES CROSSING QUARTZ GRAINS, IRELAN MINE, ADIT LEVEL

Crossed nicols.

B. LINEARLY ARRANGED VACUOLE INCLUSIONS, APPARENTLY DEVELOPED PRIOR TO INTRODUCTION OF PYRITE, RAINBOW MINE, CLINTON VEIN, LOWER DRIFT

Parallel nicols.

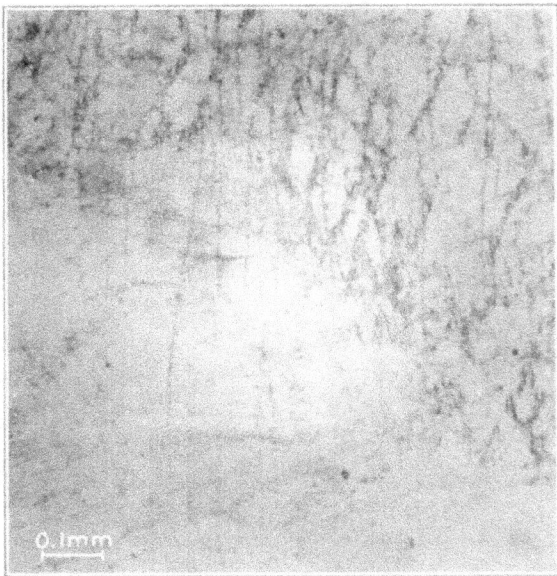

C, D. RELATION OF VACUOLES TO STRAIN TWINNING, BRUSH CREEK MINE

Strain-twinned quartz against quartz with cloudy extinction, presumably due to minor fracturing. Vacuoles in strain-twinned quartz are roughly parallel to twin lamellae, suggesting release by recrystallization of one set of lamellae. The cloudy quartz is free from vacuoles except for later lines which cross both grains. *C,* Parallel nicols; *D,* crossed nicols.

A. LINES OF VACUOLES RADIATING FROM GALENA SURROUNDING AND REPLACING PYRITE, ELDORADO MINE, ADIT LEVEL, "MILL ROCK"

Small streaks of galena follow same radiate arrangement. Radiate lines cross another set of vacuole lines (nearly vertical in the photograph). Parallel nicols.

B. GRANITE ALTERED TO FINE-GRAINED ALBITE, ORIENTAL MINE, MAIN DRIFT, WALL OF VEIN

The albite (a) is penetrated by quartz (q). Both albite and quartz are replaced by later carbonate (c). Crossed nicols.

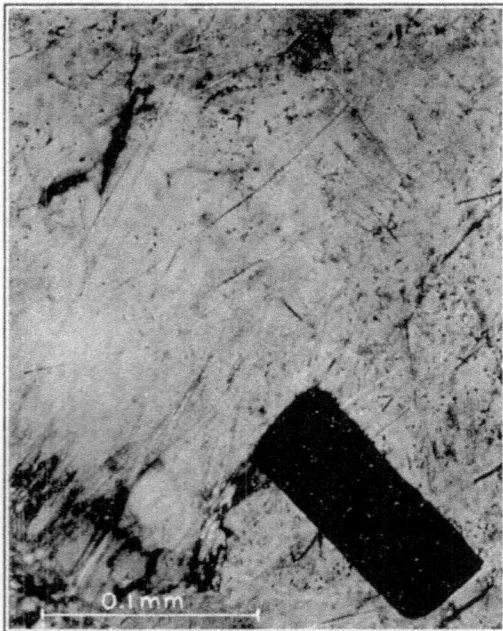

C. TUBES IN QUARTZ, RADIATING FROM GRAIN OF CARBONATE, KATE HARDY MINE, BOTTOM LEVEL

Same specimen as *D.* Crossed nicols.

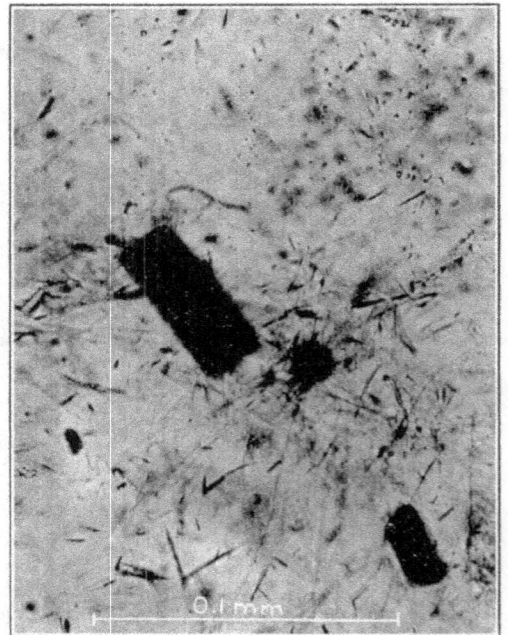

D. MINUTE TUBES IN QUARTZ, KATE HARDY MINE, BOTTOM LEVEL

Same specimen as *C.* Dark grains are later arsenopyrite. Crossed nicols.

without actual microbrecciation. In the strain-twinned grain the vacuoles are segregated into bands accordant with the twin lamellae, as if one set of lamellae were residual from the original grain and the other set the result of a new crystallization. The adjoining cloudy quartz has apparently undergone sufficient fracturing to be essentially free from vacuoles. But both the strain-twinned and the cloudy quartz are cut by lines of cavities that are clearly of later origin. On the other hand, there are numerous strain-twinned crystals in which either the earlier vacuoles are completely wiped out by recrystallization or none were originally present, and the quartz grains are crossed by numerous lines of later-formed vacuoles which bear no relation to the twinning planes.

A peculiar variation from the usual type of vacuoles was found in one specimen of quartz. Here (pl. 30, C, D) they take the form of narrow tubes as much as 0.25 millimeter in length without definite arrangement but most abundant where radiating from a grain of carbonate (pl. 30, C), although they are also found in portions of the slide that show no carbonate in their immediate vicinity (pl. 30, D).

From the foregoing description it will be seen that the vacuoles are of at least two ages, not everywhere clearly distinguishable, the first apparently contemporaneous with the present crystallization of the early quartz and possibly containing material trapped during the consolidation of the quartz, the second formed after the crystallization of the quartz and associated with later minerals that replace the quartz. The zoning shown in Plate 27, D, may indicate successive "growth rings" of the quartz, a symmetrical arrangement of the inclusions during the crystallization of the quartz, or the entrapping of the medium from which the quartz crystallized by the growth of the quartz as skeleton crystals.[49] Sorby[50] has shown that in the experimental production of crystals containing vacuoles the controlling factor was speed of crystal growth. Therefore those containing abundant vacuoles may represent conditions favoring more rapid crystallization. Where a core of bubbly quartz is surrounded by quartz that is relatively clear, either within the vein in contact with other grains (pl. 28, C) or bordering the free faces of a projecting crystal in a vug (pl. 31, A), the inference is strong that the zoning represents two periods of growth. The crystal shown in Plate 28, A, B, seems to be the result of more complex growth. Its development is interpreted as the formation of a core of bubbly quartz followed by a coating along the prism of relatively clear quartz. If this clear quartz ever completely surrounded the inner core it must have been in part removed, for the pyramid is capped by a later growth of bubbly quartz that differs in orientation from the earlier. Judd,[51] however, suggests that even vacuoles of this type may be due to the penetration of quartz by solutions working along planes of chemical weakness which may be coincident with crystallographic planes.

The later linearly arranged vacuoles, as before noted, tend to lie in planes and in many places show definite association with the later minerals, particularly the carbonate and to a less extent the sulphides and gold. The formation of these vacuoles seems to have involved actual opening of the quartz to allow the introduction of the material—gas or liquid—contained in the bubbles and later healing of the fissures through pressing together of the walls. But general lack of close parallelism (pl. 29, B) of the later rays shows that the crystallographic structure of the quartz did not ordinarily control such fracturing. The peculiar tubes shown in Plate 30, C, D, and possibly also the later bubble lines that appear to be single rays might be due to solution of the quartz. The absence of these later vacuoles in the zones of microbrecciation, although they are later than the microbrecciation, is considered to be due to the intimate fracturing of the quartz by the prior brecciation, which allowed abundant opportunity for the passage of the liquid between the broken grains.

As moderate heating, to the melting point of Canada balsam, did not perceptibly change the appearance of either bubble or liquid, it is thought likely that the liquid is principally water[52] and that liquid carbon dioxide is not present, although this has been reported from vacuoles in the quartz of California veins.[53] Similar aqueous inclusions, also probably essentially free from carbon dioxide, have been noted by Lindgren[54] in the quartz veins of Nevada City and Grass Valley and other California districts. The fluid inclusions of the quartz of the Alpine veins, analyzed by Königsberger and Müller,[55] were found to contain 85 per cent of water, 5 per cent of carbon dioxide, and 10 per cent of dissolved salts consisting of alkaline carbonates, chlorides, and sulphates, named in the order of abundance.

[49] Storz, Max, Die sekundare authigene Kieselsäure in ihre petrogenetisch-geologischen Bedeutung, Teil 1: Monographien zur Geol. und Pal., Ser. 2, Heft 4, p. 33, 1928.

[50] Sorby, H. C., On the microscopical structure of crystals, indicating the origin of minerals and rocks: Geol. Soc. London Quart. Jour., vol. 14, pp. 453–470, 1858.

[51] Judd, J. W., On the relations between solution planes of crystals and those of secondary twinning and on the development of negative crystals along the former: Mineralog. Mag., vol. 7, pp. 81–92, 1886.

[52] Chamberlin, R. T., Gases in rocks, p. 63, 1908.

[53] Courtis, W. M., Gold quartz: Am. Inst. Min. Eng. Trans., vol. 18, p. 640, 1890.

[54] Lindgren, Waldemar, The gold-quartz veins of Nevada City and Grass Valley, Calif.: U. S. Geol. Survey Seventeenth Ann. Rept., pt. 2, p. 130, 1896; Mineral deposits, 3d ed., p. 624, 1928.

[55] Königsberger, J., and Müller, W. J., Über die Flüssigkeitseinschlüsse im Quarz alpiner Mineralklufte: Centralbl. Mineralogie, 1906, pp. 72–76.

It is believed that the following inferences may be drawn from the study of observed relations of the vacuoles:

Many quartz crystals show evidence of pulsatory growth; the zones of bubbly and relatively clear quartz indicate changing composition of the solution from which they were derived or changes in the speed of growth of the crystal.

Microbrecciation involved fracturing of the crystals on a much smaller scale than is indicated by the present size of the grain in the areas of microbrecciation, for it resulted in release of the material filling the closely spaced vacuoles. This inference in turn requires that the quartz, cleared of the liquid during microbrecciation, must have been able to flow sufficiently to fill the voids. Strain twinning likewise involves actual movement of the quartz along at least one set of twin lamellae.

The development of the later vacuoles that occur in rays crossing crystal boundaries probably involved solution along actual fracture planes and partial recementation of the quartz. This fracturing did not cause rotation of the grains, as in microbrecciation; therefore the healed quartz preserves its orientation. The rare occurrence of tubular cavities in quartz is suggestive of solution.

The formation of the later series of cavities seems to have been closely associated with the introduction of minerals that replaced the quartz, particularly the carbonate.

The later quartz, though in part contemporaneous with the carbonate, is free from vacuoles, and the clear quartz is in sharp contact with the older bubbly quartz, which it veins. This feature helps to distinguish the two generations of quartz.

PLAGIOCLASE

Plagioclase feldspar is present both in the veins and as a replacement mineral in the country rock. It is not abundant as a vein mineral and where present is found only close to the walls and with quartz in little veinlets in the wall rocks near the veins. Its intergrowths with quartz (pl. 26) suggest contemporaneity of crystallization, but it is also found crystalline against the quartz and nowhere veined by the quartz. The plagioclase may in places have formed slightly later than the quartz and at its expense. The plagioclase within the veins, like the quartz which adjoins it, commonly contains abundant vacuole inclusions.

The plagioclase of this stage of mineralization occurs most abundantly in the country rocks, particularly in those which were originally feldspathic, the granite and gabbro. In the granite it apparently replaces all the original constituents, but especially the microcline and oligoclase. Here replacement began

around the edges of the older feldspars, probably favored in places by fracturing around the larger grains. Close to the veins, where replacement has been most complete, the rock is essentially a mass of fine-grained feldspar. In places small feldspar and quartz-feldspar veinlets in which the new feldspar is in optical continuity with the old cross the original feldspars of the granite. (See pls. 31, B, C; 32, C.) Where the older feldspar, particularly the oligoclase, contains abundant epidote and mica the later feldspar replaces these as well.

In the gabbro the saussurite has been replaced by clear albite (pl. 34, D), but the residual hornblende and early chlorite derived from hornblende were not attacked in this stage of mineralization.

The plagioclase is variable in composition and ranges from albite to oligoclase. The feldspar of the veins and little veinlets is rarely twinned, and consequently close determination of its composition is difficult. The average index of refraction is commonly about equal to or slightly higher than that of Canada balsam but distinctly lower than that of quartz, and the angle of extinction on twin lamellae of the rare twinned individuals indicates a composition of about $Ab_{90}An_{10}$. Pure albite with an index distinctly lower than that of Canada balsam was only rarely found.

In portions of the granite where oligoclase is the prevailing feldspar the later plagioclase has the same composition (about $Ab_{75}An_{25}$) and in places shows optical continuity with the older oligoclase. In such areas adjoining orthoclase crystals are likewise crossed and partly replaced by fine-grained later oligoclase. In parts of the granite where albite or microcline is abundant the new feldspar formed is a calcic albite. Where the original feldspar, as in the gabbro, was more calcic than oligoclase the resulting new feldspar is albite, commonly about $Ab_{90}An_{10}$.

BARITE

Barite is not an abundant mineral in the veins and was found only in thin sections of specimens from the Oriental, Kenton, and Evans veins, which are near the belt of small granite intrusions. It is everywhere closely associated with the wall rock. Most commonly it is found bordering small fragments of altered country rock within the vein. It is commonly intergrown with the quartz in such a way as to suggest contemporaneous crystallization. In a specimen from the Evans prospect barite crystals were found in albitized granite, as if here of later development than the albite. (See pl. 39, C.)

CHALCEDONY

A thin section cut parallel to the base of a large quartz crystal showed a single small spherule of chalcedony (pl. 32, D) about 0.1 millimeter in diameter.

The center is hollow, but this may be due to the tearing away of a portion during grinding. Outside the sphere through an area 0.3 millimeter in diameter are small circular segments of chalcedony within the quartz. The quartz between the outer chalcedony segments and the inner sphere contains vacuoles, but the chalcedony itself seems to be clear. There is nothing in the appearance of the surrounding quartz to suggest later introduction of the chalcedony; its annular intergrowth with the quartz implies that it is contemporaneous.

CARBONATE AND SERICITE

As already noted, carbonate and sericite are characteristic of a later stage of mineralization. Nevertheless the different stages overlapped to the extent that very small quantities of carbonate and sericite were formed contemporaneously with the earlier quartz. Several thin sections of single quartz crystals, including the large quartz crystal that yielded the spherule of chalcedony, showed isolated carbonate grains completely surrounded by quartz. There is nothing in the appearance of the quartz to suggest later introduction of the carbonate. These carbonate grains are all small, the largest found being about 0.04 millimeter in its longest dimension. A fibrous mineral that occurs in small irregular patches in some of the larger quartz crystals is probably sericite, as its refractive index is higher than that of the surrounding quartz and it is strongly birefringent. It is scattered here and there throughout the larger quartz grains and, like the chalcedony and carbonate, is completely inclosed in the quartz and therefore assumed to be contemporaneous. One of the quartz crystals examined showed small shreds of sericite arranged in zones parallel to the prism faces. (See pl. 32, *A*, *B*.) In a few of the vacuole inclusions containing liquid were seen little clusters of very minute fibers. It is suspected that these may be sericite, but they are too small to be identifiable.

APATITE

Minute grains of apatite were observed in altered gabbro from the Oriental mine. These were present both in the chlorite which replaced the hornblende and in the areas of secondary albite.

BEIDELLITE

Some of the druses in the veins are filled with a gray clayey substance which was found to be minutely crystalline and was identified by W. T. Schaller as the clay mineral beidellite ($Al_2O_3.3SiO_2.nH_2O$). Not only do the quartz crystals project into the beidellite, but prismatic crystals carrying double-pyramid terminations are completely inclosed in it. These range from a fraction of a millimeter to 2 or 3 centimeters in length and are unsupported except by the surrounding beidellite and are not corroded. It is therefore inferred that either the beidellite was formed contemporaneously with the quartz or it represents a complete replacement of solid material in which the quartz crystallized. If the first alternative is the true one, the beidellite is to be considered a primary vein-forming mineral of this stage of mineralization; on the second hypothesis its age is indeterminate. Supergene infiltration of clayey matter into the vugs is not possible to reconcile with the presence of quartz crystals within the beidellite.

CARBONATE STAGE

CHARACTER OF MINERALIZATION

After the formation of the quartz veins and the accompanying minor alteration of the wall rock came the introduction of additional minerals, chiefly carbonate, sericite, and mariposite, with sulphides, graphite, free gold, and small amounts of quartz, chalcedony, and opal. It is not intended to imply that any long interval separated these stages, and they probably overlapped to a slight extent, yet the evidence indicates not only that the crystallization of the early quartz was complete before the introduction of any appreciable amount of these later minerals, for no quartz of the early type cuts or replaces the carbonate, but that movement had taken place along and within the veins of crystalline quartz prior to the introduction of the carbonate. This is clearly shown by the fact that these minerals not only replace the quartz but are found along fault planes between strands of the quartz, in fractures crossing the zones of microbrecciation, and also as replacement minerals later than the microbrecciation. (See pl. 34, *C*.) Had the minerals of this stage not been distinctly later than the early quartz, which itself was not all deposited at one time, we should expect to find places where carbonate and sericite accompanying a strand of quartz were earlier than the quartz of an adjoining strand.

The minerals belonging to this stage are found both as replacement minerals in the wall rock and within the quartz veins. Most of these minerals occur in both positions, but certain of them, including most of the sulphides and gold, are found only within the earlier vein quartz. The wall-rock replacement is of far greater magnitude than the deposition of minerals within the quartz, though the latter, as it includes the gold, is of greater economic importance.

The replacement of wall rock by minerals of this stage, of which carbonate and mica are the most abundant, was superposed on the relatively minor wall-rock alteration that accompanied the earlier stage of mineralization. The carbonate and mica, however, have also replaced the vein quartz adjoining the altered rock or surround inclusions within the vein, but commonly they occur only as a narrow fringe at the border of the quartz. Inspection of a hand specimen showing such a fringe gives the impression that the carbonate is the earlier or is contemporaneous with the quartz; the true relations can be seen only in the thin sections. Pyrite and arsenopyrite occur in association with the carbonate and mica, but the other sulphides are confined to the veins.

Within the veins these later minerals commonly occur in small streaks through the quartz. These may consist only of the lighter-colored minerals—carbonate, mica, or quartz—but very commonly, particularly in the productive portions of the veins, there are numerous small dark streaks in which minute crystals of sulphides may be distinguished. These may follow lines of shear in the quartz, generally parallel to the walls, forming "ribbon quartz," or they may cross the vein or may be found as closely spaced irregularly curving streaks, forming the "crinkly banding" of the miners. (See pls. 40, D; 42, C; 44.)

CARBONATE

Carbonate grains belonging to this stage of mineralization examined by the aid of refraction oils almost all showed a maximum index of refraction consistent with ankerite; very rarely siderite or a mineral intermediate between siderite and ankerite was indicated. Variation in composition is in places indicated by zoning (pl. 33, A, B) and also by the occasional presence of two carbonate minerals of distinctly different age. As the mineralogic study was not exhaustive, it is thought best to use the less definite term "carbonate" in general, confining the use of more specific terms to descriptions of specimens on which tests have been carried out.

The wall rocks of the veins show everywhere more or less replacement by carbonate. This is irregular in distribution; in places there is carbonatization of the wall rock to a great distance from the vein; elsewhere the wall rock seems to be essentially unaltered. Thin sections of rock containing no carbonate visible to the eye, however, usually show more or less carbonate (pl. 33, A, B), either as sporadic crystals or in vein-like bands.

Two different habits are observable in the carbonate belonging to this stage. In most places, both in vein and in wall rock, it is fairly coarse grained and, though most commonly without sharply marked crystal outlines, shows continuity of single crystals over areas as much as several millimeters in diameter. Elsewhere it is in the form of fine-grained aggregates that show a feathery extinction. This difference is not due to

later pressure, because in places the two varieties occur together. There is some evidence that the fine-grained variety is the later of the two (pl. 33, C), but it is not known whether the difference in crystallization represents a difference in composition.

Within the quartz veins carbonate occurs both in veinlets (pl. 33, D, E) and in masses replacing the quartz (pls. 24, B; 28, D; 34, C). It is most commonly found within the veins either replacing wall-rock inclusions and also to some extent the surrounding vein quartz or in irregular replacement masses extending into the vein for a short distance from the carbonatized wall rock. The carbonate is later than the microbrecciation of the quartz, for grains of carbonate have developed at the expense of quartz within the areas of microbrecciation. (See pl. 34, C.)

The small veinlets that cut the quartz here and there contain carbonate grains. Usually the carbonate associated with such fissures is clearly not primarily a fissure filling but replaces the adjoining quartz outward from the fissure. Only in a few places where carbonate accompanies later quartz does it appear to have been deposited in fissures. Veinlets consisting entirely of carbonate (pl. 33, D, E) also cut both the quartz and the mica veinlets. In places fissures in the quartz contain angular quartz fragments cemented by carbonate (pl. 34, A); and ankerite was found coating quartz crystals in vugs.

MICA (MARIPOSITE AND SERICITE)

At least two varieties of mica occur in the same relations to the other minerals and are best considered together. One is the chromium-bearing variety mariposite, and the other may be either sericite or colorless mariposite, or both may be present. Under the microscope the green mariposite shows distinct pleochroism and high birefringence. Hand specimens containing abundant chlorite may resemble those rich in mariposite, but the green of the chlorite is of a deeper and duller shade. Under the microscope the higher birefringence of the mariposite is distinctive. The mariposite seen in thin section shows a considerable variation in color, which is presumably due to a variable chromium content. As detailed microchemical tests did not appear to be justified it is convenient and is probably essentially accurate to regard as mariposite any micaceous, highly birefringent mineral that shows a distinct green color and to refer to the colorless mica as sericite.

Both varieties occur principally in the altered country rock close to the veins but are also present in veinlets crossing the older quartz and as replacement deposits in the quartz. Sericite is the more abundant but being inconspicuous in color is easily overlooked in the hand specimen. The vivid green color of the mariposite, on the other hand, is so conspicuous that the observer tends to exaggerate its relative abundance. The bright-green altered country rock known to the miners as " blue jay " may be found, when examined under the microscope, to contain over 90 per cent of carbonate, the small content of green mariposite being of so intense a color and so evenly distributed that it colors the whole rock.

The two micaceous minerals, although of the same mode of occurrence, are not commonly found together. The controlling factor in the distribution of green mariposite is the presence of serpentine. Although mariposite is occasionally found in both wall rock and vein far from any known serpentine, it is nevertheless much more abundant in the neighborhood of the serpentine, and in veins like the Irelan, which so far as known cut no serpentine masses, it is relatively rare. It is therefore inferred that the chromium which determines the color of the mineral was derived from the serpentine.

Mica is closely associated with carbonate in the altered wall rock, and, although the two are intergrown, there is commonly a rough banding, bands a millimeter or more in thickness consisting almost entirely of carbonate alternating with those containing abundant mica. Where the mica is mariposite this gives a rock of very striking appearance, and the effect is heightened where to the close banding of white and vivid apple-green there is added the dark gray of residual streaks of only partly replaced country rock. Microscopic examination of such altered rock shows that in places such a mixture of carbonate and mica has replaced quartz which is itself the product of an earlier replacement of the original wall rock.

Mica has not migrated as far from the vein as the carbonate and is commonly found only in the rock immediately adjoining the vein or at most 10 or 20 feet distant. A partial exception to this is found where veins, such as the Oriental and Gold Canyon, fray out in the serpentine. Here mariposite may be found in the altered serpentine over a considerable area, though it is probably not present far from any one of the fissures into which the vein splits on entering the serpentine.

The close association of mica and carbonate, particularly ankerite, suggests that they are essentially contemporaneous. Apparently, however, the mica is somewhat the later of the two, for specimens were found which show an apparent replacement of carbonate by mica, though the evidence from microscopic examination is not conclusive.

Both mariposite and sericite are found within the veins, where, as in the altered wall rock, the green mariposite is more abundant in the vicinity of the serpentine. In part they occur in positions which indicate that they have replaced wall-rock residuals (pl. 35, A) and the surrounding quartz as well. More commonly, however, mica is present in little irregular veinlets crossing the quartz, accompanied by finely divided sulphides, carbon, and in places gold. In these veinlets the contact with the quartz suggests that the mica has commonly formed by replacement of the quartz along minute fissures.

There is evidence that the mica, like the carbonate, has been introduced later than the crystallization of the quartz, for mica is present as a cementing mineral of fault breccias which contain fragments of vein quartz. (See pl. 34, B.)

CHLORITE

Chlorite whose age relations show it to belong to this stage of mineralization is, like the earlier chlorite, confined to the altered wall rock and was found within the veins only at a very few places, replacing wall-rock fragments. It is not of widespread distribution and was found only in hornblendic wall rocks, particularly gabbro, where chlorite of the earlier stage has developed. (See pls. 20, A; 34, D.) There its occurrence suggests that it may be in part the product of a recrystallization of the earlier formed mineral. It seems to be of later origin than at least part of the carbonate and mica, for it replaces both. (See pl. 33, C.)

TALC

Talc occurs sporadically in the altered wall rock, where it has been developed chiefly at the expense of serpentine, and very rarely in hornblende schist, gabbro, and slate. It is not clear why serpentine next the vein should in places be altered to talc rather than as usual to a carbonate-mariposite mixture. In the hand specimen the talc is dull greenish gray and readily distinguishable from the micaceous alteration product by its more massive character, the lack of accompanying carbonate, and particularly its lower hardness. In thin sections of a rock completely altered to talc the much finer grain, botryoidal texture, and lack of schistosity are generally distinctive, but in many places it was not possible to differentiate certainly between talc and sericite.

The talc is normally confined to the wall rock and separated from the quartz by later slips, but in places (pl. 35, C, D) there is a suggestion of replacement of the vein quartz by talc. In the specimen from the Brush Creek mine chlorite, arsenopyrite, and gold are present in the talc, probably as residuals from the replacement of vein matter by talc.

CARBON (GRAPHITE)

Specks of opaque matter without metallic luster are found in the dark bands of the ribbon quartz, with both the straight and the crinkly banding. Tests by

W. T. Schaller showed these to be carbon, though, owing to the minute size of the particles, it could not be definitely determined whether or not this is present in the form of graphite. The amount present is small, but the carbon was found to occur in all the dark bands tested for it.

The carbon is almost exclusively deposited within the veins, but in specimens from the Plumbago and Oriental mines altered wall rock close to the veins showed streaks of the same graphitic material along shear planes. As far as could be determined this wall rock was originally gabbro, and only gabbro was found in the vicinity.

PALYGORSKITE

The mineral palygorskite, commonly known as "leather gouge," was found only in the Plumbago mine. It occurs principally along the hanging wall of the vein, between vein and altered country rock, at a few points in the lower levels, but was also found to penetrate the quartz for a few inches along sharply defined fissures, both parallel and normal to the wall. The mineral occurs in sheets, at most a few millimeters thick. These have a leathery texture and are capable of being folded double without breaking. This leatherlike aggregate consists of minute crystalline fibers having a maximum index of refraction not exceeding 1.52. For the most part these sheets are fairly pure, but here and there are small areas in which small amounts of very fine-grained carbonate, probably ankerite, and minute arsenopyrite crystals are intergrown with the palygorskite fibers. This intergrowth indicates that here the palygorskite is not, as has been elsewhere considered,[56] a supergene mineral formed in the zone of weathering but a primary mineral later than the quartz, which on the basis of its association with carbonate and arsenopyrite is believed to belong to this stage of mineralization.

RUTILE

Small needles of rutile were observed in the clear oligoclase feldspar which replaced the original oligoclase of the granite bordering the Oriental vein (pl. 35, B), and rutile also occurs within the altered granite as an alteration product of titanite. It appears to be later than the albitization of the granite and may be contemporaneous with the development of the carbonate. Prisms and needles of rutile were found in the chlorite in altered schist adjoining the North Fork vein, and a specimen of serpentine from the mass southeast of the Plumbago mine showed small rutile needles between the antigorite flakes.

---

[56] Fersmann, A., Ueber die Palygorskitgruppe: Acad. imp. sci. St.-Pétersbourg Bull., vol. 2, p. 271, 1908.

QUARTZ, CHALCEDONY, AND OPAL

A little silica, principally in the form of quartz, was added to the veins during the carbonate stage of mineralization. The principal feature distinguishing this later quartz from the older is its freedom from the vacuole inclusions that are almost universal in the older quartz.

The later quartz is found most abundantly in the kind of vein filling called "headcheese" by the miners. (See pls. 14, A, B; 36, C.) This consists of a breccia containing fragments, chiefly of vein quartz but in places also of altered country rock and rarely of earlier arsenopyrite cemented by fine-grained clear vein quartz. Breccia of this type is most commonly found along the wall of the vein, especially where there is a marked change in strike and dip, but in places small veins of the breccia type cross the older quartz. The later quartz in such places is not in optical continuity with the quartz of the walls but forms a very fine-grained aggregate. The inclusions of older quartz are so close together as to suggest mutual support, so that there may have been an open cavity filled with rubble of older quartz before deposition of the later fine-grained quartz. In the Eldorado mine, on the outer side of the great bend in the vein, there is "headcheese" breccia along the hanging wall (fig. 9; pl. 14, A), and extending downward into the older quartz are gash veins composed of the same material. These have a maximum width of a couple of inches at the top and taper down to a point a few feet from the wall. In a small breccia vein in the Sixteen to One mine opal cements fragments of coarser vein quartz. (See pl. 39, B.)

Narrow bands of clear quartz, commonly less than a millimeter in width, are abundant in the older cloudy quartz. (See pls. 25, C, D; 27, D; 35, C; 37, A, B.) This later quartz is generally in optical continuity with the older, but where the veinlet is above a fraction of a millimeter in width the later quartz is fine grained and not so oriented. (See pl. 38, B.) The greater sharpness of the boundary between the new clear quartz and the older bubbly quartz and the greater regularity and continuity of the veinlets distinguish the later quartz from older quartz that has been freed from vacuoles by microbrecciation.

Opal and chalcedony are also found in minute veinlets but are less common than quartz. In some veinlets all these minerals are present (pl. 36, A, B), but it is more usual to find opal veinlets crossing those of the clear quartz. Opal is more abundant than chalcedony, but a little chalcedony commonly occurs with the opal and inclosed in it. (See pls. 36, A; 37, D.) In one speciman opal and chalcedony fill a vug in the older quartz. (See pl. 37, C.)

That the clear quartz veinlets represent addition of new material and not merely recrystallization of the older quartz is shown in Plate 37, *A, B*, by the displacement of the twin lamellae of the strain-twinned quartz. The new quartz, whose orientation is controlled by the older crystal, has clearly filled an open space, for if it were merely a rearrangement of the material in the vein the twin lamellae would either be lacking or continuous across the clear veinlet without change in direction. Veinlets of later quartz also cross areas in which the older quartz has suffered microbrecciation. (See pl. 38, *A, B*.)

The mineralogic associations of the later quartz show clearly that it is contemporaneous with other minerals of this stage. Carbonate and pyrite occur in these veinlets (pl. 38, *C*), and in one specimen a clear quartz veinlet within the quartz vein was seen to pass into a cross-fiber carbonate vein in the carbonatized wall rock (pl. 25, *C, D*). Minute crystals of pyrite and an unidentified gray metallic mineral were found within the opal (pl. 38, *D*), and gold was seen in one specimen of breccia cemented by fine-grained quartz and opal (pl. 39, *B*).

Evidence of replacement of wall rock by this later silica was observed only in a single specimen, from the footwall of the Eureka vein. (See pls. 37, *D*; 39, *C*.) This shows granite largely altered to a mixture of albite and barite and later partly replaced by arsenopyrite and a little carbonate. This material is cut by small veins consisting of opal and chalcedony, and the rock adjoining the veins has been replaced, chiefly by opal but close to the veins by finely crystalline quartz. The opal of the wall rock has largely replaced the albite and remnants of the original feldspars but has left the barite untouched. The arsenopyrite crystals in the opalized area show cores of opal, though the outer edges are unbroken. This may indicate centrifugal replacement of arsenopyrite by opal.[57]

SULPHIDES

The sulphides of this stage of mineralization include pyrite, arsenopyrite, galena, tetrahedrite, sphalerite, chalcopyrite, jamesonite, and possibly pyrrhotite. Pyrite and arsenopyrite are present in both the veins and the altered wall rocks, but the other sulphides are found only in the veins and occur in irregular patches within the quartz, commonly more or less closely associated with carbonate or mica, in little veinlets that cross the quartz, and in the bands of the ribbon quartz associated chiefly with mica and graphite.

Arsenopyrite and pyrite are found in the altered wall rocks in both the carbonate and mica, and also more rarely disseminated in the less altered country

rock not in close association with either of these. Neither mineral is present in any great abundance; probably rarely are they so abundant as to form even 5 per cent by volume of the altered wall rock, and more commonly they occur sparsely scattered through the carbonate-mica rock. Pyrite is the more widespread and is found sparingly at considerable distances from the vein, commonly in minute pyritohedra or cubes. In many places a rough linear arrangement of the pyrite crystals in the country rock suggests that deposition was controlled by a fissure.

Arsenopyrite is not found far from the veins and, like the arsenopyrite of the earlier generation, seems to show some preference for the vicinity of the serpentine, though the correlation is not at all close. Arsenopyrite seems also to be rather more abundant in the gabbro than in the schists. In the altered wall rock it occurs as small crystals, rarely exceeding 5 millimeters in length. (See pl. 39, *C*.) Its common form is a combination of dome and prism, giving an octahedral appearance; less frequently small slender prisms are found.

Pyrite and arsenopyrite are more widespread within the veins than the other sulphides. Arsenopyrite belonging to this generation is not generally found in close association with other sulphides in the veins but commonly occurs either as irregular linear masses replacing the quartz, associated with mariposite (pl. 33, *D, E*), or replacing the quartz bordering the large fractured crystals of arsenopyrite of the earlier stage (pl. 39, *A*). It also occurs in small crystals in little veinlets in the quartz and in the dark streaks of the ribbon quartz. Pyrite is present in the ribbon quartz and is also found in crystals and crystalline masses replacing the quartz, most commonly near altered wall-rock inclusions. In many places it has formed a nucleus for the deposition of other sulphides, which surround and partly replace it. (See pl. 40, *A*.)

The other sulphide minerals occur only in the quartz, generally connected definitely with later fissuring. Tetrahedrite, which carries a little arsenic, is probably the most abundant and was found principally in small irregular spots replacing pyrite and also, though rarely, in masses as much as 2 or 3 centimeters in diameter. Chalcopyrite is less abundant and is closely associated with tetrahedrite. Minute needles of a metallic mineral found in clusters in small vugs of the quartz (pl. 40, *B*) were identified as jamesonite by M. N. Short, and it is thought probable that similar needles within the quartz (pl. 40, *C*) are the same mineral. These are found scattered apparently at random within the quartz, though usually near areas containing mariposite or sericite.

Wherever the relations of the galena to other sulphides can be observed it seems to be the latest metallic

[57] Lindgren, Waldemar, Mineral deposits, 3d ed., p. 210, 1928.

mineral except the gold and has replaced all the earlier sulphides, especially the pyrite. Locally even the minute slender jamesonite needles are replaced by galena (fig. 15), but galena also occurs in acicular form, probably as an original filling of minute fissures. (See pl. 30, *A*.) It is apparently contempora-

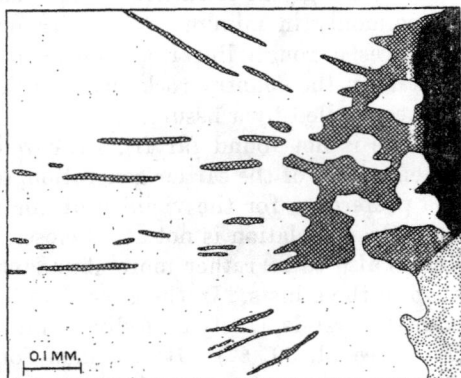

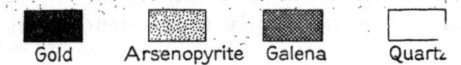

FIGURE 15.—Radiate needles of galena, probably replacing jamesonite

neous with the gold, and gold and galena are commonly closely associated. (See pl. 41, *A*.) Like the other sulphides it is found most commonly in association with the dark threads of the ribbon quartz, though usually not within them.

Pyrrhotite has been found in a vein only in a single locality, on the tenth level of the Tightner (Sixteen

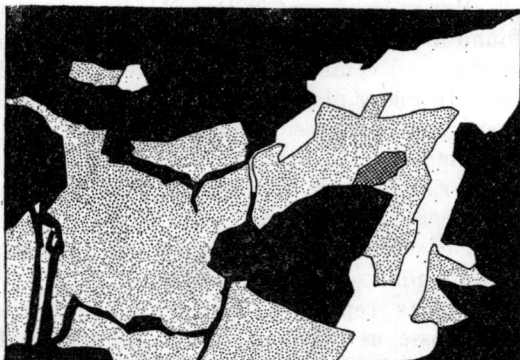

FIGURE 16.—Gold and galena in arsenopyrite. Gold in part replaces quartz which veins the arsenopyrite

to One) mine, where small grains were observed in the quartz, not in association with other sulphides. It has not been found in recent deeper work on the same vein. Its relation to the quartz and earlier sulphides is unknown, and it may belong to the preceding stage of mineralization. Small patches of pyr-

rhotite are fairly common in the hornblendized gabbro throughout the district, apparently not in any way associated with the veins.

GOLD

The spatial occurrence and distribution of the gold are considered in detail on pages 52–60. Here only a few features bearing on its paragenesis and relations to other minerals will be noted.

The gold is clearly later than the early quartz and is also later than the other sulphide minerals except possibly galena, with which it may be contemporaneous. Free gold is found only within the veins themselves, though rarely, as in a specimen of rich ore from the narrow Clinton vein of the Rainbow mine, threads of gold extend from the quartz into the adjoining altered county rock for distances of less than a centimeter. The pyrite and arsenopyrite of the altered

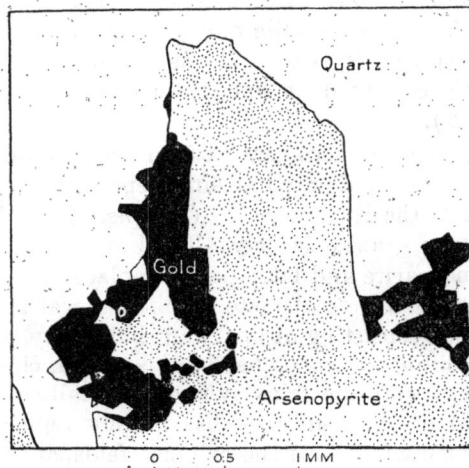

FIGURE 17.—Gold replacing quartz and arsenopyrite

wall rocks are, however, slightly auriferous, and assays of the sulphide-rich portions of the wall rock show a low tenor of gold. Assays of the relatively heavily pyritized granite in the Oriental mine, for instance, are said to indicate a considerable body of rock carrying $5 or less to the ton, but here there are also a large number of small quartz veins, so the relatively high tenor may be due to gold in the quartz.

Free gold is commonly found in association with the coarse-grained early arsenopyrite. (See pls. 21, *A*, *B*; 41, *A*, *B*; figs. 16, 17.) It has replaced the arsenopyrite but even more abundantly has replaced quartz adjoining the arsenopyrite. Where there is abundant gold associated with arsenopyrite a little galena is usually to be found in the same relation to the arsenopyrite as the gold. Sericite or mariposite is a nearly constant associate of gold. The veinlets of later quartz and the later quartz cementing breccias are commonly not gold bearing, but gold has been observed here and there in this later quartz. (See pl. 39, *B*.)

The minute veinlets of later minerals that cross the quartz or form the ribbons of the ribbon quartz in places contain gold in the vicinity of other sulphides, especially where the later arsenopyrite is abundant. Generally where associated with these veinlets the gold occurs as a replacement mineral, usually of the older quartz but rarely of the carbonate also. It has replaced the later pyrite and tetrahedrite but not the galena. Where found in areas of microbrecciation the gold is in the zones containing the most minutely divided quartz. (See pl. 42, A.) Although much of the gold is extremely coarse, thin sections examined under the microscope show individual grains down to the limit of vision (about 0.001 millimeter). (See pl. 42, B.) Many of the individual particles show definite crystal faces. The contacts of larger masses of gold with other earlier minerals commonly show a peculiar fuzzy outline due to the presence of minute projections from the gold into the host mineral. Sometimes crystalline gold, in small octahedra, is found in vugs in the quartz.

The sulphides contain little invisible gold. Concentrates in general show a tenor of less than $20 a ton. In one or two mines assays of concentrates, particularly where the older generation of arsenopyrite is abundant, may be as high as $100 a ton, but the minute threads of gold that cross the arsenopyrite suggest that such concentrates may owe their unusually high tenor to the presence of free gold not separated from the sulphide matrix in crushing. No tellurides have been found in any of the Alleghany mines.

In all the mines particles and even small nuggets of gold that resist amalgamation are occasionally encountered. As described to the writer, the substance coating the gold is not released by hammering but yields to weak hydrochloric acid.

No analysis has been made of the native gold to determine the exact percentage of silver and base metal, but a review of the production returns of the district as a whole shows that gold averages 84.74 per cent of the total quantity of gold and silver, which is somewhat higher than the fineness of the bullion, as the content of base metal is not considered. The fineness shows a wide variation, however, even within a single stope. Annual returns from individual mines show a percentage ranging from 79.61 to 86.96, but the variation can not be correlated with any of the geologic or mineralogic peculiarities of the different mines. The gold obtained from the Clinton vein of the Rainbow mine was consistently of higher than average grade, the bullion being valued at $18.50 an ounce, as against about $17.50 for the district as a whole. The average seems to be not greatly different from that of other California districts,[58] nor is the

wide range in fineness exceptional, for, according to Lindgren,[59] the fineness of the gold in most of the veins of Nevada City and Grass Valley ranges from 0.800 to 0.860, but extremes show fineness of the vein gold as high as 0.950 and as low as 0.730 and of the bullion as low as 0.650.

FINAL STAGE

The minerals of the last stage of mineralization are inconspicuous in total amount and of no economic importance. Here and there small veinlets of calcite and pyrite, either separately or together, cross both the veins and the altered country rock. The calcite is the only carbonate mineral of this stage and, so far as can be judged from its indexes of refraction and its response to weak acid, is nearly pure calcium carbonate. The pyrite is so minutely crystalline that it requires a strong lens to distinguish the individual crystals. It is commonly of cubic or octahedral habit, in contrast to the dominant pyritohedra that characterize the older pyrite. It occurs not only in little veins crossing the older minerals but in places as a coating, which may be several millimeters thick, on drusy quartz crystals. Marcasite has been noted by local observers in similar situations, and specimens collected in 1913 were found, when unpacked some months later, to have oxidized extensively—a characteristic tendency of marcasite.

Although no definite evidence was obtained, it is thought possible that the minerals of this stage are the result of deposition by supergene waters. The calcite and pyrite occur together and seem to be contemporaneous. Pyrite of similar habit occurs in the Tertiary gravel and clay. Although this later pyrite is not confined to the upper levels of the mines, none of the mines have reached a sufficient depth to make the presence of minerals of supergene origin inherently unlikely.

OXIDATION

The products of oxidation are inconspicuous. The period of erosion which closed with the end of Eocene time left a country of subdued topography in which, it is to be presumed, the outcrops of the veins were but little above the ground-water level. As the veins have been uncovered in the present cycle of erosion, deep canyons have been cut below the Tertiary outcrops, but the work has been too rapid to permit oxidation to keep pace with erosion, and the total volume of sulphides is so small that their oxidation at the outcrop was not sufficient to furnish appreciable amounts of acid to aid materially in the solution and transfer of metals by ground water.

[58] Lindgren, Waldemar, Tertiary gravels of the Sierra Nevada of California: U. S. Geol. Survey Prof. Paper 73, p. 68, 1911.

[59] Lindgren, Waldemar, The gold quartz veins of Nevada City and Grass Valley, Calif.: U. S. Geol. Survey Seventeenth Ann. Rept., pt. 2, p. 116, 1896.

There are small remnants of the results of oxidation accomplished in Tertiary time. The alteration of serpentine at the vein contacts to iron-rich carbonates accompanied by more or less pyrite has yielded in places a massive limonitic gossan. Huge boulders of gossan of this type occur in the old gravel workings of Chips Flat. The partial oxidation of the coarsely crystalline sulphides found in the upper tunnels of veins outcropping under the lava also must date from the period of Eocene erosion. Such oxidized sulphides are not found more than a comparatively short distance below the old surface. Apparently there has also been some very slight solution and redeposition of gold close to the old surface. Specimens of high-grade ore from the old upper tunnel of the Tightner mine, close beneath the lava, show minute specks of reddish spongy gold resting upon the surfaces of crystals of light-yellow gold in the quartz. Gold of this type may have been released from oxidizing sulphides and redeposited on the primary gold. The total amount present is so minute that any such enrichment is without economic significance.

Oxidation initiated during the present cycle has produced inconspicuous results except at the outcrops, but here and there throughout the mines are small patches of sulphides partly oxidized to limonite with a little scorodite and deposits of iron oxide or limonitic sludge along watercourses.

There is nothing either in the general geologic relations of the ore bodies or in the mineralogy of the ores to indicate that the high-grade ore is in any way related either to the present surface or to that existing at the end of the Eocene epoch.

ORE SHOOTS

CHARACTER

The characteristic ore of the Alleghany district is of "high grade"—that is, it shows coarse gold in considerable amounts. As a rule such ore carries from $2,000 a ton in free gold up to many times this amount.

The following recorded occurrences of especially rich ore show how rich some of the high-grade ore has been. A single shipment from the Rainbow mine weighing 1,935 pounds yielded $116,337, and a single specimen, weight unstated, was calculated to contain gold to the value of $20,468.[60] In recent work on the Clinton vein in the same mine $30,000 was obtained from 600 pounds of ore. In the Sixteen to One mine one lot of 80 pounds mined in 1924 yielded $5,000, and in 1928 a single piece weighing 160 pounds was found to contain $28,000 in gold.

The high-grade shoots range in size from "bunches" yielding a few hundred dollars to those from which many thousand dollars' worth of gold has been taken. The largest shoot in the Sixteen to One mine yielded a total of nearly $1,000,000. The richest ore in this shoot was confined to an area less than 40 feet square on the vein and came mostly from the 2 feet of quartz next to the hanging wall. A shoot that yielded $736,000 from an area of 14 by 22 feet on the vein was mined in the Oriental mine. In several mines single shoots have produced over $200,000 each. "Mill rock" shoots, with a tenor of perhaps $10 to $20 a ton, in which gold accompanied by mixed sulphides is disseminated through a considerable area in the vein, are rare in the district, though a few such shoots have been mined, and in several mines there have been areas in which small patches of high-grade ore, separated by barren quartz, were sufficiently close together to permit stoping of a considerable area of the vein.

The quartz outside the high-grade shoots may be practically barren, with a tenor of less than $1 a ton, or may contain enough gold to nearly repay the cost of mining. The tenor of the quartz outside the shoots varies considerably in different mines and in different parts of the same mine. In the Plumbago mine, particularly on the lower levels, the gold was in part so widely disseminated that in places sampling of a blocked-out area gave an approximation of the gold content of the block. In the work in the upper levels of the Sixteen to One mine removal by hand of all quartz showing visible gold kept the mill heads as low as $1.50 a ton, but in recent work (in the summer of 1928), principally below the 1,000-foot level, with similar close sorting, the quartz without visible gold, which alone went to the mill, had a tenor of slightly over $5 a ton, about enough to pay the cost of mining and milling, and amounted to about 30 per cent of the production of the mine. The gold content of this low-grade quartz is in part due to gold too finely divided to be easily visible but also in some degree to coarse gold either overlooked by the "specimen bosses" or not visible on the surface of the quartz fragments.

Although barren quartz may adjoin a high-grade shoot, it is usually found that in one or more directions the change is less abrupt than in others, and along such a course for a considerable distance from the high-grade shoot the quartz may contain more gold than the average vein quartz. Nevertheless, the change is in places very sharp, and a single round of shots may bring a drift or raise from almost barren quartz into rich ore.

The high-grade ore rarely extends completely across the vein. It is usual for high-grade ore to be confined to a single strand, with the gold either ceasing abruptly at the edge of the strand or visible for a few inches into the next. High-grade ore is more commonly near

⁶⁰ Hanks, H. S., Second report of the State mineralogist, p. 149, Sacramento, 1882.

the hanging wall than the footwall, especially in the vicinity of lenticular swellings or outward bends of the vein either in strike or dip.

The attempt to determine the factors that control such unusual concentration of gold as that at Alleghany was the prime object of the present studies in both field and office. Naturally a district in which such concentrations exist is a favorable one for the study of the factors that determine the presence of ore shoots, but even the quantities of gold in the high-grade shoots represent relatively small concentrations of the gold considered in terms of percentages. For instance, in high-grade ore with a value of $6,000 a ton the gold represents only about 1 per cent of the ore by weight and less than 0.1 per cent of its volume.

Although the authors were able to see high-grade ore in place in several of the mines, it is inevitable that a study of this kind must depend largely on information furnished by those who have had more constant opportunities for observation. In this respect the authors were particularly fortunate. For many years all those associated with the mines of the district have observed closely the features of the veins concomitant with the occurrence of high-grade ore, and such observations have greatly aided the studies here reported. To a large extent this section will be devoted to a rationalization of these empirical guides and a consideration of their relative importance.

In the development of mines containing ore bodies of this type small high-grade shoots are, of course, easily missed unless the veins are thoroughly explored; on the other hand, it has been found by experience that mining large amounts of nearly barren quartz in order to avoid missing high-grade bunches is not profitable. If, therefore, intensive development can be confined to the portions of veins in which high-grade ore is likely to occur the chances of profitable operation will be greatly increased. It is probable that future operations will involve exploration of the undeveloped territory beneath the lava. This work will be more likely to be successful if development of the unproductive portions of veins is confined to the minimum of necessary drifts, excessive crosscuts are avoided, and only the parts of the veins that appear to show promise are prospected closely.

The sporadic distribution of the high-grade shoots has at times led operators to attempt to stope and mill whole sections of the vein in the hope of encountering " bunches " that would otherwise have been missed. Even where this procedure has been rewarded by the discovery of high-grade ore it has usually not been profitable, though partial exceptions to this generalization should be noted for the Sixteen to One and Plumbago mines, as stated on page 52. Moreover, as patches of high-grade ore tend to occur near together, separated by more or less barren quartz, it is, of course, only common prudence to stope the barren quartz surrounding a high-grade shoot. It is also possible that in portions of a vein which show abundant indications of promise or along which " prospects " are found in the drifts, stoping as a means of prospecting might pay for itself by the discovery of " bunches " that would otherwise be missed.

PRODUCTIVITY OF VEINS OF DIFFERENT TYPES

In the descriptions of the individual mines (pp. 87–135) the writer has attempted to set forth in some detail the features of the high-grade shoots. In the accompanying table are summarized what appear to be the major features of the principal shoots. As in describing ore deposits of this kind any attempt to express the average grade of the ore in dollars to the ton would be meaningless, the attempt has been made to obtain a rough measure of the productivity of the different veins by comparing the amount of work done with the production. In making this ratio no account has been taken of crosscuts and blind exploration work, and on account of lack of data only the principal raises and winzes have been included. It will be readily seen that the chances for error in such a procedure are large, but it is believed that at least in order of magnitude this ratio is significant.

The veins with easterly dip that follow reverse faults have yielded about 90 per cent of the total production of the district, and the ratios of production to development work show that on the whole these have been richer than the other veins. The average production per foot of development work for all the mines of this group is over $150. It is likely that the Kenton, Docile, and Oriflamme veins also follow reverse faults. If so, the proportion of the total production assignable to the veins of this group is greater, but the average production per foot is less. The other eastward-dipping veins have had a relatively small production, and such data as are available indicate that for those following normal faults the output per foot of development has been much less than for the veins of the first group.

The Kate Hardy and Brush Creek mines, which follow the fissure system that borders the district on the west, have both had a considerable production, but the ratio of production to development from the Brush Creek is not known and that for the Kate Hardy is less than the average for the veins that follow the eastward-dipping reverse faults. Most of the other westward-dipping veins have not proved highly productive, and in those mines for which data are available the ratio of production to development is less than for either group of eastward-dipping veins.

Productivity of veins of different types

Veins with easterly dip, following reverse faults

Mine	Production to the end of 1928		Principal features of the high-grade shoots
	Total	Per foot of development on the vein	
Oriental (Oriental vein)	Over $2,000,000, of which $1,260,000 was obtained below third level.	Nearly $400 for mine as a whole; about $250, exclusive of work above third level; $40 for work on adit level alone.	Vein productive where strike changes at serpentine contact. Very rich ore in upper levels, where vein has a steeper than average dip, probably controlled by vein junction. Gold associated with arsenopyrite.
Rainbow (Rainbow vein)	About $2,200,000, of which $990,000 came from Rainbow vein below Groves tunnel.	$190 for Rainbow vein below Groves tunnel.	Old stopes above Groves tunnel show serpentine footwall. Productive area below Groves tunnel follows swelling of vein near junction with minor branches. Gold associated with arsenopyrite.
Rainbow (Parallel vein)	None		
Rainbow Extension	About $5,000	$3 to $4	High-grade ore at vein junction.
Sixteen to One (including workings on vein in Twenty-one and Tightner mines).	$9,000,000	$230	Vein faults 2 serpentine dikes, so serpentine forms either footwall or hanging wall over a considerable portion of the productive area. Productive portion of mine has flatter than average dip. Ore occurs in 2 broad productive zones with flat southerly rake, separated by a barren zone. Quartz in productive portions thicker than average, but thickest swells in quartz are not productive. Junctions of vein with branches control position of much of the ore; also high-grade ore at intersection of vein with westward-dipping faults in upper levels. Several very productive shoots not at junctions but are crossed by small faults. Arsenopyrite less abundant than in other mines with large production.
Eldorado (Eldorado vein)	$325,000	$70 for work on Eldorado vein alone.	Serpentine on either footwall or hanging wall in productive area. High-grade ore associated with sharp change in strike of vein and with splits in vein. Arsenopyrite not abundant. Mill ore at change in strike of vein. Dip of vein in productive area probably steeper than average.
Yellow Jacket (Yellow Jacket vein).	Reported $20,000 from outcrop sluicing.		Vein, the northward continuation of the Eldorado, is here very weak.
Eastern Cross	$2,500	Small	High-grade ore from surface close to serpentine.
Irelan	About $300,000	Probably less than $100	No serpentine cut in workings. Productive area close to junction of 2 branches. Change in strike and dip. Gold largely associated with arsenopyrite.
Plumbago	Over $2,500,000	$86	Serpentine near footwall in upper workings and near hanging wall on lowest level. High-grade ore, generally in small bunches, may occur wherever there is a greater than average thickness of quartz, but thickest lens of quartz was not productive. Some of the high-grade shoots are clearly associated with junctions of minor branches; for others no definite control is evident. Gold generally associated with coarse arsenopyrite.
German Bar	$200,000	About $80	High-grade ore associated with arsenopyrite at vein junction. No serpentine. Considerable mill ore reported to have been mined.
Gold Canyon	$735,000	More than $100	All production derived from region where vein curves toward serpentine contact. Individual shoots in large part controlled by small faults that carry quartz. Gold associated with coarse arsenopyrite.

Veins with easterly dip following normal faults

Mine	Production to the end of 1928		Principal features of the high-grade shoots
Ophir	Over $60,000	About $20	Productive area coincides with crossing of serpentine by vein and with sharp change in strike close to junction with Sixteen to One vein. No high-grade ore along minor intersections.

Productivity of veins of different types—Continued

Veins with easterly dip following normal faults—Continued

Mine	Production to the end of 1928		Principal features of the high-grade shoots
	Total	Per foot of development on the vein	
Eclipse	$20,000 to $50,000	Probably less than $40	No serpentine. Ore shoot at vein junction. Quartz thicker than average.
Morning Glory	$80,000 to $100,000	About $30	No serpentine in workings. Productive area coincides with change in strike. Arsenopyrite not abundant.
Osceola	About $100,000	Less than $50	High-grade ore at junction of 2 branches, also close to serpentine. Arsenopyrite present.
Mariposa	Over $50,000	Probably less than $40	Southward continuation of Ophir vein. Probably principal production derived from area in which dip is flatter and quartz thicker than average.
Eldorado (Bullion vein)	None		Probably on same normal fault as Colorado and Osceola.
Yellow Jacket and Colorado (Colorado vein).	$8,000	$8	Probably on same normal fault as Bullion and Osceola.

Veins with easterly dip, direction of faulting unknown

Mine	Total	Per foot of development on the vein	Principal features of the high-grade shoots
Tomboy	Small		
Eureka	About $2,000		Small shoot of rich arsenopyrite at vein junction. Vein in gabbro near serpentine.
South Fork	Small		Vein thickest where high-grade ore was encountered; also numerous stringers. Gold associated with coarse arsenopyrite.
Amethyst	$6,000	Unknown	Close to serpentine.
Kenton	$90,000	About $50	Principal production from ore close to serpentine contact, where vein changes in strike and breaks up into stringers. Also high-grade ore at distance from serpentine, where quartz is thicker than average and much sheared. Gold said to have been associated with coarse arsenopyrite. Vein probably follows a reverse fault.
Oriental (Alta vein)	Small		A little high-grade ore with arsenopyrite close to serpentine. Conditions in old workings unknown.
Minnie D	do		
Red Star	$80,000	About $40	Reported to have been close to serpentine.
Eldorado (Middle vein)	None		
Docile	$100,000 to $200,000	About $140	High-grade ore close to serpentine. Change in strike and dip of vein.
Eastern Cross (Eastern Star vein).	$500	Small	Close to serpentine.
Oriflamme	None		
Arcade	Small		

Veins following fault along eastern border of district

Mine	Total	Per foot of development on the vein	Principal features of the high-grade shoots
Mammoth Springs	Small		Gold in slate close to vein. Near serpentine.
Kinselbach	$8,000		Surface pockets.
Independence	Unknown		Serpentine close to footwall.

Veins with westerly dip

Mine	Total	Per foot of development on the vein	Principal features of the high-grade shoots
Brush Creek	About $1,000,000	Unknown	Serpentine footwall. In recent workings small high-grade shoots in widest part of vein, ore associated with change in dip. Arsenopyrite present but not abundant in recent workings. Relations in old workings unknown.
Kate Hardy	About $300,000	About $100	High-grade ore in small bunches with erratic distribution. Serpentine near footwall in productive portion. Gold associated with coarse arsenopyrite.
Mugwump	About $5,000	About $4	Change in strike and dip. Vein widest at ore shoot. Arsenopyrite present. No serpentine.
Gold Bug	None		
Diadem	$20,000	Unknown	At serpentine.

Productivity of veins of different types—Continued

Veins with westerly dip—Continued

Mine	Production to the end of 1928		Principal features of the high-grade shoots
	Total	Per foot of development on the vein	
North Fork	$125,000	About $25 for recent work. Possibly over $100 for old work.	High-grade shoot of old workings in contact with serpentine. Recent shoot at junction of main vein and minor branches. Sharp change in dip. Gold associated with coarse arsenopyrite.
Roye-Sum	None		
General Sherman	----do		
Wyoming (Dead River)	Unknown, probably not large.		Probably near serpentine.
Rainbow (Clinton vein)	$50,000	$22	Gold associated with fault that cuts the quartz. No serpentine or arsenopyrite. Gold of more than average fineness. Productive only near junction with Rainbow vein.
Sixteen to One (westward-dipping faults).	None		
Extension of Minnie D	Less than $10,000	About $12	High-grade ore in main vein near intersection but not in fault veins. Gold limited by fault, which cuts the quartz.
Eastern Cross (Western Cross vein).	None		
Eastern Cross (Eastern Star vein).	$500	Small	At serpentine.
Rising Sun	$58,000	Less than $100	Ore largely of mill grade. In part at split in vein.
Dreadnaught	$50,000 to $100,000	Unknown	In part at split in vein.
Belmont	Under $1,000	Small	Probably near vein junction. Ribbon quartz.
Golden King	$100,000 to $250,000	Unknown	Ore largely of mill grade. Details of occurrence unknown.
Mountain View	None		
Maryland	----do		
Cooper	----do		

FAVORABLE STRUCTURAL FEATURES

In all the veins that have been productive there are certain structural features which appear to be generally associated with the high-grade shoots, and within such favorable areas there are closer guides furnished by the mineralogy of the veins.

The gold reached its present position later than the crystallization of the quartz, and it is believed that the major control of the deposition of gold in the high-grade shoots lay in the presence in the vein of conditions which favored the local shattering of the quartz and consequent ready entrance of the solutions that deposited the gold.

It is well recognized that any irregularity of the vein is a favorable indication. Such irregularities may be caused by changes in strike and dip, sudden swellings in the vein, junctions of veins or splits in the vein (the same thing looked at from opposite points of view), and minor faults. The stresses that caused shearing along and in the veins were operative during as well as before and after the stage of mineralization in which the gold was deposited. Where the veins follow straight courses shearing has for the most part taken place along the walls rather than within the veins. The most favorable localities for fracturing of the quartz are naturally where some initial irregu-

larity of the vein tended to favor later fracture. The veins with gentle easterly dips have much more sinuous courses than the steeply dipping veins and, as shown above, have proved far more productive.

The influence of the large masses of serpentine on the courses of the veins has already been noted. The veins do not persist in the larger serpentine masses, but nevertheless these control the courses of the veins close to their contacts. The Gold Canyon (fig. 46), Oriental (pl. 50), and Kenton (fig. 28) veins, within the gabbro but close to the serpentine, all curve sharply toward parallelism with the serpentine contact, and the chief production from each mine came from the part of the vein within the bend.

The veins that cut the smaller serpentine bands also tend to carry high-grade ore in the vicinity of the serpentine. This may be due to deflection of the veins, particularly on the dip, caused by the serpentine, as, for instance, the Eldorado vein (secs. H–H' and I–I', pl. 10, and pl. 56) and less obviously the Sixteen to One (pl. 53). The Rainbow vein was most productive above the Groves tunnel, where the footwall was serpentine. In other mines high-grade ore, though not close to serpentine, was sufficiently near it to suggest a causal relation. In the upper workings of the Plumbago, which were very rich, serpentine occurs close to the vein in the footwall, and an upward projection of the vein would meet the projected serpentine contact at a short distance above the present surface. Similarly the high-grade shoots of the Kate Hardy mine were found a short distance below a probable contact of vein and serpentine, now eroded. It is estimated that 80 per cent of the production of the district has been derived from those portions of the veins where serpentine either formed one wall or was less than 100 feet from the ore. This is believed to be due to at least two causes—the effect of serpentine in causing bends in the vein, both in strike and dip and in branching of the vein, all of which favor later fracturing; and the greater abundance of the early arsenopyrite, which, as shown below, is an effective precipitant of gold, near the serpentine. As is suggested on page 75, there is a possibility of even more direct connection between the serpentine and the gold.

A knowledge of the position of the serpentine in relation to the veins may, therefore, be of value in planning development work. To some extent the extension of the serpentine belts beneath the lava can be predicted from the geologic map that forms Plate 1. The valley of Kanaka Creek, however, contains many small masses of serpentine, and it is to be supposed that similar conditions prevail beneath the lava to the north and south. It is important, therefore, that the position of all serpentine belts encountered in workings, whether drifts or crosscuts, should be accurately recorded and their possible extension relative to the veins carefully studied. The only visible mineralogic guide to the probable presence of serpentine in the vicinity of a vein is the increase in the amount of the mariposite-carbonate mixture (" blue jay ") in the altered wall rock. Therefore the presence of abundant " blue jay " should lead the operator to consider whether the vein may not be near a serpentine mass. As the serpentine contains sufficient magnetite to affect the compass needle, it is possible that magnetic surveys, both on the surface and underground, may in places yield information of value as to the position of the serpentine.

The minor eastward-dipping veins that follow normal faults tend to bend toward the controlling major veins. A good example of this is the change in strike of the Ophir vein close to the Sixteen to One. (See figs. 31, 32.) Such curving, presumably due to control of the minor fissure by the major, gives a favorable locality for gold deposition in the minor veins. The principal production from the Ophir vein was derived from material mined within the curve formed by its bending toward the Sixteen to One. A similar though less pronounced bend seems to have been the controlling feature of the ore shoots of the Morning Glory. (See pl. 54.)

Junctions of veins, particularly of minor branches, naturally favor the fracturing of the quartz in the main vein, owing to the resistance offered by the minor veins to movement parallel to the major vein, and are therefore favorable factors for the localization of high-grade shoots. It is likely that more high-grade shoots occur at or near junctions than elsewhere in the veins. The production from the Clinton vein in the Rainbow mine was derived from the neighborhood of its intersection with the Rainbow vein. Where two veins join at an acute angle, as the Ophir and Sixteen to One, not only is the bending of the minor vein favorable for the introduction of gold, but the major vein seems also more subject to fracture, presumably owing to deflection of stresses from the wall of the minor vein. In many places in the vicinity of vein junctions there are an unusually large number of stringers in the wall rock. This is particularly noticeable on the side of the junction within the acute angle, where " crossover " stringers are likely to be particularly abundant. Even small quartz stringers branching out from the major veins seem in places to have been effective in localizing ore. Much of the high-grade ore of the Plumbago mine seems to have been due to localization by such junctions.

There was much rich ground in the Sixteen to One vein close to the junction of the vein with the quartz-bearing faults that displace it in the upper levels of the mine. Owing to later movement along the steep fault the quartz is not now continuous between the two segments of the main vein, but earlier the quartz may have had a Z-shaped form, with the arms of the Z canted to an angle of 30° and the upright about perpendicular to the arms. Such a rigid body surrounded by the relatively plastic schist would yield to stresses by fracturing at the junctions of the upright and the arms, thus permitting the introduction of the gold-bearing solutions.

It is possible that in some places where fracturing of the quartz and later introduction of gold were dependent on the junction of veins there may have been later faulting along the major vein, which has displaced the junction relative to the ore shoot that is dependent upon it. If so, ore originally formed at the junction of veins may now be a short distance away, owing to later movement along the vein. Where this is suspected the direction of movement shown by the fault striae may be a guide for exploration.

As the gold is later than the quartz, it follows that faulting later than the quartz, if prior to the introduction of the gold, may have given the gold-bearing solutions access to the veins.

A description of the minor thrusting in the veins is given on pages 34–36. The small faults of this type that are most in evidence are those which are later than the mineralization. But there is evidence that faulting of this type was in progress before the deposition of the earlier quartz was finished and it doubtless continued intermittently during the whole period of mineralization. Ribbon quartz, in which the ribbons contain minerals of the carbonate, is in part caused by such faulting, nearly parallel to the vein walls, prior to or during the period in which the gold was deposited, for minerals nearly contemporaneous with the gold were deposited along the shear planes thus formed. It was noted that the effects of postmineral thrusting were generally most conspicuous in the productive parts of the veins, and it is to be supposed that the portions of the veins most affected by this late faulting were also most susceptible to faulting of the same type during the stage in which the gold was deposited. Even the veins that follow normal faults show evidence of postmineral faulting in the areas in which high-grade ore has been mined. The westward-dipping veins, on the other hand, are relatively much more free from faults that cross the veins. In these the postmineral gouge as a rule follows the walls of the quartz.

There are places where minor faulting later than the quartz and more directly transverse to the veins seems to have been associated with the formation of high-grade shoots. A small vertical fault with displacement of about 4 feet crosses the very rich shoot of the Sixteen to One mine above the 1,300-foot level but also extends into the unproductive portion of the mine. At the edge of the stope in the Sixteen to One mine, above the 800-foot level, from which the largest high-grade shoot was mined, there could be seen a small vertical fault which displaced the vein less than a foot, but which was itself cut by the hanging-wall slip. Another fault believed to be later than the gold limits the ore of this shoot. Likewise at the Extension of the Minnie D (fig. 33) and in the Gold Canyon mine (fig. 46) and parts of the Plumbago mine small faults transverse to the vein seem to have exercised some control over the position of the high-grade shoots. The influence of the " crossings " on the position of the ore shoots at Grass Valley has been noted by Hoover [61] and Howe.[62] At Grass Valley, as at Alleghany, the gold is later than the fracturing of the quartz. Howe notes that faults along the crossings, where present, are small.[63]

On the other hand, many small slips cross the veins in unproductive portions of the mines, and as faulting later than the introduction of the gold was far more widespread than the faulting which took place in the period between crystallization of the quartz and introduction of the gold, these transverse breaks in the veins do not in themselves furnish a very reliable guide to favorable areas. But the portion of a vein where evi-

dence of later faulting is prominent is worthy of investigation, because the later movement, the results of which may alone be evident, may have been determined by movement prior to the introduction of the gold. In certain ore shoots the high-grade ore has stopped (or begun) sharply against a later fault, and fault planes have been found that were gilded by the drag of the gold. It is evident that under such conditions the continuation of the high-grade shoot is to be looked for in the faulted segment, with the striae on the fault plane for a guide. Most of the later faulting produced small reverse faults within the vein and closely parallel to it, of not more than a few feet throw. (See fig. 10.) Where a strand carrying high-grade ore ends abruptly against a fault that makes a small angle to the vein, the continuation of the ore should be sought in the direction of reverse movement.

The throw of most of these thrusts does not exceed a few feet, and the discovery of the faulted segment offers little difficulty. Probably in most of the places where the gold stops at the boundary of a single strand this is due to original deposition of gold in one strand and not in others, but where the boundary of the strand is marked by slickensides, particularly if polished sulphides or gold are present and the change from high-grade ore to barren quartz is abrupt, there is the possibility that the high-grade strand has been " telescoped " upon its barren neighbor, or vice versa.

The features noted in the preceding paragraphs comprise the general physical characteristics of the veins that are believed to indicate areas which may be favorable for the presence of high-grade shoots. Of course only a very small area within such favorable portions of the vein as those indicated above actually carries high-grade ore, and in veins that are elsewhere very productive there may be areas in which for no apparent reason the high-grade ore fails entirely. In the Sixteen to One mine such a barren zone extends from the 800-foot to the 1,100-foot level, with extremely profitable ground immediately above and below. Moreover, such guides are necessarily only of a general nature, and certain mineralogic features may be better indicators within the favorable areas outlined by the broader structural features.

MINERALOGIC INDICATIONS

The most reliable mineralogic guide is the actual presence of free gold, though not necessarily in amount sufficient to make even milling ore. In a shoot of high-grade ore there is commonly one direction in which the change to barren quartz is not as sharp as in others, so that the ore body has a " tail " consisting of quartz from which a little free gold can be panned. Therefore quartz that shows even small " prospects " may indicate the proximity of high-grade ore.

[61] Hoover, H. C., Some notes on crossings: Min. and Sci. Press, vol. 72, pp. 166–167, 1896 ; reprinted in vol. 120, pp. 743–744, 1920.

[62] Howe, Ernest, The gold ores of Grass Valley, Calif.: Econ. Geology, vol. 19, p. 600, 1924.

[63] Howe, Ernest, op. cit., p. 597.

A. QUARTZ CRYSTAL SHOWING TWINNING IN INNER AND OUTER ZONES, ELDORADO MINE, SOUTH RAISE

Same large crystal as shown in Plates 27, *B, C,* and 32, *D.* Shows lack of parallelism between edge of core containing abundant vacuoles and outer rim of clear quartz. The twinning planes also have a different direction. The white specks are small crystals of quartz with different orientation. Crossed nicols (upper nicol turned 6° to bring out twinning).

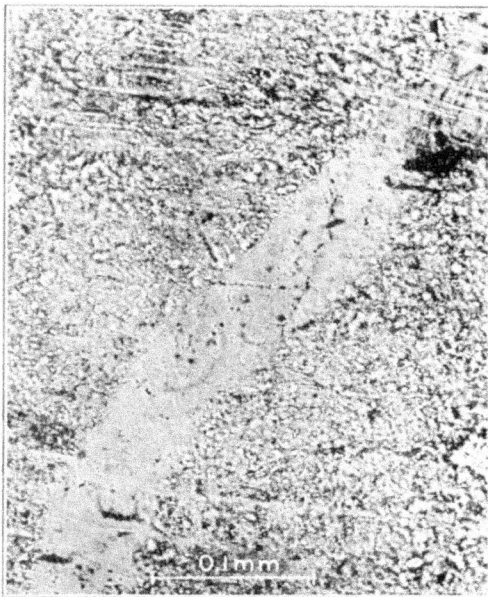

B, C. VEIN OF ALBITE CROSSING OLDER ALBITE CRYSTAL

Detail of Plate 32, *C.* Original albite altered in part to epidote and sericite (?) veined by later clear albite. *B,* Parallel nicols; *C,* crossed nicols.

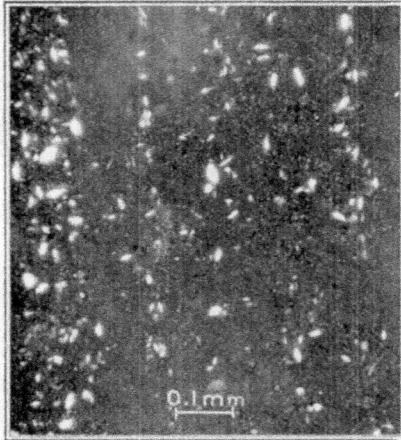

A. ZONAL DISTRIBUTION OF SERICITE
IN QUARTZ CRYSTAL

Portion of large quartz crystal showing zones parallel to the prism face containing shreds of a
fibrous mineral, probably sericite. Crossed
nicols.

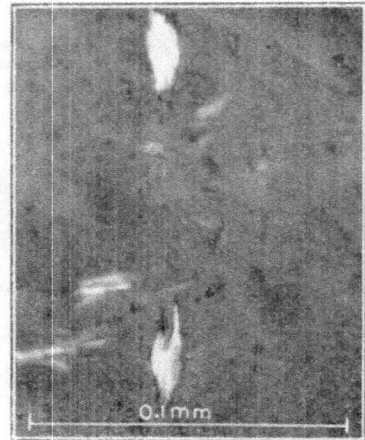

B. DETAIL OF A
Crossed nicols.

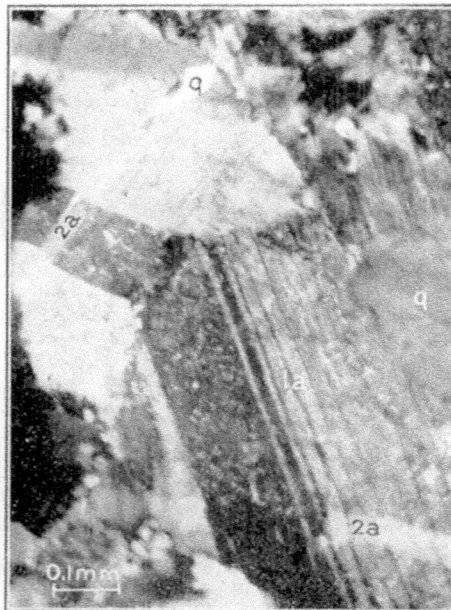

C. ORIGINAL ALBITE OF GRANITE VEINED BY
QUARTZ AND LATER ALBITE, ORIENTAL
MINE, MAIN LEVEL, 75 FEET FROM VEIN

Albite is speckled from development of epidote (?) and
sericite (?) and is partly replaced by quartz (q) and cut
by two veinlets, the upper of which contains quartz and
albite (2a) and the lower albite (2a) only. These veinlets represent the introduction of new material rather
than recrystallization, for in the upper one (left) not
only is quartz as well as albite present, but the boundaries of the twinned crystal are displaced, and in the
lower one there is an apparent bending of the twinning
lamellae across the veinlet, due to the alignment of new
material according to the crystal structure of the old.
Crossed nicols. See also Plate 31, *B, C.*

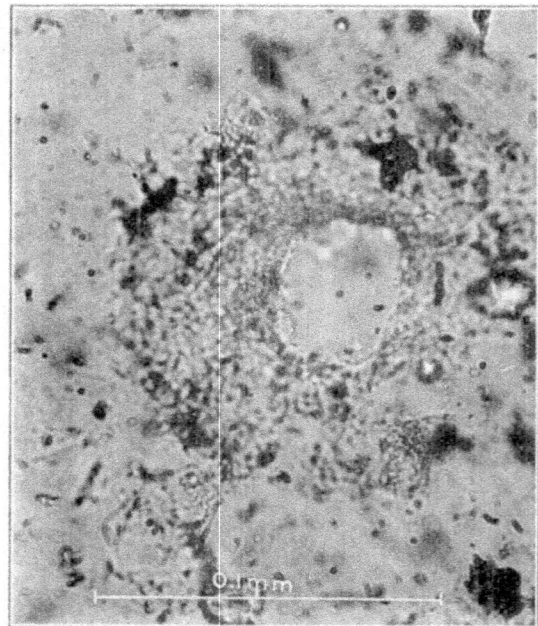

D. SPHERULE OF CHALCEDONY IN LARGE QUARTZ
CRYSTAL, ELDORADO MINE, SOUTH RAISE

Dark specks are vacuoles, mostly containing bubbles. Parallel
nicols. Same crystal as shown in Plates 27, *B, C,* and 31, *A.*

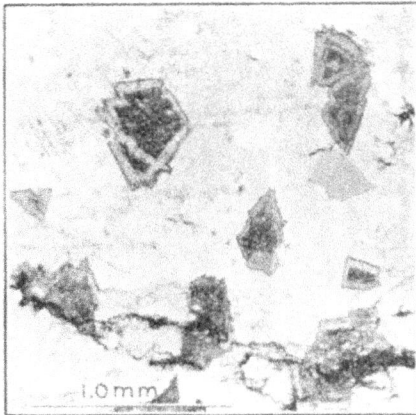

A

B

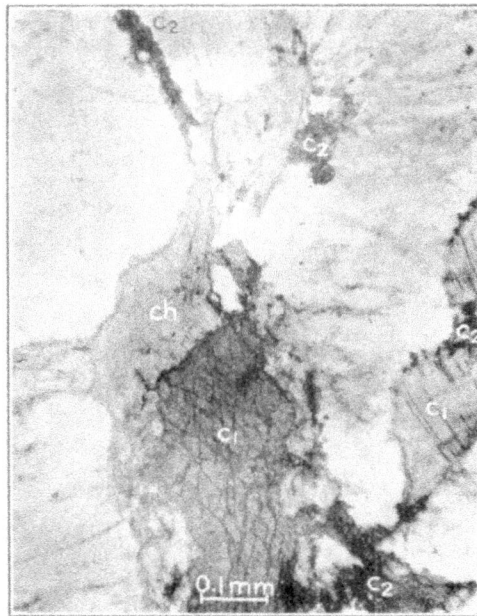

C. QUARTZ REPLACED BY CHLORITE AND TWO GENERATIONS OF CARBONATE, BRUSH CREEK MINE

Coarse carbonate (c_1) replacing quartz and partly replaced by chlorite (ch). Fine-grained carbonate (c_2) appears to be later than the chlorite. Rays of vacuoles (dark lines) seem to be dependent on earlier carbonate or chlorite. Parallel nicols.

A, B. FINE-GRAINED ALBITE REPLACING GRANITE AND ITSELF PARTLY RE-PLACED BY ZONED CARBONATE, ORIEN-TAL MINE, MAIN DRIFT

A, Parallel nicols; *B,* crossed nicols.

D

E

D, E. COARSE EARLY ARSENOPYRITE (a) BROKEN AND VEINED BY QUARTZ (q) AND FRINGED WITH MINUTELY CRYSTALLINE LATER ARSENOPYRITE (a_2) INTERGROWN WITH MARIPOSITE (m), GOLD CANYON MINE

A streak of crinkly banding containing mariposite and later arsenopyrite crosses the upper part of the picture. Veinlets of carbonate (c) cross the slide, and carbonate in part replaces the quartz that veins the older arsenopyrite. *D,* Parallel nicols; *E,* crossed nicols. See also Plate 39, *A.*

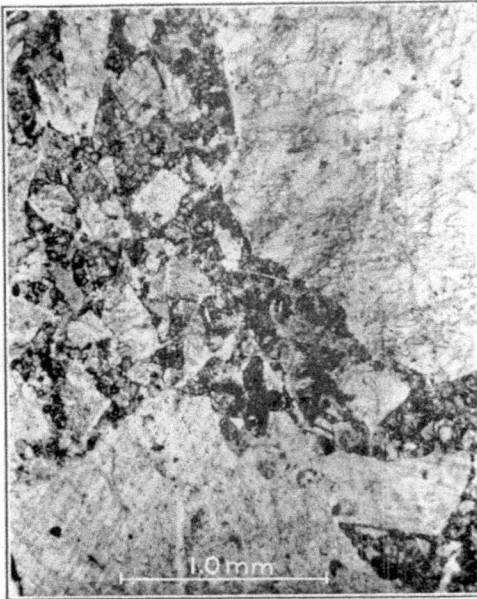

A. BRECCIA OF QUARTZ FRAGMENTS CEMENTED BY CARBONATE (DARK), PLUMBAGO MINE, MAIN LEVEL

Parallel nicols.

B. FAULT BRECCIA ALONG WALL OF VEIN, CONTAINING FRAGMENTS OF VEIN QUARTZ ALMOST COMPLETELY REPLACED BY CARBONATE (ANKERITE), PLUMBAGO MINE

Matrix consists of mariposite and ankerite with opaque material which is in part metallic.   Parallel nicols.

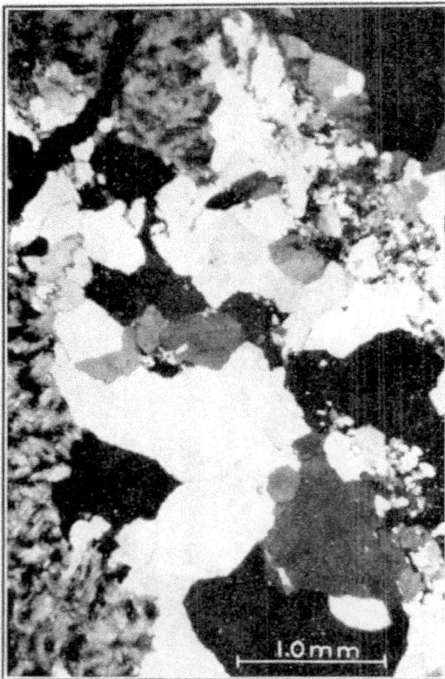

C. ZONE OF MICROBRECCIATION IN QUARTZ, WITH LATER REPLACEMENT OF QUARTZ BY FEATHERY CARBONATE, ORIENTAL MINE, WINZE FROM MAIN LEVEL

Crossed nicols.

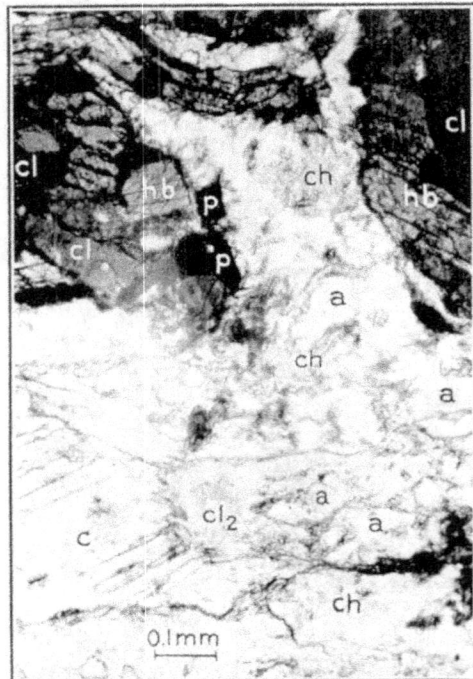

D. ALTERED GABBRO SHOWING REMNANTS OF HORNBLENDE, PLUMBAGO MINE, LEVEL 10, 5 FEET FROM VEIN

Original feldspar replaced by albite (a), which is partly replaced by chlorite (ch). Hornblende (hb) in part replaced by mixture of opaque material and chlorite (ch). Later stage of alteration involved development of carbonate (c) and replacement of earlier minerals by later chlorite (cl₂) and pyrite (p).   Parallel nicols.

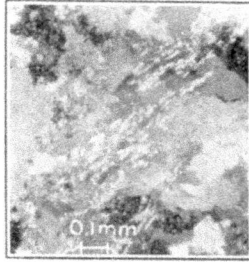

A. MARIPOSITE IN PAR-
ALLEL ARRANGEMENT
IN QUARTZ, SIXTEEN
TO ONE MINE, 1,300-
FOOT LEVEL

Parallelism of fibers suggests
that the mariposite has
replaced wall-rock residuals.
Crossed nicols.

B. RUTILE IN OLIGOCLASE, ORIENTAL
MINE, MAIN LEVEL, 75 FEET FROM
VEIN

Primary oligoclase of granite (o₁) with speckled
appearance due to partial alteration to
minute grains of epidote and sericite, replaced
by clear oligoclase (o₂) which contains
needles of rutile. Crossed nicols.

C. BOTRYOIDAL TALC REPLACING QUARTZ, BRUSH
CREEK MINE

Veining of later clear quartz across older quartz and beween talc
and quartz. Parallel nicols.

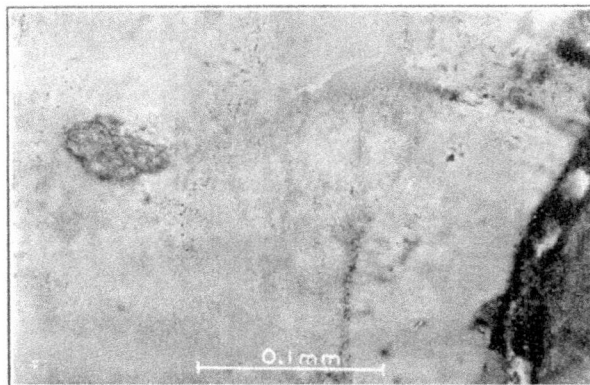

D. DETAIL OF *C*

Showing veining of later quartz around margin of larger talc mass and
"island" of talc in earlier quartz. Crossed nicols.

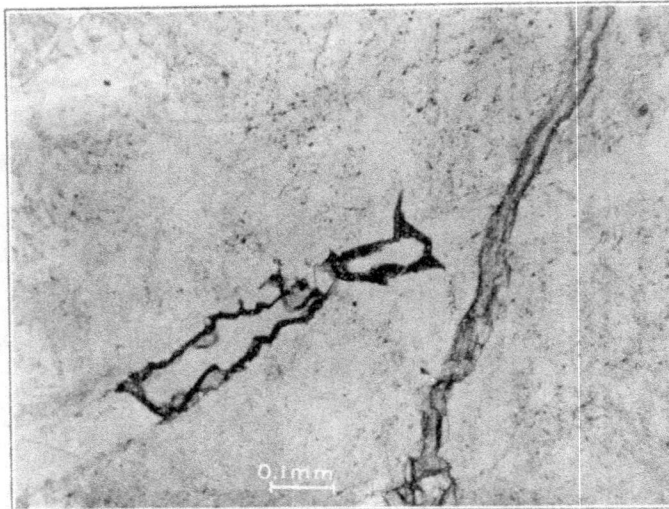

A. VEINLET OF LATER QUARTZ WITH OPAL AND CHALCEDONY, ROYE-
SUM PROSPECT DUMP

Vein of clear quartz passes into opal (dark) and chalcedony where it crosses an area of
fine-grained quartz and is cut by later veinlet of opal.   Change of grain in older
quartz is probably original, as inclusions are not cleared and fine-grained quartz
shows a few crystal boundaries.   Parallel nicols.

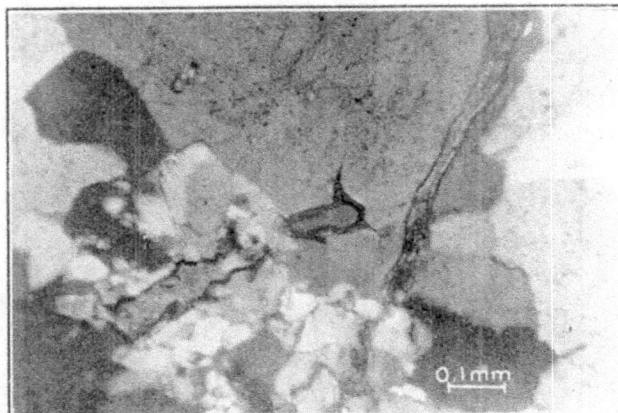

B. SAME AS *A*, SHOWING COINCIDENCE OF CHALCEDONY WITH
FINER GRAIN OF OLD QUARTZ

Crossed nicols.

C. "HEADCHEESE" BRECCIA, PLUMBAGO MINE, MAIN LEVEL

Later fine-grained quartz cementing breccia of coarse-grained vein quartz.   With the
fine-grained quartz are a little carbonate and specks of a metallic mineral, probably
pyrite.   Crossed nicols.

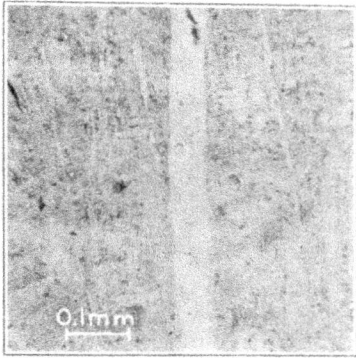
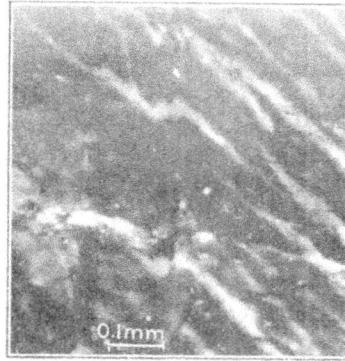

A, B. VEINLET OF CLEAR QUARTZ CUTTING STRAIN-TWINNED CRYSTAL OF OLDER QUARTZ AND CRYSTALLIZED IN OPTICAL CONTINUITY, RAINBOW MINE, CLINTON VEIN, LOWER DRIFT

Deflection of the light bands indicates fissure filling.   *A,* Parallel nicols; *B,* crossed nicols.

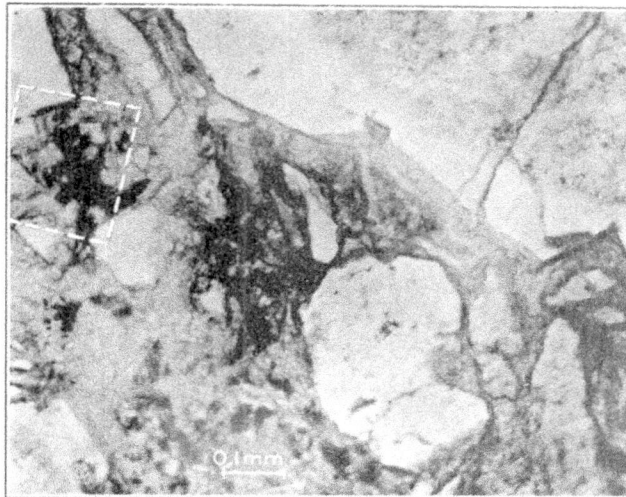

C. VUG IN OLDER QUARTZ, FILLED WITH OPAL AND CHALCEDONY, ROYE-SUM PROSPECT DUMP

Opal also veins quartz.   Drusy crystals of older quartz have cloudy centers and clear margins.   Opal contains pyrite and unidentified gray metallic mineral. Area within dotted lines shown in larger scale in Plate 38, *D.*   Parallel nicols.

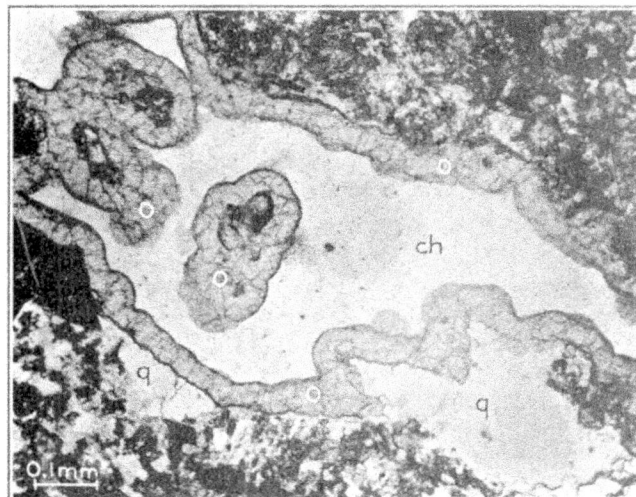

D. VEINLET CONTAINING QUARTZ (q), OPAL (o), AND CHALCEDONY (ch) CUTTING OPALIZED GRANITE, EUREKA PROSPECT, FOOT-WALL OF VEIN

Same specimen as Plate 39, *C.*   Parallel nicols.

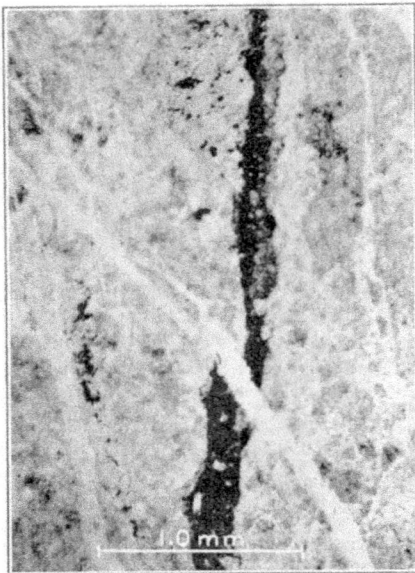

A   *B*

A, B. VEINLET OF LATER QUARTZ CROSSING AREA OF MICROBRECCIATION, TIGHTNER MINE, KNICKER-
BOCKER TUNNEL DUMP

Older quartz microbrecciated with clearing of inclusions in fine-grained portions.   Later introduction of gold and arsenopyrite following and crossing zone of microbrecciation. Vein of clear quartz, carrying chalcedony in places, crosses brecciated zone and streak
of gold and arsenopyrite. *A*, Parallel nicols; *B*, crossed nicols.

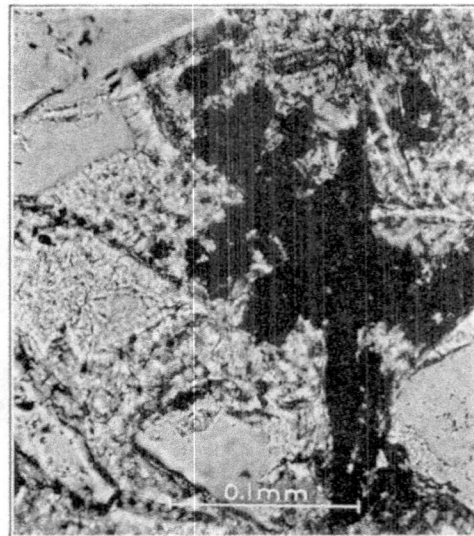

C. VEINLET OF CLEAR QUARTZ CUTTING
OLDER CLOUDY QUARTZ, PLUMBAGO
MINE, LEVEL 10

Contains apparently contemporaneous carbonate (high
relief). Pyrite (dark) replaces older quartz and is
cut by the veinlet of later quartz. Crossed nicols.

D. OPAL CONTAINING METALLIC MINERALS,
ROYE-SUM PROSPECT DUMP

Detail of Plate 37, *C*. Parallel nicols.

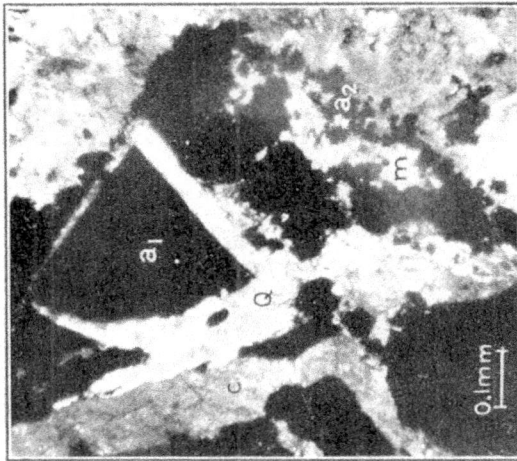

A. LATER ARSENOPYRITE FRINGING EARLIER, GOLD CANYON MINE

Detail of Plate 33, E. Early arsenopyrite (a₁) veined by quartz (q). Later veining and replacement of quartz by carbonate (c). Fringe of later arsenopyrite (a₂) with a little mariposite (m) developed around edges of early arsenopyrite crystals. Crossed nicols.

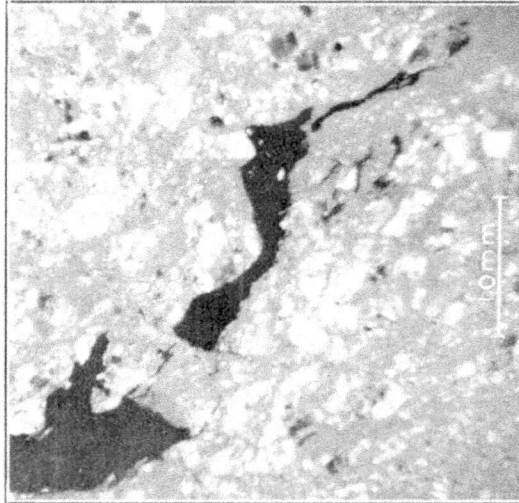

B. GOLD IN BRECCIA CONTAINING QUARTZ FRAGMENTS IN CEMENT CONSISTING LARGELY OF OPAL, SIXTEEN TO ONE MINE

Crossed nicols (upper nicol rotated slightly).

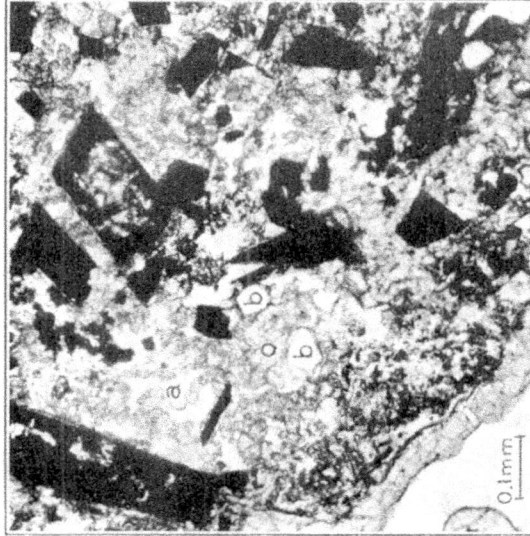

C. ARSENOPYRITE AND BARITE IN ALBITIZED GRANITE, EUREKA PROSPECT, FOOTWALL OF VEIN

Opal (o) has almost completely replaced the fine-grained albite (a) and attacked the center of the arsenopyrite crystals but has left the barite (b) untouched. Vein of opal and chalcedony in lower left corner. Same specimen as Plate 37, D. Parallel nicols.

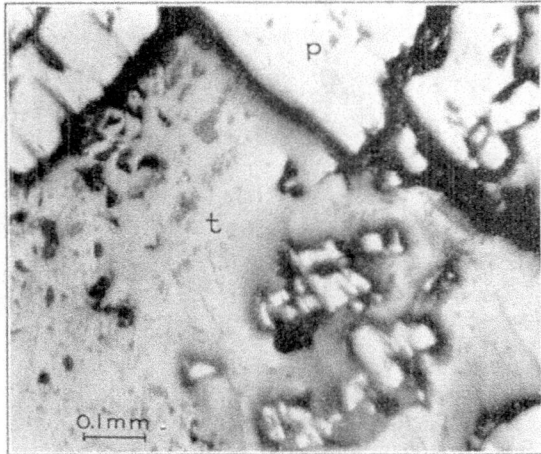

A. PYRITE REPLACED BY TETRAHEDRITE, SIXTEEN TO
ONE MINE, 600-FOOT LEVEL

Polished section.

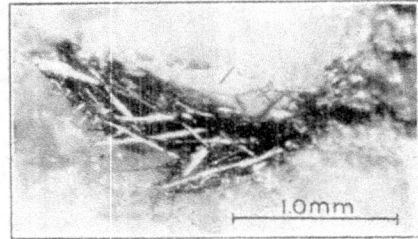

B. CLUSTER OF JAMESONITE CRYSTALS IN
SMALL VUG IN QUARTZ, RAINBOW MINE,
CLINTON VEIN, LOWER DRIFT

C. NEEDLES OF JAMESONITE IN QUARTZ RADIATING FROM
FRAGMENT OF COUNTRY ROCK REPLACED BY SERICITE
AND PYRITE, PLUMBAGO MINE, LEVEL 5

The needles appear to have no connection with the lines of vacuoles but
are for the most part associated with small clear streaks in the quartz.
Parallel nicols.

D. ARSENOPYRITE ALONG CRENULATIONS IN QUARTZ, ELDORADO MINE

Natural size.

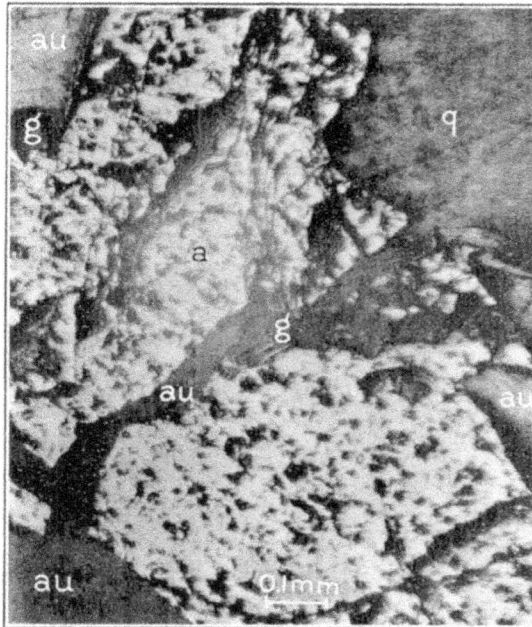

A. COARSE ARSENOPYRITE (a) VEINED BY QUARTZ (q),
WITH LATER GOLD (au) AND GALENA (g), SIXTEEN
TO ONE MINE

Detail of Plate 21, *B.* The preservation of crystal outlines of arsenopyrite and the presence of gold and galena along the quartz veins
crossing the arsenopyrite imply that the gold, though deposited
against the arsenopyrite, did not replace it to any great extent.
Polished section.

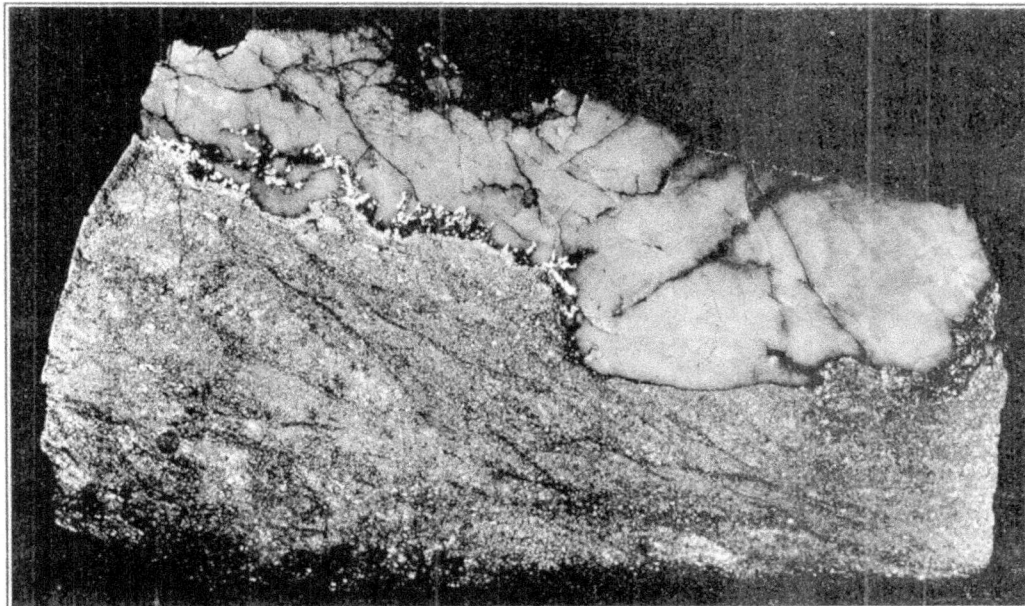

B. CRUSHED ARSENOPYRITE FRINGED BY GOLD, IRELAN MINE

The gold (white) penetrates the crushed arsenopyrite but chiefly replaces the quartz. Twice natural size.

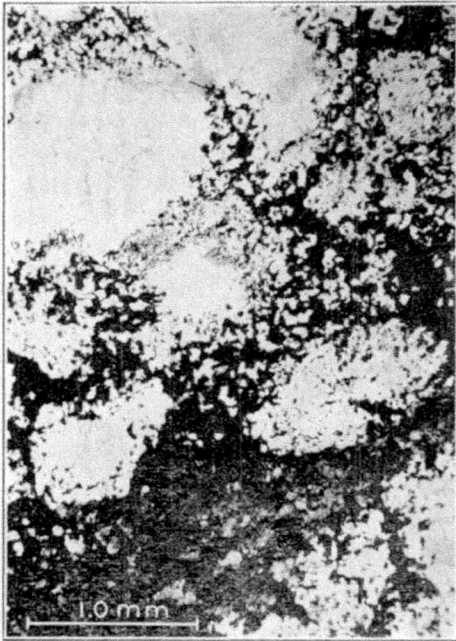

A. GOLD (BLACK) WITH A LITTLE CARBONATE (GRAY) REPLACING QUARTZ IN ZONE OF FRACTURING, RAINBOW MINE

Parallel nicols.

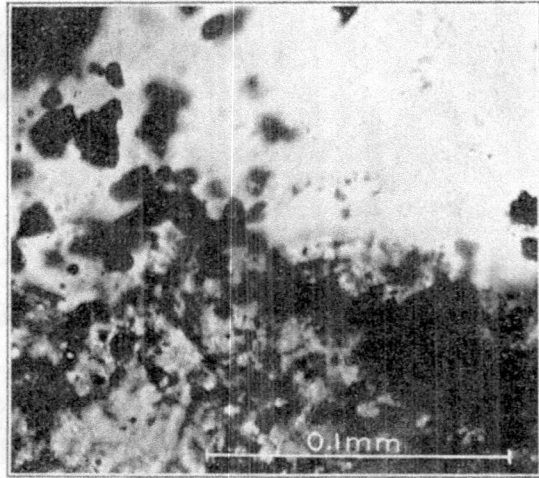

B. MINUTE CRYSTALS OF GOLD (BLACK) IN BOTH QUARTZ AND CARBONATE

Detail of A.   Parallel nicols.

C. SPECIMEN SHOWING CRINKLY BANDING, ELDORADO MINE

Dark bands contain graphite.   About 3/4 natural size.   See also Plate 44, C.

In the Alleghany district the mill is a valuable adjunct to prospecting. It is the usual custom to mill all quartz, however barren, mined in the course of development work, and whenever there is an increase, no matter how slight, in the gold recovered, the working faces are carefully investigated. As shown above (p. 52), the production from the low-grade quartz milled, though not necessarily profitable in itself, may furnish a considerable fraction of the total production.

In one mine in which samples were constantly taken it was found that if the vein was sampled strand by strand an assay value of $2 or more to the ton was a guide to ore.[64] In such sampling the object was not to get a fair sample of the quartz, but to try to find the part richest in gold. It is the custom to pan the drillings where the vicinity of high-grade ore is suspected. This is not only an aid to prospecting but a precaution against "high grading," because if gold is found in the drillings only trusted men are allowed to work at that face. It is thought likely that more panning, particularly wherever banded quartz is present, might be of use as a guide to ore. Galena, which is the later sulphide most closely associated with gold, should also be looked for in the concentrates.

Other than the actual presence of gold, the best mineralogic indication of high-grade ore is the coarsely crystalline arsenopyrite, which has crystallized prior to the quartz. This mineral seems to have been effective in causing the deposition of the gold and is in places partly replaced by gold (pl. 21, A), though more commonly gold is found to have replaced the quartz in contact with the arsenopyrite (pls. 21, B; 41, A, B). It is the impression of the writer that the greater portion of the gold obtained from the high-grade shoots of the Alleghany district was found in close association with arsenopyrite though not necessarily in immediate contact with it. Arsenopyrite was abundant in the high-grade ore from the Oriental (particularly the adit level), Kate Hardy, Kenton, Plumbago, Irelan, German Bar, and Gold Canyon mines. In the Rainbow mine it was present in the high-grade ore from the Rainbow vein but lacking in that mined from the Clinton vein. In the Sixteen to One mine there is a larger proportion of the high-grade ore not in association with the early arsenoyprite than in most other mines, but the late general manager, Mr. T. Bradbury, stated that the lower portions of all high-grade shoots contained more or less arsenopyrite. This would suggest that the arsenopyrite, even where not in close association with the gold, may have been the direct cause for the location of most of the high-grade shoots, but in 1913 gold was observed in the Gold Canyon vein below arsenopyrite. (See fig. 18.) The small bunches of very rich ore of the

thin Clinton vein of the Rainbow mine contained no arsenopyrite, and arsenopyrite is not commonly present in the most spectacular specimens of high-grade ore, though it may have been abundant in the shoots from which such specimens were obtained. The efficiency of arsenopyrite as a precipitant of gold is well known.[65] It is therefore natural that where other conditions are favorable the presence of large masses of arsenopyrite should have an important effect in localizing the deposition of gold.

On the whole the early arsenopyrite is most abundant in those portions of the veins in or near the smaller serpentine bands and in the neighborhood of the serpentine contact in those veins which fade out

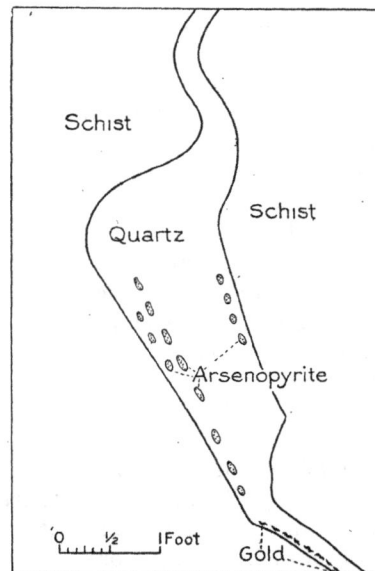

FIGURE 18.—Sketch of Gold Canyon vein, showing occurrence of arsenopyrite and gold

close to or against the serpentine. This may be due to interaction of the vein-forming solutions with the wall rock, or, as inferred for the chromium content of the mariposite, abundant iron may have been derived from the serpentine and combined with sulphur and arsenic of the vein-forming solutions to form the arsenopyrite.

The presence within the quartz vein of minerals that belong to the same stage of mineralization as the gold is naturally a favorable indication, though by no means a certain one. Readily visible evidences of the introduction of later materials are the later banding, either the ribbon quartz or the crinkly banding with dark lines of graphite and sulphides standing out sharply against the white quartz, and the breccia in which the fragments of earlier coarse white quartz in a matrix of fine-grained dark quartz give the "head-

---

[64] Simkins, W. A., The Alleghany district of California: Pacific Mining News of Eng. and Min. Jour.-Press, vol. 2, p. 291, 1923.

89277—32——5

[65] Palmer, Chase, and Bastin, E. S., Metallic minerals as precipitants of silver and gold: Econ. Geology, vol. 8, p. 156, 1913.

cheese" effect. Naturally as the gold in many places accompanies the minerals found in the dark bands and is only rarely associated directly with the later quartz that forms the cement of the "headcheese" breccia, the banding is the more reliable guide, though both indicate localities at which the vein has been reopened. It is not likely that the graphite in the dark streaks has had any direct influence on the deposition of the gold. The gold in these bands is commonly found in association with sericite, ankerite, and the sulphides, especially galena and the finely crystalline later arsenopyrite, and only rarely near or adjoining the opaque specks of graphite. Moreover, the carbon present in the country rock seems to have had no effect on the localization of the ore shoots. The Rainbow vein crosses alternate belts of greenstone and carbonaceous slate, but the distribution of its ore shoots seems to have been entirely independent of the country rock other than serpentine. Of the sulphides found in the banded quartz, galena appears to be most commonly associated with gold, and therefore may be the best indicator of possible proximity of high-grade ore.

The presence of mariposite ("blue jay") is regarded as a favorable indication, but its value is believed to lie merely in the fact that it is most abundant near the serpentine and that it is therefore an indication of the presence of that rock, which for the reasons stated above exercises a considerable amount of control over the location of the high-grade shoots.

Experienced miners of the Alleghany district distinguish between "live" and "dead" quartz and consider the "live" quartz a favorable but by no means certain indication of high-grade ore. The distinction between the two varieties is very faint and may often be warped by the optimism of the observer. Nevertheless it appears to be real, though the writer's limited experience did not enable him to make it with any certainty. The "live" quartz has a milkier and less lustrous appearance that the "dead" quartz. Under the microscope it is apparent that this difference is due to the greater degree of microbrecciation in the "live" quartz. It follows, therefore, that, on the assumption that portions of the veins in which later fractures are prevalent are the more favorable, this distinction has some validity as an indication. Plate 42, *A, B*, shows how gold may follow zones of microbrecciation. Probably the value of this distinction is negative; that is, portions of the vein in which the quartz is entirely of the more vitreous, "dead" type are less favorable.

PROBABLE PERSISTENCE IN DEPTH

The question most frequently asked us by those interested in the district was, "Will the high-grade ore continue in depth?" To this question it is impossible to give a categorical answer. The Alleghany district stands alone among the districts of the California gold belt in its peculiar type of ore shoots, so that no analogy based on the persistence of ore to great depths in such districts as Grass Valley and the Mother Lode region is pertinent. It may be said, however, that the fear often expressed that the rich shoots are of supergene origin is without foundation. The gold has been deposited from hypogene solutions and is in no way dependent upon either the present surface or the surface as it existed prior to the outflow of the lava, and there is no reason to anticipate impoverishment in depth on this account. It is possible, however, that there may be a change in the character of the ore with greater depth. The wider distribution of gold in the lower levels of the Sixteen to One mine, as shown by the data given on page 52, is perhaps an indication that with greater depth the gold will tend to occur in larger shoots of lower grade. But, on the other hand, ore of the Gold Canyon mine, whose workings are at about the same altitude as the lower productive levels of the Sixteen to One, was of the same high-grade type as has been found elsewhere in the district at higher altitudes, and though "mill rock" was mined at the German Bar mine most of the production was derived from high-grade ore. The mill-rock shoots of the Tightner, Plumbago, and Eldorado were at higher levels than either of these. It has been found that barren zones exist even where a combination of favorable features is present, and it is possible, though the depth to which work has extended in any one mine is not sufficient to say with certainty, that there may be a rough horizontal distribution of barren zones and zones containing high-grade shoots.

In conclusion, the authors are of the opinion that there is no reason indicated by the present study why the mines should not prove equally productive to a much greater depth than had been reached at the time the field study was made. This opinion is, of course, not meant to be taken as an encouragement to deep exploration of a vein in which shallow development has revealed nothing of promise.

ORIGIN OF THE DEPOSITS

The preceding description of the veins and ore deposits has laid emphasis on those features which have some bearing on the localization of the gold in the ore shoots. Any visitor to the district whose imagination is at all active must be intrigued by the number of interesting problems without apparent direct economic bearing which present themselves in even a cursory study of the mines. The following pages will be devoted to a consideration of some of these problems. Such problems are not only interesting in themselves, but even where no direct economic gain can be seen and where no final solution can be attempted, the record of inferences drawn, with a statement of the data on which they are based, may be of value in leading to a more complete knowledge of the origin of deposits of this type and hence to future rules and generalizations of economic value. The subjects here considered—the age of the veins, their depth of formation, the origin of the fissure system, the pressure and temperature at the time of ore deposition, the source of the minerals, and the methods of vein formation—are to a large extent mutually dependent, and all have a bearing on the major problem—the mode of origin of the ore deposits.

As such conclusions can not be clearly formulated until all the data derived from field, laboratory, and office study have been assembled and correlated, it is evident that the responsibility for this discussion rests solely upon the senior author.

AGE OF THE VEINS

Granite is the youngest of the pre-Tertiary rocks of the Alleghany district, and it is cut by the Oriental vein, which is identical in character with the other veins of the district. Therefore the direct evidence from this district is that the veins were formed prior to the deposition of the Eocene auriferous gravel and after the intrusion of the granite. There is, however, no reason to consider that the Alleghany veins, in spite of certain features peculiar to this district, are of different age from the auriferous veins of the neighboring districts. Over all the California gold belt veins are found to be in rather close association with but later than granitic intrusives, particularly the satellitic batholiths west of the main mass. The intrusion of the granite and allied rocks took place at or near the end of the Jurassic period, for the intrusives and their associated veins cut the Mariposa slate, of Upper Jurassic age, but are older than the Knoxville formation of the valley, which is chiefly if not wholly of Lower Cretaceous age but may contain some Upper Jurassic beds. The veins must therefore have been formed at about the end of Upper Jurassic time.

DEPTH OF VEIN FORMATION

It is clear from the general geology of the region and the association of the veins with granitic rocks that the portions of the veins now being mined were formed at considerable depths below the surface. Plate 3 shows that the Alleghany district lies in the midst of areas that show large surface exposures of granitic rocks. These areas of granite themselves imply a great depth of erosion, for the granite now exposed must have crystallized at considerable depth beneath the surface. Therefore the portions of the veins now visible must have been formed at a depth below the surface measurable in thousands of feet.

Lindgren [66] gives a range of 4,000 to 12,000 feet for depth below the actual surface at the time of formation of deposits of this general type, and speaking particularly of the California veins he says: [67]

The exposures by unequal erosion or by mining operations show, in many districts, that the vertical range of gold deposition without notable change in richness of shoots was over 4,000 feet; the relations in some districts lead to the conclusion that the deepest parts now mined were formed 7,000 feet or more below the surface.

He estimates the amount of erosion in the near-by Nevada City district at 3,000 to 4,000 feet.[68] Knopf [69] considers that the veins of the Mother Lode system were formed at a depth of several thousand feet below the surface.

In the following pages the attempt has been made to reach, by as many different methods as possible, a quantitative estimate of the amount of cover removed by erosion during the interval from the formation of the veins after the intrusion of the granite to their burial beneath the lavas at the end of the Eocene epoch. In so doing it will be necessary to present in considerable detail the data on which the estimates are based.

The time was probably ample for several erosion cycles. Barrell's estimates [70] for the time from the beginning of the Cretaceous period to the end of the Eocene, in years, are as follows:

	Minimum	Maximum
Eocene	20,000,000	26,000,000
Upper Cretaceous	40,000,000	50,000,000
Lower Cretaceous	25,000,000	35,000,000
	85,000,000	111,000,000

The more recent estimates of Holmes [71] are less detailed and for this reason are not used here but agree closely with Barrell's minima for these periods.

The following methods have been employed:

A. An estimate for Eocene erosion based on present topography is applied to the ratio of Eocene time to the total of Cretaceous and Eocene.

B. Estimates for the present rate of erosion are applied, with corrections, to the time scale given above.

C. An estimate is based on the thickness of the Cretaceous and Eocene sediments.

[66] Lindgren, Waldemar, Mineral deposits, 3d ed., p. 598, 1928.
[67] Idem, p. 626.
[68] Lindgren, Waldemar, The gold quartz veins of Nevada City and Grass Valley districts, California: U. S. Geol. Survey Seventeenth Ann. Rept., pt. 2, p. 173, 1896.
[69] Knopf, Adolph, The Mother Lode system of California: U. S. Geol. Survey Prof. Paper 157, p. 10, 1929.
[70] Barrell, Joseph, Rhythms and the measurements of geologic time: Geol. Soc. America Bull., vol. 28, pp. 884–885, 1917.
[71] Holmes, Arthur, Age of the earth, p. 78, New York, Harper Bros., 1927.

D. An estimate based on the amount of erosion as represented by the probable original content of Alleghany gold in the Eocene gravel is extrapolated for the entire period considered.

A. The first method involves an estimate of Eocene erosion, derived from the present topography, and extrapolation of this figure to the whole period of erosion.

Some idea of the depth of erosion can be obtained from the difference in altitude between the outcrops of the Alleghany veins and the high peaks to the west and southwest. The altitude of Alleghany as given on the Colfax topographic map is 4,500 feet. To the east and southeast is a line of high peaks formed of pre-Tertiary rocks. (See pl. 43.) These include the Sierra Buttes (8,615 feet), English Mountain (8,404 feet), Grouse Ridge (7,800+ feet), Snow Mountain (8,048 feet), and McKinstry Peak (7,918 feet). The difference in altitude, ranging from 3,300 feet on Grouse Ridge to 4,115 feet on the Sierra Buttes, is, however, not a direct measure of erosion, for there was a tilt toward the southwest in late Tertiary time, estimated by Lindgren [72] to have been about 70 to 80 feet to the mile. Using these figures, assuming the direction of tilt to have been S. 66° W. (normal to the direction of Lindgren's inferred pre-Tertiary axis of the range [73]), and multiplying by the distance in this direction from a line parallel to this axis passing through Alleghany gives corrected figures which vary from 2,410 feet on McKinstry Peak to 3,040 feet on the Sierra Buttes, on the basis of an assumed tilt of 80 feet to the mile, and 2,540 to 3,170 feet on the basis of an assumed tilt of 70 feet to the mile. These figures clearly represent only a small fraction of the total erosion from the beginning of the Cretaceous to the end of the Eocene. To them must be added estimates for the depth of erosion of the granite and for the original depth of cover of the batholith. There are no definite data for estimating these unknown items, but each of them can hardly be of less magnitude than the figures given above.

If Lindgren is correct in considering that this line of peaks represents the position of the pre-Tertiary divide,[74] then the figures of approximately 2,400 to 3,200 feet given in the preceding paragraph afford a rough measure of Eocene erosion. If the ratios (not necessarily the total lengths of time) of Eocene and Cretaceous time given on page 61 are approximately correct, Eocene time is to the total of Eocene and Cretaceous time as 1 to 4.25. Hence the amount of erosion during Cretaceous and Eocene time as estimated by this method is about 10,000 feet for the minimum and 13,000 feet for the maximum. But there must have been some grade to the original surface of which these

peaks are a remnant, and there should be a deduction to allow for it. The grade of the Tertiary river channels is given by Lindgren as 20 feet to the mile, so the slope of the pre-Tertiary surface, presumably a region of subdued topography, could not have exceeded this amount and was probably somewhat less. The correction to be made therefore can not exceed 1,000 to 2,000 feet. The estimate for total erosion by this method accordingly lies between 8,000 and 12,000 feet, but as no allowance is made for reduction in height of the peaks since Eocene time these figures should be increased by an unknown though probably small amount.

B. The second method involves an estimate of the probable rate of degradation during the entire time the veins were exposed to erosion. The studies of Dole and Stabler [75] furnish estimates of the present rate of denudation in different parts of the country based on the amounts of dissolved and suspended material in the streams. For the whole United States this figure is 0.00138 inch a year. This is equivalent to 114 feet in a million years. Expressed in this way the rates for the different large divisions of the country range from 21 feet per million years for the Hudson Bay drainage area within the United States to 189 feet per million years for the Colorado River drainage basin. For the San Francisco drainage area the rate is slightly lower than for the country as a whole and amounts to 107 feet per million years, but this includes not only the mountain area, undergoing active erosion, but also the valley region, where erosion is at a minimum. These estimates refer to present erosion, which is increased by the uncovering of the soil through agriculture, but do not take into account the tractional load—the material rolled along the stream bottom. In exceptional cases this may be as great as the suspended load [76] or it may be as little as 11 per cent of the suspended load, as estimated for the Mississippi.[77] Probably 125 feet in a million years is not far from the present rate for this region, taking into account the normal tractional load but not the increased load, which has resulted from the movement of placer-mining débris.

Evidently figures derived from the estimates of the present rate of denudation can be used as a measure of Cretaceous and Eocene erosion only with the aid of many assumptions as to the physiographic conditions prevailing throughout these periods. The present height of the land is probably greater than that which continued for any long interval during the time under consideration, and there is an additional amount of

[72] Lindgren, Waldemar, The Tertiary gravels of the Sierra Nevada of California : U. S. Geol. Survey Prof. Paper 73, p. 41, 1911.
[73] Idem, pl. 1.
[74] Idem, p. 38 and pl. 1.

[75] Dole, R. B., and Stabler, Herman, Denudation (in Papers on the conservation of water resources) : U. S. Geol. Survey Water-Supply Paper 234, pp. 84–93, 1909.
[76] Gilbert, G. K., Transportation of débris by running water : U. S. Geol. Survey Prof. Paper 86, p. 230, 1914.
[77] Humphreys, A. G., and Abbott, H. L., Physics and hydraulics of the Mississippi, p. 149, 1867.

material derived from the exposed soil of cultivated areas. Therefore the average rate of denudation in the past was presumably less than the present rate for the mountain regions. On the other hand, the large proportion of clastic material in the Cretaceous and Eocene sediments shows that the land was subject to active erosion during the greater part of these periods, and that the rate of denudation was, therefore, probably nearer to the present rate of 125 feet per million years than to the minimum cited above, 21 feet per million years. The physiographic history so far as it can be deduced gives evidence of only one period of even approximate peneplanation in the time under consideration. It is thought, therefore, that a fair assumption would be that the average rate of denudation was of the order of magnitude of 90 or 100 feet per million years. This is about the present rate for the North Atlantic drainage area as given by Dole and Stabler [78] plus a moderate allowance for tractional load. This figure applied to Barrell's time scale (p. 61) gives an estimate for erosion of the veins ranging from about 7,500 feet to 11,000 feet.

To some extent the assumptions used in this estimate serve as a check against the first estimate, for if, as assumed, the amount of Eocene denudation is of the magnitude of 2,500 to 3,000 feet, the rate of erosion during the Eocene was of the order of magnitude of 96 to 125 feet per million years, which is closely in accord with the figures already used.

C. The third method is based upon estimates of the thickness of the sedimentary rocks, in large part derived from the Sierra region. The data used are entirely independent of those of the first two methods but are so fragmentary that the estimate is less satisfactory.

There is a great thickness of Cretaceous and Eocene rocks in the region west of the Sierra Nevada in this general latitude. The maximum thicknesses and ages of these formations are summarized by Smith [79] as follows:

	Feet
Upper Eocene: Tejon formation (including Meganos formation, 3,000 feet)	10,000
Lower Eocene: Martinez formation	4,000
Upper Cretaceous: Chico formation	4,000
Lower Cretaceous: Horsetown formation	6,000
Lower Cretaceous (?): Knoxville formation	20,000
	44,000

The Ione formation, which is of the same age as the Eocene auriferous gravel, is equivalent to the upper part of the Tejon formation, according to Dickerson. [80]

During the period in which the Knoxville, Horsetown, and Chico formations accumulated sediment was furnished not only from land occupying roughly the present site of the Sierra Nevada but also from a land mass to the west of the present coast line. [81] The Eocene sediments were for the most part derived from land to the east, but the shore line varied in position at different times from nearly its present position eastward to the western edge of the present range. [82] A rough though arbitrary quantitative estimate may be obtained by assuming that half of the Cretaceous sediments were derived from the present Sierran region and the area of deposition was roughly equivalent to the area then subject to erosion and that all the Eocene material was derived from the east. It is assumed that within the Eocene the average area of erosion in the Martinez epoch was about double the present area and in the Tejon one-third greater, as the shore line was generally a considerable distance west of the present base of the Sierra Nevada. In addition there should be deducted a certain amount, say 20 per cent, [83] to compensate for increased volume and porosity in the unmetamorphosed Cretaceous and Eocene sediments as compared with the compact metamorphic and crystalline rocks from which they were derived. On the basis of half the thickness of Cretaceous and Martinez sediments (17,000 feet) and two-thirds that of the Tejon [and Meganos] (6,667 feet), less 20 per cent, an estimate for depth of erosion of about 19,000 feet is obtained, which may be in excess of the true figure, inasmuch as the maximum thicknesses of the sediments as exposed on the edges of the basin of deposition were used rather than the average. The writer is not sufficiently familiar with the stratigraphy of the California Cretaceous and Eocene to determine the order of magnitude of this excess.

D. The fourth method is based on estimates of the amount of gold derived from Alleghany veins originally present in the Eocene auriferous gravel and of the gold content of the Alleghany lodes within a given vertical range.

The methods used in the three previous estimates are applicable to northern California veins in general, without particular reference to the Alleghany district. In making an estimate based on lode and placer production, however, the writer found it necessary to confine himself to a consideration of the gold derived from the Alleghany district alone, as an attempt to

[78] Dole, R. B., and Stabler, Herman, op. cit., p. 84.
[79] Smith, J. P., Geologic formations of California: California State Min. Bur. Bull. 72, pp. 34–35, 1916.
[80] Dickerson, R. E., Stratigraphy and fauna of the Tejon Eocene of California: California Univ. Dept. Geology Bull., vol. 9, p. 406, 1916.

[81] Schuchert, Charles, Sites and nature of North American geosynclines: Geol. Soc. America Bull., vol. 34, pp. 227–229, 1923. Diller, J. S., and Stanton, T. W., The Shasta-Chico series: Idem, vol. 5, pp. 435–464, 1894.
[82] Dickerson, R. E., op. cit., pp. 467–474.
[83] Leith, C. K., and Mead, W. J., Metamorphic geology, p. 70, New York, 1915.

make a general estimate for a larger area involved the use of factors to which it was impossible to give a reasonable quantitative expression. This method, though involving greater detail than the others, requires an even larger number of assumptions. It will be necessary to use the estimates of different writers for the gold content of the placer gravel, both mined and remaining, in different parts of the Tertiary channels in which gold from the Alleghany district was deposited as a basis for estimation of the total gold in the Eocene gravel and then to make estimates of the proportion of this gold that was probably derived from the Alleghany district and the time required for its accumulation. The data for these estimates are given in considerable detail in order that the reader may have opportunity to judge of their validity.

As the mines in many California districts have been developed to great depths, the excess of placer production over lode production is in itself an indication of a considerable amount of erosion. Hill's estimate of total gold production from California to 1923 [84] is as follows:

Placer production_____ $1,210,083,000
Lode production from siliceous ores_____ 543,708,000
Lode production from base-metal ores_____ 9,236,000

1,763,027,000

Lindgren's estimate of the California production to 1909,[85] which is in close agreement with Hill's estimate of production to that date, gives a total production for California of $1,200,000,000 to $1,500,000,000, of which one-fifth is estimated to have been derived from quartz veins, about $300,000,000 from the Tertiary gravel, and the remainder from Quaternary placer deposits.

The courses of the principal Tertiary streams of this part of California are shown in Plate 43, which is derived from Plate 1 of Professional Paper 73 and from the Colfax and Smartsville folios. Gold from the Alleghany district was carried to the Tertiary Yuba River by a branch stream that flowed southward and joined the Eocene river between Moores Flat and Snow Point, on the ridge south of the district. This stream, as shown by the depth of the valley cut in the bedrock and the relatively small amount of gravel, was a small one, not comparable in magnitude with the main river, but owing to erosion of the exposed high-grade ore in the Alleghany veins its gravel was exceedingly rich. Another stream flowing westward, following about the course of the present Oregon Creek, tapped the northwestern part of the district but carried a much smaller amount of gold.

The following table gives an estimate of the gold eroded from the Alleghany district during the Eocene, and in the succeeding correspondingly numbered paragraphs is given a description of the method by which the figures were obtained.

Estimate of gold eroded from lodes of the Alleghany region originally present in the Eocene gravel

	Minimum	Maximum
1. Originally present in the Eocene gravel of the Alleghany district and in branch channels as far south as the Eocene river_____	$20,000,000	$35,000,000
2. Originally present in the Eocene gravel of the Colfax quadrangle outside the Alleghany district_____	40,000,000	144,000,000
3. Originally present in the Eocene gravel of the Smartsville quadrangle_____	18,000,000	52,000,000
4. In marine sediments (Ione formation)_____	0	5,000,000
5. Derived from oxidation of sulphides_____		
6. Enrichment from solution of silver_____		
7. Intervolcanic gravel not otherwise accounted for_____		
8. Gold on valley sides and in minor channels_____	2,000,000	14,000,000
9. Gold in quartz pebbles_____		
10. Excess ratio of fine gold in placer estimate over that for lodes_____		
	80,000,000	250,000,000

1. On page 26 is given an estimate of $12,000,000 to $14,000,000 as the production from the Tertiary gravel of the Alleghany district. All descriptions agree that the gold of the Tertiary gravel was coarse and showed little evidence of transportation. The placers of the district have been famous for the large nuggets produced. Therefore, although the streams that deposited the gravel entered the district from the north, the gold is believed to be largely of local derivation. Probably the gold still remaining in unworked portions of the channels amounts to as much or more than the material brought in from outside the district, and $12,000,000 may be taken as a minimum estimate of the amount of gold derived from Alleghany veins in the portions of the gravel not yet eroded. On the basis of relative areas, $8,000,000 may be added for the gold contained in the parts now eroded. Thus $20,000,000 may represent the gold derived from Alleghany veins in the Tertiary gravel of the region, both existing and eroded, as far south as the junction of the Tertiary stream that drained the Alleghany district with the main Tertiary stream near Snow Point.

The maximum estimate is based on data derived from Alling,[86] who states that the principal Tertiary

[84] Hill, J. M., California gold production, 1849–1923: Econ. Geology, vol. 21, p. 174, 1926.
[85] Lindgren, Waldemar, The Tertiary gravels of the Sierra Nevada of California: U. S. Geol. Survey Prof. Paper 73, p. 81, 1911.

[86] Alling, M. N., Ancient auriferous gravel channels of Sierra County, Calif.: Am. Inst. Min. Eng. Bull. 91, p. 1721, 1915.

channel crossing the Alleghany district, known as the "Great Blue Lead," was immensely rich and that "wherever found in Sierra County it has been worked with large profits, yielding, as a rule, $400 to $500 per linear foot breasted." These figures applied to the total length of the channel from the edge of the Colfax quadrangle to its point of junction with the Tertiary Middle Fork of the Yuba River (5½ miles) give totals of $11,600,000 and $14,500,000 for the gold originally present in the channel that would have been recoverable by drift mining. This includes only the gold in the rich layer along bedrock and takes no account of the gold in the overlying poorer gravel, so that it is presumably less than half of the total content of gold. Alling also lists a number of minor channels in the ridge between Downieville and Forest, several of which derived the principal part of their gold content from the Alleghany veins. If the gold content of these channels and also the gold not recoverable by drift mining are taken into account, a maximum estimate for the total would probably be about $35,000,000.

2. The Eocene stream (pl. 43) that drained the major portion of the Alleghany district joins the main stream just above Orleans, in Nevada County. Eastward from Orleans the main stream received gold from the Sierra Buttes and American Hill districts. From Snow Point to the entrance of the next tributary, at Columbia Hill, however, the Alleghany gold in the gravel must form a large proportion of the total, inasmuch as this district was the nearest source of supply. In the lower section, from Columbia Hill westward, other tributaries draining auriferous areas to the south, entered the main channel and reduced the proportion of Alleghany gold, but Alleghany was still the nearest important source. The Tertiary stream that followed the course of the present Oregon Creek could have drained only a small portion of the Alleghany district.

The minimum estimate for the major stream is based on the assumptions that for the upper section above Columbia Hill the Alleghany gold is about equal in amount to that estimated for the Alleghany district, say $22,000,000, for the lengths of channel are about equal, and, though farther from the source, the total gravel is several times greater; and that for the lower section the amount must be somewhat less, say, $15,000,000, giving a total of $37,000,000.

The maximum estimate is based on estimates of the amount and gold content of gravel already mined and the total remaining susceptible of mining by the hydraulic process and the following estimate of the proportion of Alleghany gold in the gravel.

Between Orleans and Snow Point the greater portion of the gold, perhaps over two-thirds, must have been derived from Alleghany, for it was the nearest known important source of supply. Gold from the

Sierra Buttes district must have traveled a distance of about 13 miles to reach this section of the stream. The veins of the American Hill district are nearer, about 6 miles, but these have never yielded coarse gold comparable to that from the Alleghany veins. The channel joining the main stream at Columbia Hill drained an auriferous region to the east and southeast, but as far as the junction of the next stream below this Alleghany remained the nearest important source of gold, and the Alleghany gold of the gravel may amount to over half of the total. Westward from the stream junction near North Columbia to the western border of the Colfax quadrangle, a distance of about 1½ miles, the proportion of Alleghany gold must have been greatly reduced, perhaps to one-third of the total, by the gold brought in by the large stream entering from the south. The gravel of this stream, however, where mined at Blue Tent, Quaker Hill, and Scotts Flat does not seem to have been particularly rich.[87] Therefore it is estimated that on the average five-eighths of the total gold in the gravel between Orleans and the western border of the Colfax quadrangle was derived from the veins of the Alleghany district.

Lindgren[88] estimates that of the Tertiary gravel near Moores Flat, North Columbia, and North Bloomfield, 81,000,000 cubic yards has been removed, but Gilbert's measurements, quoted by Lindgren,[89] give a total of 191,000,000 cubic yards. Lindgren gives 310,000,000 cubic yards for the gravel still available. Jarman[90] estimates the total gravel remaining in the area between Badger Hill (in the Smartsville quadrangle) and North Bloomfield as 886,000,000 cubic yards. Of this about a quarter lies within the Smartsville quadrangle, leaving 665,000,000 cubic yards for the area under consideration. This does not include the small Moores Flat and Snow Point gravel areas, included in Lindgren's estimate. The estimate of the board of engineers, quoted by Jarman,[91] gives 14,200,000 cubic yards for Moores Flat, Snow Point, and Orleans Flat, so 10,000,000 cubic yards may be allowed for the first two. The total of auriferous gravel both mined and available in the Colfax quadrangle is, therefore, taken as between 290,000,000 cubic yards, on the basis of Lindgren's estimates, and 866,000,000 cubic yards, on the basis of Gilbert's estimate for the gravel removed and Jarman's estimate for that remaining.

As measured on the map accompanying the Colfax folio the portions of the channels containing gold from Alleghany and other sources show gravel exposed for a distance of 6.5 miles, capped by lava for 5 miles, and lost by erosion for 4.7 miles. On the

[87] Lindgren, Waldemar, op. cit. (Prof. Paper 73), p. 143.
[88] Idem, pp. 139–141.
[89] Idem, p. 20.
[90] Jarman, Arthur, Report of the Hydraulic Mining Commission upon the feasibility of the resumption of hydraulic mining in California, pp. 74–75, Sacramento, 1927.
[91] Idem, p. 71.

basis of this ratio (6.5 to 16.2) the total gravel originally present was between 970,000,000 and 2,157,000,000 cubic yards.

Jarman [92] gives the yield of gravel worked at the North Bloomfield mine as 9.8 cents to the cubic yard and at the Boston mine at Moores Flat at 14.5 cents, and quotes an estimate of 10.3 cents for 408,000,000 cubic yards of the unworked gravel in the North Bloomfield area. He estimates a probable yield of 10 cents a cubic yard for the 800,000,000 cubic yards of workable gravel of the total of 886,000,000 cubic yards of the main area. For purposes of this estimate the total gold originally present may be considered to have been between $97,000,000 and $220,000,000, including a small amount allowed for the 86,000,000 cubic yards considered by Jarman to be below the workable grade, and that from Alleghany between $61,000,000 and $138,000,000.

The Tertiary stream that followed the course of the present Oregon Creek drained a small portion of the Alleghany district. Small patches of the gravel deposited by this stream still remain and have been worked,[93] but no data are available as to their gold content. As most of the Alleghany district must have drained toward the south, the amount of Alleghany gold present here was probably not great and might be assumed to have been between $3,000,000 and $6,000,000. These figures added to the minimum estimate given on page 65 and the maximum derived from Jarman's data give a range of $40,000,000 to $144,000,000 for the area considered in this section.

3. In the Smartsville quadrangle, adjoining the Colfax quadrangle on the west, several branch streams joined the main Tertiary river and reduced the proportion of Alleghany gold. One of these, entering from the north, above North San Juan, carried gold derived from districts in the Downieville quadrangle and also a small amount of Alleghany gold brought in by the Tertiary stream that followed the course of Oregon Creek. Another, which joined the main stream near French Corral, brought in gold from the vicinity of Nevada City; and a third, which followed approximately the present course of Deer Creek and joined the main stream above Mooney Flat, drained the Grass Valley district. It is therefore assumed that for the portion of the main stream above the junction of the first mentioned branch one-third of the gold may be credited to Alleghany, and for the remainder smaller proportions, decreasing with the distance from the source and the dilution by the gold brought in by the branch streams.

Only small remnants of Tertiary gravel remain in the Smartsville quadrangle, and, though the production from these is known to have been large, the records are so fragmentary that Lindgren has not ventured an estimate of past production. Consequently estimates for the Smartsville quadrangle have a much greater probable error.

On the basis of Jarman's estimate 221,000,000 cubic yards may be allowed for the available gravel in the extension of the North Bloomfield area into the Smartsville quadrangle. (Even if this proportion is in error, the total for the two quadrangles is not seriously affected.) Lindgren [94] estimates that in the same area 22,000,000 cubic yards has been excavated and 121,000,000 cubic yards remains. Gilbert's estimate for the excavated gravel, quoted by Lindgren,[95] is 41,000,000 cubic yards. Therefore, the total for gravel mined and remaining is taken as between 143,000,000 and 262,000,000 cubic yards.

For this section of the Tertiary streams, from the eastern border of the quadrangle to its junction with the first branch near North San Juan, the gravel remains over a distance of 3.3 miles and has been eroded for 3.1 miles. Therefore the gravel eroded may have amounted to between 134,000,000 and 246,000,000 cubic yards and the total between 277,000,000 and 508,000,000 cubic yards. On the basis of Jarman's estimate of 10 cents a yard, the total gold in this section may be taken as from $28,000,000 to $51,000,000, of which from $9,000,000 to $17,000,000 is credited to the Alleghany district.

For the stretch between North San Juan and French Corral, Lindgren's estimates [96] are 52,500,000 cubic yards for gravel excavated and 24,500,000 cubic yards for gravel available, but Gilbert [97] gives figures of 105,500,000 cubic yards for the gravel excavated. Therefore the total may be between 77,000,000 and 130,000,000 cubic yards. Jarman does not consider this area in detail. The geologic map of the Smartsville quadrangle shows gravel remaining over 5 miles and eroded along 3 miles of this section of the course of the Tertiary river. Therefore the total original gravel may have been between 123,000,000 and 208,000,000 cubic yards. But if the volumes of the gravel were equivalent for equal distances the ratio derived from the maximum yardage for the next higher stretch would give 601,000,000 cubic yards for this section. Lindgren [98] quotes figures of 10 to 35 cents a cubic yard for the gravel mined in this stretch, but as the areas are remnants from which the lava cover and a portion of the low-grade top gravel have been eroded, the total average content was probably nearer the lower figure. This would give an estimate for total gold originally

[92] Jarman, Arthur, op. cit., pp. 77–84.
[93] Lindgren, Waldemar, op. cit. (Prof. Paper 73), pp. 138–139.

[94] Idem, p. 122.
[95] Idem, p. 20.
[96] Idem, p. 122.
[97] Idem, p. 20.
[98] Idem, p. 122.

present of $12,000,000 to $60,000,000. The proportion of Alleghany gold may be taken as between one-twelfth and one-fifth, giving minimum and maximum figures of $1,000,000 and $12,000,000.

From French Corral to the mouth of the Tertiary river, a distance of 17 miles, the gravel has been eroded, except for an area near Smartsville and two smaller areas below.

For these areas Lindgren [99] gives 46,500,000 cubic yards excavated and 115,000,000 cubic yards remaining. Gilbert's figure [1] for excavated gravel is about 59,000,000 cubic yards. Applying the lower figure, of 161,500,000 cubic yards, for the uneroded gravel covering 3.5 miles of the course of the Teritary stream to the total distance of 17 miles below French Corral gives an original total of 784,000,000 cubic yards for this stretch. On the basis of ratios of length of stream the maximum would be 1,278,000,000 cubic yards.

Much gold was brought into this stretch of the Tertiary stream by the branch streams from the vicinity of Nevada City and Grass Valley, and Lindgren [2] quotes an estimate of 37 cents a cubic yard for the yield of the gravel near Smartsville. Presumably here, as farther upstream, some of the low-grade upper portion had already been eroded, and therefore the original average gold content may be taken as 15 cents a cubic yard and the total gold as between $118,000,000 and $192,000,000. On account of the increasing distance and the probable large amount of gold brought in from the region of Nevada City and Grass Valley the proportion of Alleghany gold may have been only from one-tenth to one-twentieth, or between $6,000,000 and $19,000,000.

Probably the amount of Alleghany gold brought into the Smartsville quadrangle by the Tertiary Oregon Creek was less than estimated for the same stream in the Colfax quadrangle, or from $2,000,000 to $4,000,000.

The estimate for the total of Alleghany gold in the Smartsville quadrangle is then as follows:

Estimate of gold eroded from lodes of Alleghany region and carried into Smartsville quadrangle

	Minimum	Maximum
Main Tertiary stream between eastern border of quadrangle and point 1½ miles northeast of North San Juan	$9,000,000	$17,000,000
Same from point 1½ miles northeast of North San Juan to end of gravel near French Corral	1,000,000	12,000,000
Same from French Corral to mouth of river	6,000,000	19,000,000
Brought in by Tertiary Oregon Creek	2,000,000	4,000,000
	18,000,000	52,000,000

Compared with the estimates for the main stream in the Colfax quadrangle of $40,000,000 to $144,000,000 from 16.2 miles of channel as against 31.4 miles in the Smartsville quadrangle, these figures seem in fair agreement when the decrease in the amount of Alleghany gold with increasing distance from the source is taken into account.

4. Gold was carried out to the Eocene sea and in places deposited in the beaches in such amount that ancient beach placers have been workable.[3] The amount contributed by a district as far inland as Alleghany must have been a very small proportion of the total. The method used to obtain a quantitative estimate for this gold involves a consideration of the deposition of gold in the Pleistocene gravel laid down during the glacial epoch,[4] which forms the present dredging ground in the valley. The production of gold from dredging in Yuba County to 1925 has been nearly $70,000,000.[5] To this may be added $10,000,000 for ground worked since that date and yet to be worked[6] and another $20,000,000 for gold in areas of too low grade for working.

The time in which this gravel was deposited was much shorter than the time during which the Eocene stream was depositing its load, but, on the other hand, the land was higher and the volume and velocity of the glacial streams and consequently their tractional load much greater, and the major portion of their gold content was derived, not from the veins but from the more easily eroded Eocene gravel. It is therefore assumed that $100,000,000 represents about the right order of magnitude for the gold deposited in Eocene marine gravel. The proportion contributed by the distant Alleghany district may have been so small as to be negligible or may conceivably have amounted to a twentieth of the total.

5-10. Several other items believed to be of minor importance are grouped together and given an arbitrary estimate. No single one of these is believed to be large, but together they may amount to a significant proportion of the total.

5. If the problem were to make an estimate of the source of all placer gold in the Yuba drainage basin, the amount of gold released from oxidizing sulphides at the vein outcrop would be an important factor. In the Alleghany district, however, the proportion of sulphides is small and their gold content low, so probably this item may be neglected.

6. The average higher fineness of placer gold over lode gold indicates that there has been a loss of silver in the placer gold, and as the estimates for both placer and lode gold are given in dollars instead of

[99] Idem, p. 122.
[1] The writer is unable to identify all the localities given in Gilbert's table of estimates.
[2] Op. cit., p. 122.

[3] Lindgren, Waldemar, op. cit., p. 122.
[4] Idem, p. 220.
[5] U. S. Bur. Mines Mineral Resources, 1925 and previous years.
[6] Hill, J. M., California gold production: Econ. Geology, vol. 21, p. 177, 1926.

ounces this difference should be considered. The total figure for Alleghany gold in the table, however, is equivalent to between 4,000,000 and 14,000,000 ounces, and so the silver lost must have ranged from less than 1,000,000 to not over 3,000,000 ounces.

7. The estimate already made includes the gold of the intervolcanic channels in the Alleghany district, but no data exist for making an allowance for such channels elsewhere. As the gold of these channels was principally derived from the Eocene gravel, there is an unknown but probably rather small amount which should be added. But in the Alleghany district the intervolcanic streams flowed on bedrock and received some gold directly from the veins. This factor therefore, at least in part, balances the other.

8. Undoubtedly some gold eroded from the Alleghany veins during the period of Eocene gravel accumulation never reached channels that were of sufficient size to form minable deposits but remained on the hillsides and in small rills. This gold could hardly, however, be more than a small percentage of the amount estimated for the Alleghany district and has to some extent already been taken into account.

9. Vein quartz is abundant in all the Eocene gravel deposits, and occasionally pebbles are found that show free gold. Attempts have even been made, though without success, to mill the quartz pebbles of the placers. The Alleghany gold thus locked up is probably only a very small proportion of the total of the free placer gold, and with increasing distance from the source it becomes less.

10. So far no attempt has been made to take into account the fine gold that can not be recovered in mining. This must in all amount to a considerable figure. But in the estimate for the gold content of the lodes described below consideration of the loss in mining and milling and the loss of the gold contained in low-grade quartz is likewise omitted. If these two factors bear the same ratio to the figures used in the two estimates no error is involved in their omission, because comparable estimates only are sought; but it is thought likely that, as the estimates involve gold that has been transported for considerable distances, the percentage of loss for placer gold is higher than that for lode gold. The amount can not be large, however, for by far the greater part of the Alleghany gold is coarse, and the change from "high-grade" material to essentially barren quartz, which would yield only a little fine gold to the placers, is sharp.

The figures of $2,000,000 to $14,000,000 taken to cover items 5 to 10 are chosen less from any confidence in their accuracy than to get figures of about the right order of magnitude to round out the totals. Items 5, 6, and 7 are not of enough importance to consider, except for the sake of a complete summary. Items 8,

9, and 10 are more important, but the errors involved in the estimate can hardly be sufficiently great to affect the result seriously.

It now remains to consider the gold content of the lodes.

In the area covered by Plate 1 the summit of the bedrock ranges in altitude from 4,250 to 5,000 feet. Over most of the area, however, especially where the veins pass beneath the lavas, the altitude is close to 4,500 feet. The 3,700-foot level represents the depth to which extensive exploration has been carried in most mines. The estimate for gold originally present in the Alleghany veins to an average depth of 800 feet below the lava cap is as follows:

Gold content of Alleghany veins between altitudes of 3,700 and 4,500 feet

1. Production from veins above 3,700 feet	$19,000,000
2. Estimate for veins concealed beneath lava	10,000,000
3. Estimate for new discoveries in and near veins now being developed	6,000,000
4. Amount lost by erosion	7,000,000
	42,000,000

1. The first item is easily arrived at and fairly accurate, because the total production of the district can be closely estimated, and except for the Gold Canyon and German Bar mines and recent development in the Sixteen to One mine there has been no notable production below the 3,700-foot level.

2. The second item is obtained by considering the whole lava-covered area, roughly 12 square miles (including a small strip north of the area mapped), and deducting 6 square miles for underlying areas consisting of serpentine and for the area east of the eastern fault. Of the remaining 6 square miles, 3 square miles may be considered sufficiently explored to preclude the likelihood of very important new discoveries at a distance from veins now being developed, and the remaining 3 square miles may be considered virgin territory. The production from veins outcropping beneath the lava and gravel, including most of that from the Sixteen to One and Tightner, all from the Rainbow and Irelan, and a part from the Plumbago, amounts to about $10,000,000. It is estimated that the gold contained in the veins of the unexplored territory is equal to this.

3. In several of the larger mines there are still blocks of undeveloped ground above an altitude of 3,700 feet; many of the smaller mines are not yet developed to this level, and there are many known veins on which development is not far advanced.

4. The estimate for the fourth item takes into account the depth of erosion of veins below the 4,500-foot contour and the productivity of such veins as shown by development of the uneroded portion. It is based on the following subordinate estimates:

Gold lost by Pleistocene and recent erosion

Brush Creek, Kate Hardy, and neighboring minor veins	$1,000,000
Oriental, Alta, and neighboring veins	500,000
Kenton and minor veins in lower part of Kanaka Creek Valley	100,000
Sixteen to One vein	2,000,000
Veins in Kanaka Creek Valley above Sixteen to One	1,000,000
Plumbago	500,000
Irelan and minor veins in Yuba drainage basin	100,000
Gold Canyon and German Bar	1,800,000
	7,000,000

No estimate is made for loss in mining or for gold contained in the low-grade quartz, for it is believed that such gold would bear about the same ratio to the total as the fine gold of the placers, which was likewise neglected in the estimate, bears to the placer total.

This estimate gives a figure of $50,250 per foot for the gold of the entire Alleghany district below the lava cap. Applied to the figures of $80,000,000 to $250,000,000 obtained for the Alleghany gold in the Eocene gravel the amount of erosion represented by this gravel can be estimated as between about 1,800 and 5,000 feet. But clearly only a portion of Eocene time is represented by the accumulation of the gold. Manifestly it took far longer than merely the time required for the deposition of the gravel, for the immensely greater richness of the gravel resting on bedrock than of the gravel above bespeaks a long period for the erosion of the bedrock gold.

The physiographic history of the Eocene epoch [7] includes the reduction of the region to a rolling plateau slightly below the height reached by the peaks of the old divide. Such a topographic condition was doubtless favorable for deep disintegration,[8] which favored the release of the gold when, in the later part of the Eocene time, renewed erosion began. The Eocene rivers cut valleys in this plateau to considerable depths.[9] Erosion continued until a postmature topography had been attained. The trunk streams flowed southwestward across the strike of the bedrock formations, but the tributary streams were in part controlled by the structure of the underlying bedrocks. The valleys, moreover, had gently sloping sides, and the grade of the major Eocene streams, estimated by Lindgren[10] at about 20 feet to the mile, was less than one-third that of the present Middle Fork of the Yuba River (about 73 feet to the mile from the Alleghany district to the edge of the valley below Smartsville). The period of erosion was followed by the shorter period in which the auriferous gravel was deposited. This

is correlated[11] with the transgression of the sea in Ione time. The end of the Eocene epoch is marked by the recession of the waters, with renewal of erosion for a short time, and the beginning of volcanic eruptions.

Clearly the portion of Eocene time represented by the accumulation of the Ione formation is too short for a measure of the erosion represented by the auriferous gravel, for the much greater richness of the gravel near bedrock shows that the gold there accumulated was eroded from the veins prior to the beginning of gravel accumulation. Nor, on the other hand, can the gold represent the whole period of Eocene erosion, for the gold eroded from the veins during the formation of the plateau and the early stages of valley cutting could not all have remained in the later Eocene gravel.

Therefore it is fair to say that the gold content of the gravel represents a time much longer than required for the deposition of the Ione formation, which, according to Dickerson,[12] is equivalent to the uppermost zone of the Tejon (upper Eocene) formation, but, on the other hand, considerably less than the total length of Eocene time. It is perhaps a reasonable assumption that the uplift represented by the break between the Martinez and Tejon formations began the release of the gold that was eventually deposited in the gravel. As the Tejon (10,000 feet including Meganos) is two and a half times as thick as the Martinez[13] (4,000 feet), five-sevenths of Eocene time is allowed for the accumulation of the gold. Then the estimate for Eocene erosion lies between 2,500 and 7,000 feet, and applying the factor (4.25) previously used gives as the estimate for the total erosion in Cretaceous and Eocene time between about 11,000 and 30,000 feet.

To summarize, the various methods outlined in the preceding pages give the following results:

Amount of material removed by erosion since the formation of the Alleghany veins

	Feet
A. Based on estimate for Eocene erosion, derived from present topography	8,000–12,000
B. Based on present rate of erosion	7,500–11,000
C. Based on thickness and extent of Cretaceous and Eocene sediments	19,000
D. Based on gold content of Eocene gravel	11,000–30,000

Even the smallest of these estimates is in excess of those quoted at the beginning of this section, and the writer realizes that in spite of attempts to maintain a judicial attitude, the personal equation has probably entered into the study, particularly because these estimates confirm the order of magnitude which he had in

[7] Lindgren, Waldemar, op. cit., p. 37.
[8] Idem, p. 45.
[9] Idem, p. 38.
[10] Idem, p. 41.

[11] Idem, p. 45.
[12] Dickerson, R. E., Stratigraphy and fauna of the Tejon Eocene of California: California Univ. Dept. Geology Bull., vol. 9, p. 417, 1916.
[13] Smith, J. P., The geologic formations of California: California State Min. Bur. Bull. 72, pp. 34–35, 1916.

mind when he began the consideration of the problem. Nevertheless, although the different estimates are interlocked to some extent, as many different factors as possible were used, the results obtained are for the most part roughly equivalent, and to some extent each method employed serves as a check on the validity of the factors involved in the others. The writer therefore believes that it is safe to say that the thickness of the cover removed from the veins is of the order of magnitude of 10,000 feet and that this figure is likelier to be below than above the truth.

It is interesting to speculate as to how far above the present surface the California gold-bearing veins were of the same type as those now being mined. Clearly there was no essential change within the distance through which erosion extended during the period of accumulation of the auriferous gravel. The pebbles of quartz and altered country rock within the auriferous gravel show material of exactly the same sort as the veins now being mined. A suggestion that the quartz veins retained their auriferous character to heights far above the present surface is given by Lindgren in his description of auriferous gravel in a watercourse of pre-Chico age.[14] It is therefore apparent that in Lower Cretaceous time erosion had reached the auriferous zone of the veins. It follows that the depth of auriferous veins removed by erosion is of the order of magnitude of several thousand feet. As lode mining in California has now reached a depth of nearly 5,000 feet, it is evident that the conditions which permitted the deposition of gold must have prevailed over a very great vertical range.

ORIGIN OF THE FISSURE SYSTEMS

The principal veins of the Alleghany district follow reverse faults with gentle easterly dip along which the displacement does not exceed 900 feet, and this is reduced to about 200 feet in a distance of 1,500 feet along the strike. On the hanging-wall side of at least one of these faults are faults with small normal throw. Other veins follow fissures with steep westerly dip. For most of these the displacement is unknown, but where measurable it is found to be less than 200 feet and also in the reverse direction. Along the eastern border of the district is a fault with steep easterly dip and probable large reverse displacement. The dip of the schistose country rocks is generally steep to the east, and the smaller intrusive masses are roughly parallel to the schistosity of the older rocks. The district lies midway between large areas of granitic rocks (pl. 3), and a belt containing small granite intrusions passes through the district, just to the west of the area containing the principal veins.

The fissure system at Alleghany is in many respects similar to that of the Mother Lode region.[15] In both areas a major fault with easterly dip lies along the eastern border, and it is suggested (p. 22 and pl. 6) that this fault may be continuous between the two areas. On the Mother Lode many of the veins also follow reverse faults with easterly dip which are about parallel to the major fault and lie a short distance away on its footwall side. Knopf's explanation of the Mother Lode fissures as auxiliary to the major fault[16] is therefore equally applicable to the eastward-dipping reverse faults of Alleghany.

No fissures analogous to the westward-dipping veins of Alleghany appear to be present in the Mother Lode region. In the Nevada City and Grass Valley districts,[17] as at Alleghany, the veins follow two systems of reverse faults, but the strikes instead of being nearly parallel are at right angles. The fissures of the north-south system are older than most of the east-west fissures and are faulted by them.[18] The major reverse fault, that of the Merrifield and Ural veins, in which the throw exceeds 1,000 feet,[19] has a northerly strike and dips to the east, indicating that the principal thrust was from the east. Lindgren[20] says: "It is most probable that the fissure systems have been produced by a succession of compressive stresses applied in different directions, chiefly from east to west and from north to south." The explanation of the Grass Valley fissure system given by Howe[21] supposes that at the time of intrusion of the batholith the expansive force of the intrusion caused tensional cracks. When settling took place after the consolidation of the granodiorite, "the arched roof of the new batholith would be compelled to support the weight of the overlying rocks, and the tension fissures would be closed." With further shrinkage lateral pressure on the flanks "would tend to force inward the invaded rocks and the upper portion of the batholith, producing strains that would find relief in reverse faults of moderate throw." It seems doubtful whether the loss of volume consequent upon consolidation of the granite would be sufficient to account for all of this compression, but the same result would be obtained by the withdrawal of a portion of the still liquid granite beneath an outer shell. If at Alleghany the belt of granite dikes can be considered the top of a larger mass of granite beneath,

---

[14] Lindgren, Waldemar, op. cit., p. 23.

[15] Knopf, Adolph, The Mother Lode system of California: U. S. Geol. Survey Prof. Paper 157, pp. 45–46, 1929.
[16] Idem, p. 46.
[17] Lindgren, Waldemar, The gold quartz veins of Nevada City and Grass Valley districts, California: U. S. Geol. Survey Seventeenth Ann. Rept., pt. 2, pp. 167–170, 259, 1896; U. S. Geol. Survey Geol. Atlas, Nevada City folio (No. 29), 1896.
[18] Idem (Seventeenth Ann. Rept.), p. 254.
[19] Idem, p. 168.
[20] Idem, p. 170.
[21] Howe, Ernest, The gold ores of Grass Valley, Calif.: Econ. Geology, vol. 19, pp. 602–603, 1924.

such a process may have played a part in the formation of the fissure system, and the westward-dipping fissures may be due to such collapse, whereas the larger displacement on the eastward-dipping fissures may be the net result of movement auxiliary to that of the major fault to the east and collapse consequent on consolidation and probable deep-seated withdrawal of the granite.

The parallelism of the two systems of Alleghany fissures as contrasted with the perpendicular relation at Grass Valley and Nevada City might be due to difference in form of the granitic intrusion; that at Grass Valley, as far as revealed by erosion, is not markedly elongate, but at Alleghany the distribution of the small dikes suggests that the deeper-seated mass from which there are offshoots is elongate in a direction parallel to the strike of most of the veins.

It might be supposed that the westward thrust on the eastern vein and the auxiliary fractures came first, followed by the collapse that caused the formation of the westward-dipping fractures, which in general fault the earlier series.

On the other hand, it is possible to discard the collapse theory with reference to the Alleghany veins and suppose that the westward-dipping fissures originated from resistance offered to a westward thrust by the inferred elongate mass of granite underlying a portion of the district. Against this hypothesis is the fact that the westward-dipping faults are as a rule later than those with easterly dips.

The major fault to which the principal eastward-dipping faults are thought to be auxiliary is probably a structural feature which long antedates vein formation, as it lies along the belt of serpentine intrusions. Renewal of movement on this fault, however, took place after the intrusions of both the basic and the granitic rocks, for quartz has been introduced along it. Such renewed movement may have been caused by the pressure exerted by the intrusion of the great Sierra batholith, a few miles to the east.

The small normal faults, such as those followed by the Ophir and Osceola veins, may be due to fracture of the hanging-wall block of the thrust faults, consequent upon irregularities of the fault plane, or, as the greatest displacement along the Plumbago fissure is not in line with that of the Sixteen to One, they may be due to relief of torsional strain. Their position relative to the two en échelon reverse faults suggests "flaws," but the surface geology gives no indication of a greater horizontal than vertical component.

The feathering out of the Oriental, Kenton, and Gold Canyon veins into stringers within the gabbro but close to the serpentine contact may be explained on the supposition that owing to volume expansion accompanying serpentinization the gabbro was under

strain at the time the fissure system was developed and that the relief of this strain had begun when the fissures were formed. The change in strike of these veins is explained on the assumption that strains due to the pressure of increasing volume were set up, whose relief diverted the original fissures close to the serpentine contacts. The fact that the smaller serpentine belts similar to those cut by the Sixteen to One vein have not caused similar diversion of the fissures may be explained by their smaller volume and consequent less pressure exerted during serpentinization. It may be that a part of the movement along the fissures is also due to increase in volume during serpentinization. Formulas for the formation of serpentine [22] imply an increase of volume of 34 to 40 per cent on the assumption that the resulting silica and carbonate remain in place:

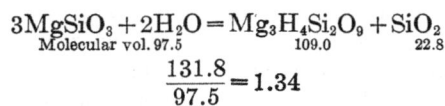

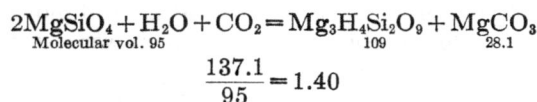

$$\underset{\text{Molecular vol. 97.5}}{3MgSiO_3} + 2H_2O = \underset{109.0}{Mg_3H_4Si_2O_9} + \underset{22.8}{SiO_2}$$

$$\frac{131.8}{97.5} = 1.34$$

$$\underset{\text{Molecular vol. 95}}{2MgSiO_4} + H_2O + CO_2 = \underset{109}{Mg_3H_4Si_2O_9} + \underset{28.1}{MgCO_3}$$

$$\frac{137.1}{95} = 1.40$$

But there is no field evidence for the presence of any great amounts of quartz or magnesite (or allied carbonate) within the mass of the serpentine. If silica and carbonate were thus formed, they have largely migrated, and the volume increase is essentially only that of the serpentine over the original basic rock. The increase then becomes 15 per cent for the first formula given above and 12 per cent for the second.

A line drawn on the map along the valley of Kanaka Creek between the points where the creek enters and leaves the mapped area represents a distance of 18,000 feet, of which 3,700 feet crosses the outcrops of serpentine masses. This line crosses the reverse faults followed by the Sixteen to One and Eldorado veins, as well as minor steeply dipping veins. It is possible that if the pressure caused by the increase in volume of the serpentine over the original basic rock were largely relieved by movement along these faults this process is competent to account for a minor part of the reverse faulting.

That movement along the fissures continued after ore deposition had ceased is shown by the numerous "walls" and thrust faults that cut the quartz and later minerals, but there is also evidence of movement along and in the veins during the process of mineralization, both during the introduction of the quartz and between the filling of the quartz veins and the intro-

[22] Clarke, F. W., The data of geochemistry, 5th ed.: U. S. Geol. Survey Bull. 770, p. 613, 1924.

duction of carbonate, mica, and gold. The separation of different strands of quartz by septa of country rock implies that the introduction of the quartz was a pulsatory process. Relations such as are shown in Figure 10, *d*, where the curved strand of quartz follows a minor fault formed after the earlier quartz had been deposited, indicate that there was movement in the vein during the period of quartz introduction. It is impossible to determine closely how much of the total displacement on the fissures is due to movement after the beginning of quartz deposition. Most of the thrusts later than the quartz whose displacement can be determined indicate movement of only a few feet on any one fault (fig. 10 and pl. 17, *A*, *B*), but Mr. Gannett's notes on the Sixteen to One mine show in one place a probable movement in the plane of the vein of as much as 40 feet.

On the other hand, the steeply dipping veins are not displaced by the later thrusts along the gently dipping veins, and there seems to be in places a deflection of the postmineral faults from one system to the other (figs. 6, 13; pl. 7, *C*), as if the postmineral movement were due to purely local causes and were small in amount. It is therefore thought probable that the major displacement on the fissures occurred prior to the introduction of the quartz. If, as suggested above, serpentinization, though initiated prior to the quartz, proceeded concomitantly with the vein filling, a motive force for these later movements is indicated.

PRESSURE AND TEMPERATURE

Few data are available as to the pressure prevailing in the veins or country rock at the time of vein formation. At a depth of 10,000 feet the pressure due to the weight of rock is about 11,000 pounds to the square inch, and the weight of a column of water of this height is about 4,200 pounds to the square inch. Additional pressure may have been supplied by strains originating within the earth's crust, other than simple gravity, or this pressure may have been decreased by tensional stresses. The evidence of recurrent reverse faulting along the veins shows that compressive stresses continued at least intermittently during the entire period of ore deposition. But the introduction of the vein minerals may have taken place during intervals in which the compressive stress was relaxed, as is suggested by the Osceola and Ophir veins, which occupy normal faults of small displacement.

It seems probable that if the depth below surface at the time of vein formation was of the magnitude of 10,000 feet or more the schistose and crushed rocks adjoining the fissure could not have long permitted the existence of large open spaces. Similarly, although it is possible that meteoric water under favorable conditions may penetrate to such a depth, it is doubtful whether any appreciable circulation could have been maintained.[23]

Neither does the depth alone give any clue to the temperature at the time of vein formation, for as the vein filling probably took place not long after the intrusion of the granite it is to be supposed that a temperature greater than normal prevailed. Also the vein-forming solutions presumably were hotter than the rock they traversed. The normal temperature gradient of 1° F. to 63 feet applied to the assumed depth of 10,000 feet would give a temperature of about 220° F. (110° C.). The gradient observed by Knopf[24] in the Mother Lode was only 1° F. to 150 feet, which would give a rock temperature at a depth of 10,000 feet of only about 125° F. (52° C.).

Little can be deduced as to temperature from the association of minerals in the deposits. The presence of garnet in the veinlets of serpentine crossing the chromite may indicate a relatively high temperature at the time of its formation, presumably an early stage in the process of serpentinization. On the other hand, the barite of the quartz veins suggests a lower temperature during vein filling. Lindgren[25] has estimated that in deposits of this type the temperature may have ranged from 175° to 300° C.

Attempts have been made to estimate the temperature of the crystallization of quartz by the relations of liquid and vapor in the vacuoles. As shown above (pp. 42-43), it seems certain that a part of these vacuoles were formed later than the quartz. Others seem to be contemporaneous with the final crystallization of the quartz, and for these such deductions may be valid. But if, as suggested on pages 77-80, the quartz has undergone recrystallization, only the temperature at the time of such recrystallization, which, however, must have been prior to the carbonate stage, is indicated.

The relative proportion of vapor and liquid in the vacuoles of quartz has been used by Spencer[26] as a means of estimating the temperature of the formation of the quartz, on the assumption that the liquid completely filled the cavity at the time of formation and, therefore, that the present volume relations of liquid and bubble are the result of contraction of the liquid with fall of temperature. In the vacuoles of the Alleghany quartz that contain both liquid and bubble (pl. 27, *B*, *C*) the proportion of gas is estimated at not over 5 per cent, and this volume applied to the curve given by Spencer would imply a temperature, at

[23] Meinzer, O. E., The occurrence of ground water in the United States: U. S. Geol. Survey Water-Supply Paper 489, pp. 42-50, 1923. Lindgren, Waldemar, Mineral deposits, 3d ed., pp. 36-44, 1928.

[24] Knopf, Adolph, The Mother Lode system of California: U. S. Geol. Survey Prof. Paper 157, pp. 22-23, 1929.

[25] Lindgren, Waldemar, Mineral deposits, 3d ed., p. 598, 1928.

[26] Spencer, A. C., The geology and ore deposits of Ely, Nev.: U. S. Geol. Survey Prof. Paper 96, pp. 63-64, 1917.

the time of crystallization, possibly as low as 100° C. It is uncertain, however, whether the vacuoles contain only water and water vapor, and if other substances are present there may have been a considerable variation. Spencer [27] observed cubes of possible sylvite in the vacuoles of the Ely quartz and inferred that the original temperature of the liquid may have been slightly higher than indicated for the curve of contraction of pure water. Only very rarely was a crystal found in the vacuoles of the Alleghany quartz, and here the fibrous form was suggestive of sericite. If so, the liquid may have been of complex composition.

Königsberger [28] has deduced the temperature range of quartz crystallization from observations of the temperature at which the bubble disappeared on heating. It was found that the size of the bubble decreased steadily on heating, nearly in proportion to the increase in temperature. For vacuoles in which the volume of the gas was from one-ninth to one-fourth of the cavity, the liquid expanded to fill the entire cavity at temperatures of 160° to 260° C. Although Königsberger's analysis shows that the liquid contained a considerable amount of dissolved salts, the results obtained fit fairly well with the theoretical curve given by Spencer for pure water. Nacken,[29] however, doubts the reliability of Königsberger's inferences, owing to the complexity of the temperature-pressure relations, particularly where carbon dioxide and water are both present, and Larsen [30] considers that the vapor content of vacuoles in the minerals of igneous rocks has no relation to the temperature at which the rock consolidated. But Bowen [31] admits that " at temperatures up to perhaps 300° C. the compressibility of water is sufficiently small so that lack of knowledge of the pressure does not introduce great uncertainty into the volume of the liquid as fixed by temperature alone."

Long prior to Königsberger's work on the Swiss veins, Phillips [32] had applied the same method to a study of the vacuoles of California quartz veins, with rather discordant results. He says: [33]

In every instance, however, they were found to require very different degrees of temperature to become full, since in the same specimens some vacuities disappeared at 180° F (82° C.), others filled at temperatures slightly above that of boiling water, whilst many, though much reduced in size, remained perfectly visible at 365° F. (185° C.).

The whole process of mineralization clearly continued over a considerable period, as is shown by the successive veinings by quartz, the recurrent faulting in and along the veins, and the progressive change in the character of mineralization. If on the analogy of the recurrence of earthquake shocks on a present-day active fault it may be supposed that the movements within the vein took place at intervals of several years, the evidence of the recurring movements indicates that the process of vein formation may have extended over hundreds or even thousands of years. If so, the temperature may have varied greatly, though it seems doubtful if, at the depth inferred, it could have been as low as 82° C.

It is, however, concluded that if the evidence afforded by the relations of the vapor and liquid in the vacuoles is of any value, it indicates that the final crystallization of the quartz took place at a relatively low temperature, probably nearer the minimum than the maximum of Lindgren's hypothetical temperature range. Also the great vertical range through which the quartz was formed as well as the duration of the process suggests that a very high temperature is unlikely.

SOURCE OF THE MINERALS

POSSIBILITY OF DERIVATION FROM THE WALL ROCK

As the filling of the quartz veins followed the intrusion of the granite, and as the association of granitic rocks and quartz veins of this type is observable not only over all the Sierra Nevada region but throughout the world, it is logical to conclude that the components of most of the minerals of the ore deposits were derived, if not from the granitic magma, at least from the same source. Yet there is abundant evidence of extensive alteration of the country rock at the time of ore deposition, and it would be reasonable to assume that, beginning with the first escape of the emanation from the parent magma, reaction with the wall rock caused an interchange of constituents that altered to some extent the constitution of the original solution. This interchange would presumably have continued until the composition of the solution was such that it was stable with respect to the wall rocks. It may be, therefore, that only a part of the material now in the quartz vein is directly derived from a magmatic source, and the remainder is the result of the reaction of magmatic solutions and country rock.

Studies of the Alpine veins by Königsberger and Parker [34] present convincing evidence that the minerals of those veins were formed at moderately high temperatures from material derived from the immediately

[27] Spencer, A. C., op. cit., p. 63.

[28] Königsberger, J., Die Minerallagerstätten im Biotitprotogin des Aarmassivs: Neues Jahrb., Beilage-Band 14, p. 111, 1901. Königsberger, J., and Müller, W. J., Über die Flüssigkeitseinschlüsse im Quarz Alpiner Mineralklüfte: Centralbl. Mineralogie, 1906, pp. 72–76.

[29] Nacken, R., Welche Folgerungen ergeben sich aus dem auftreten von Flüssigkeitseinschlüssen in Mineralien?: Centralbl. Mineralogie, 1921, pp. 12–20. 35–43.

[30] Larsen, E. S., The temperature of magmas: Am. Mineralogist, vol. 14, p. 93, 1929.

[31] Bowen, N. L., Geologic thermometry in Fairbanks, E. E., The laboratory investigation of ores, p. 192, 1928.

[32] Phillips, J. A., Note on the chemical geology of the gold veins of California: Philos. Mag., 4th ser., vol. 36, pp. 321–336, 422–433, 1868.

[33] Idem, p. 333.

[34] Königsberger, J., Über alpine Minerallagerstätten: Schweizer Min. pet. Mitt., Band 5, Heft 1, pp. 67–127, 1925. (Contains references to earlier publications.) Parker, R. L., Alpine Minerallagerstätten: Idem, Band 3, Heft 3, pp. 298–348, 1923. For an abstract of both these papers in English see Campbell, Ian, Alpine mineral deposits: Am. Mineralogist, vol. 12, pp. 157–167, 1927.

adjacent wall rock. In the Alleghany veins, however, the evidence bearing on this point is obscure and conflicting, so that the following paragraphs offer little in the way of definite conclusion as to the material contributed to the veins by the wall rocks in distinction to that derived from the inferred magmatic source. But it may be that geologists have been too prone to refer the contents of veins to a magmatic source without giving sufficient weight to the possible contributions by the country rock, and therefore speculation on this subject, even though inconclusive, may be desirable.

MINERALS OF THE CHLORITE STAGE

In the description of minerals of the earliest or chlorite stage of mineralization it has been shown that the formation of these minerals did not involve the accession of material other than water and carbon dioxide. It is reasonable to suppose that, in this first stage, the formation of the new minerals, principally chlorite from hornblende and epidote and mica from feldspar, and at least a part of the serpentinization of the peridotite was effected by emanations that may have had their origin in the granite magma and were not contaminated by reaction with the wall rocks.

QUARTZ AND CARBONATE

Knopf[35] considers that the quartz veins of the Mother Lode were formed by the action of carbon dioxide of magmatic origin on the wall rocks of the fissures, through the agency of heated meteoric waters. Ankerite was formed in the wall rocks, and the silica set free was deposited in the veins. So far as concerns the Alleghany veins, the fact that the carbonate is later than the quartz of the veins and replaced the vein quartz after it had undergone microbrecciation seems to the writer an objection to such a theory of origin of the quartz. For although the quartz was deposited in successive pulses, the carbonate is everywhere later than the bulk of the vein quartz, although it may be contemporaneous with the little veinlets of later clear quartz. But the extensive development of carbonate in rocks containing original silicates of calcium, magnesium, iron, and aluminum must have released silica and alumina in so far as these were not retained by the relatively small amount of mariposite and sericite. Here is presented a possible source for the silica of the later veinlets of clear quartz, opal, and chalcedony and the cementing silica of the "headcheese" breccias. These are contemporaneous with at least a part of the carbonate, and it may well be that they consist of silica freed from the adjacent wall rock during carbonate replacement, in the manner which

Knopf has suggested, though if the writer's estimate of 10,000 feet or more for the depth of vein formation at Alleghany is valid, it seems doubtful whether waters of the meteoric circulation could have played any effective part. Quantitatively, the amount of later silica introduced in these veinlets is less than that which must have been removed from the wall rocks as a result of replacement by carbonate.

But the abundant basic wall rocks of the veins, including both the basic intrusives and the schists, might, as Knopf has suggested, have furnished the necessary iron, magnesia, and lime to combine with carbonic acid of the vein-forming solutions to form ankerite. This suggestion gains weight from a consideration of relations over a wider area. In the districts in California in which quartz veins are abundant basic rocks are also present and the same type of carbonate alteration of the wall rock has been prevalent. In western Nevada, on the other hand, where basic rocks are not prominent, altered rock of this type does not commonly accompany the deep-seated quartz veins which occur in association with similar granitic batholiths that are also satellitic to the Sierra batholith.[36] If, however, the wall rocks yielded the bases necessary for the ankerite, the bulk of the alumina remains unaccounted for, because the only minerals containing alumina—mariposite and sericite—are present in far less amount than the carbonate.

If any alteration of the peridotite and pyroxenite to serpentine took place according to the first formula quoted on page 71, there must have been a release of silica; according to the second formula, a release of magnesium carbonate. At first sight this suggests a source of both quartz and ankerite. The alteration to serpentine was already under way before the introduction of the quartz, but the process may have continued during the entire period of mineralization. In view of the large masses of serpentine present, such an origin seems quantitatively possible. The gently dipping veins projected downward in the direction of the dip would all meet serpentine at no great distance. It might therefore be inferred that the silica and carbonate given off during serpentinization were deposited in the fissures. But quartz veins of this type exist in many districts in which serpentine is lacking, and, conversely, there are great masses of serpentine in regions that show no prominent quartz veins. The carbonate is distinctly younger than the quartz and has reached its present position in the veins, not only after the crystallization of the quartz but after its shearing and brecciation, whereas serpentinization, as is shown by the relation of quartz veins to the serpentine masses, was certainly well advanced prior to the deposition of the quartz. The fact that serpentine

[35] Knopf, Adolph, The Mother Lode system of California: U. S. Geol. Survey Prof. Paper 157, pp. 32, 46, 1929.

[36] Ferguson, H. G., The mining districts of Nevada: Econ. Geology, vol. 24, p. 127, 1929.

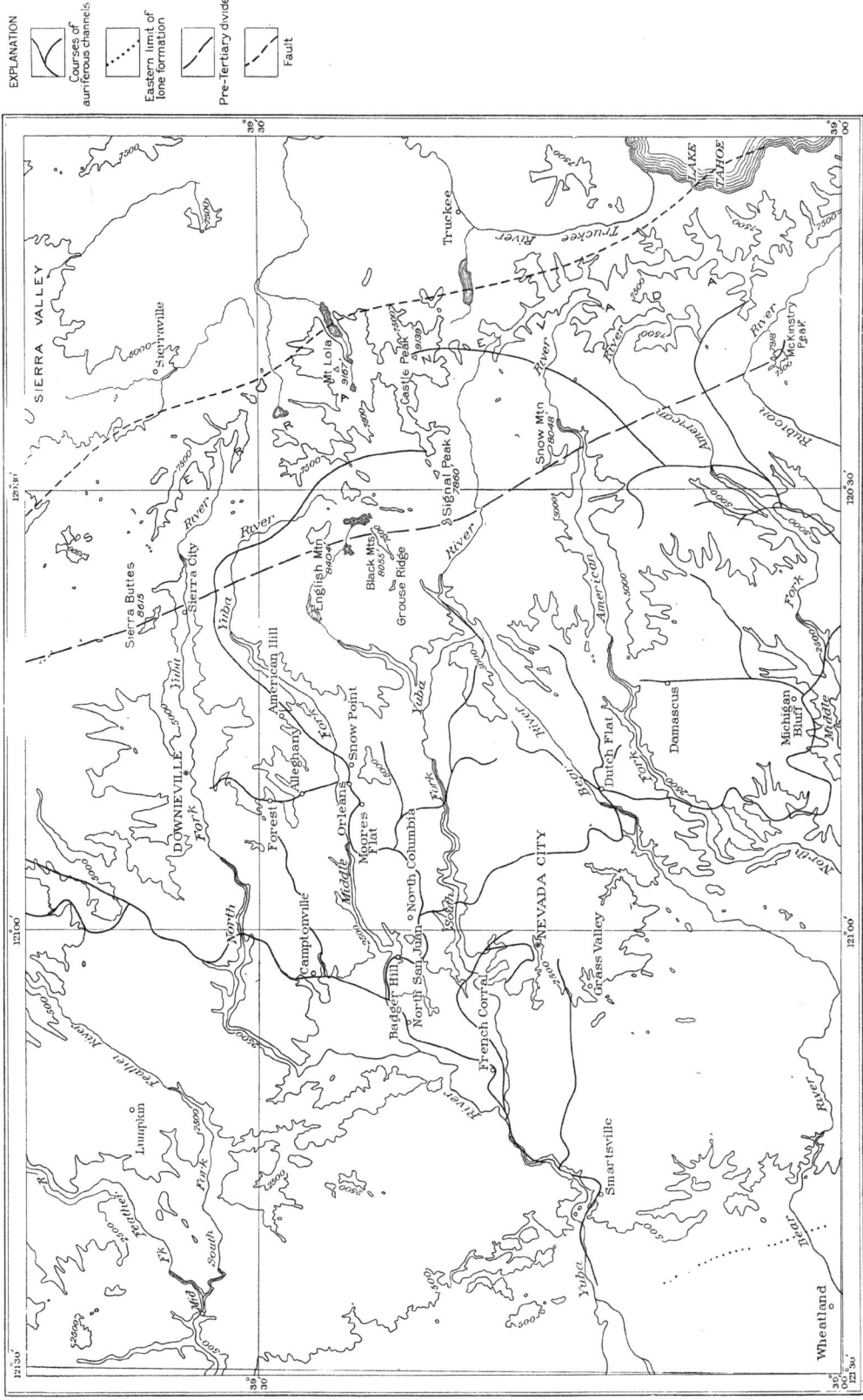

EXPLANATION

Courses of auriferous channels

Eastern limit of Ione formation

Pre-Tertiary divide

Fault

MAP OF A PORTION OF THE SIERRA NEVADA SHOWING MAJOR FEATURES OF TERTIARY DRAINAGE

Data from Professional Paper 73 and Downieville, Colfax, and Smartsville folios of the Geologic Atlas of the United States.

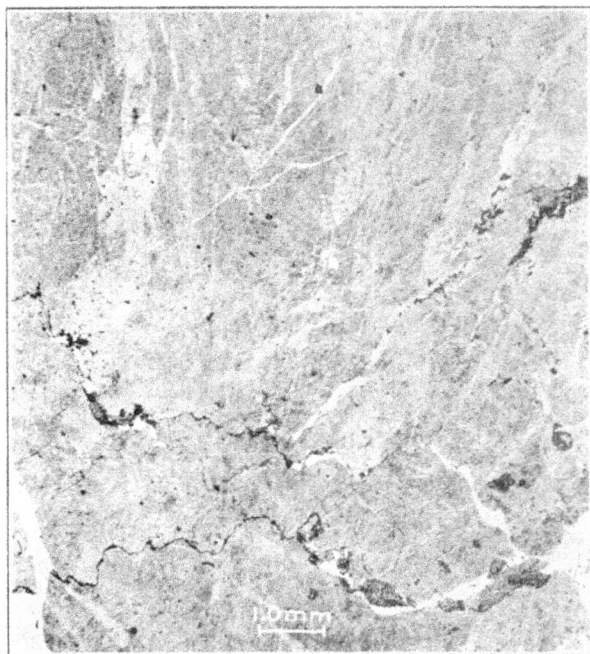

A, B. CRINKLY BANDING IN QUARTZ WITH LATER MICROBRECCIATION, PLUMBAGO MINE

The dark bands contain carbonate, mica, sulphides, and graphite, which are also present in the microbrecciated areas. The clear streaks of microbrecciated quartz do not cross the dark bands. *A,* Parallel nicols; *B,* crossed nicols. In *B* only a few quartz grains show strain twinning. See also Plate 46, *A, B.*

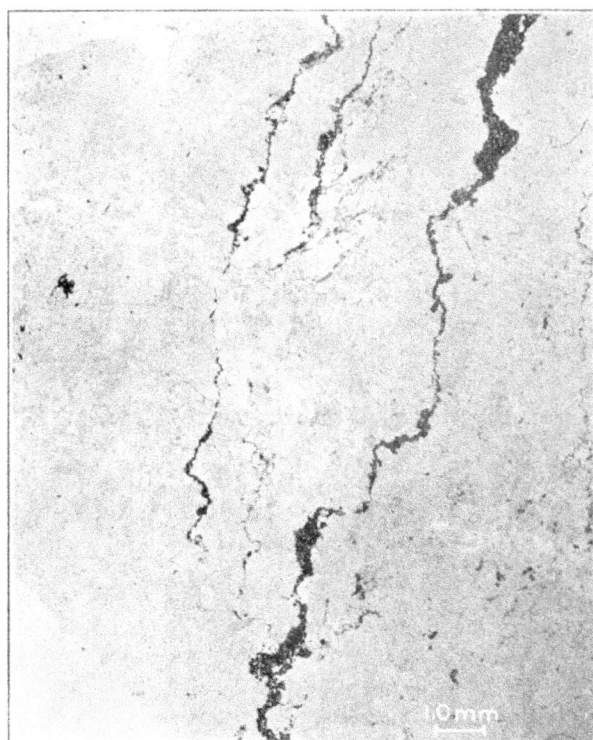

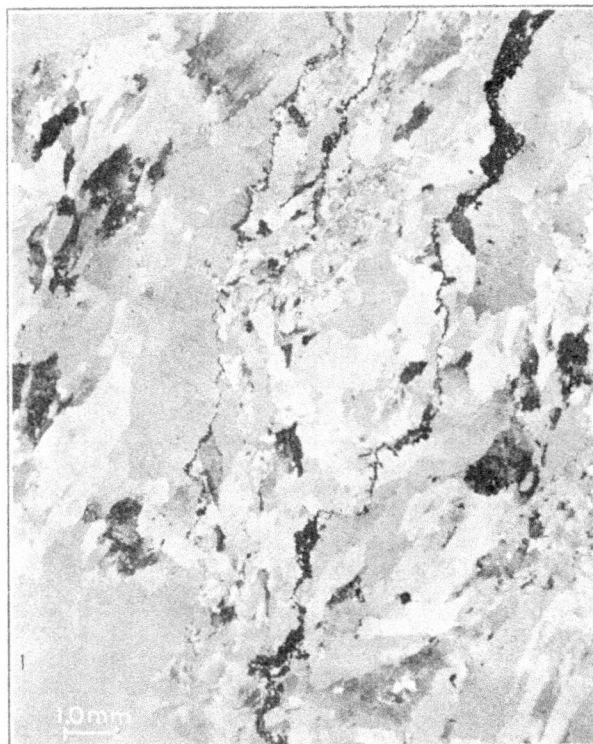

C, D. QUARTZ WITH CRINKLY BANDING, WITHOUT MICROBRECCIATION, ELDORADO MINE

Same specimen as Plate 42, *C.* *C,* Parallel nicols; *D,* crossed nicols (upper nicol rotated slightly to reduce contrast). *D,* shows parallel texture of quartz, believed to be due to recrystallization.

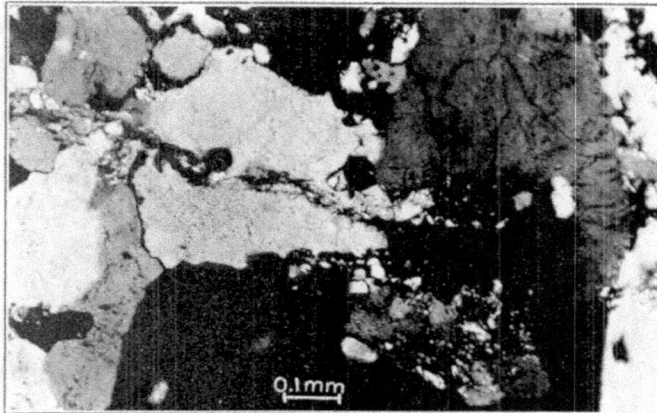

A. SERICITE BAND CROSSING SINGLE QUARTZ GRAIN, ELDORADO MINE

The band, which shows minor crenulations, contains sericite and carbonate and passes through one quartz grain and part way through the adjoining grain. Crossed nicols.

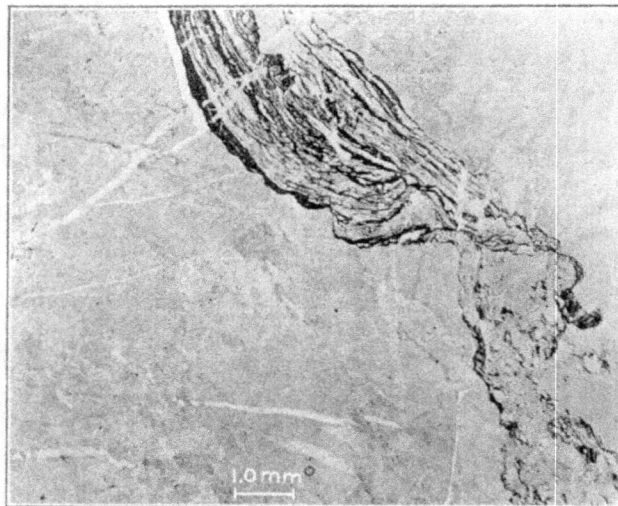

B

C

B, C. FOLDED SEPTUM OF PARTLY REPLACED SLATE BETWEEN QUARTZ STRANDS, BRUSH CREEK MINE, DUMP OF OLD SHAFT

The septum grades into crinkly banding. Both quartz and slate cut by veinlet of later clear quartz. No relation between crenulation and microbrecciation. *B,* Parallel nicols; *C,* crossed nicols.

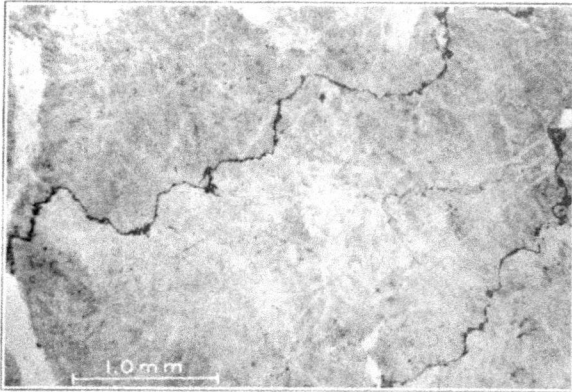

A, B. CRINKLY BANDING CROSSING STRAIN-TWINNED QUARTZ, PLUMBAGO MINE

Same specimen as Plate 44, *A, B.*   *A,* Parallel nicols; *B,* crossed nicols.

C. GROWTH OF CRYSTALS OUTWARD
FROM ARSENOPYRITE WITH ADJOIN-
ING ZONE OF LAMELLAR QUARTZ.
ELDORADO MINE

Crossed nicols.

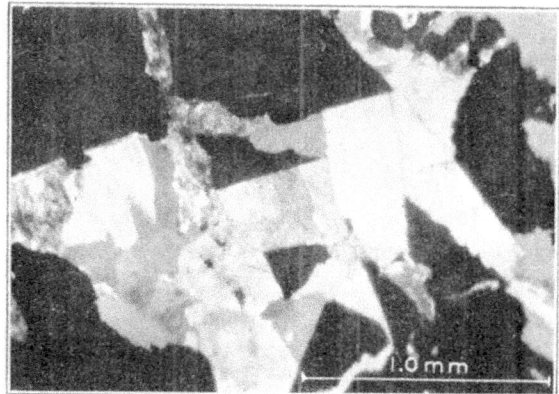

D. ARSENOPYRITE VEINED BY QUARTZ, GERMAN BAR
MINE

Shows also growth of quartz crystals on face of arsenopyrite crystal.
Fragment of arsenopyrite in center has broken off from larger
crystal above.   Crossed by veinlet of carbonate which also replaces
the quartz.   Crossed nicols.

E. FRAGMENTS OF ARSENOPYRITE IN QUARTZ, SIXTEEN TO ONE
MINE

About ¾ natural size.

F. COARSELY CRYSTALLINE ARSENOPYRITE IN BLADED
CRYSTALS GROWING OUT FROM WALL, VEINED BY
QUARTZ, GERMAN BAR MINE

About natural size.

has been replaced by carbonate and mariposite also shows that, at least along the veins, serpentinization was earlier than the development of these minerals. It is possible to suppose simultaneous deposition of quartz and carbonate with later solution and redeposition of the carbonate, but no evidence for this was found. In the Mother Lode region serpentines form a smaller proportion of the wall rock than at Alleghany, but carbonatization of the wall rocks has apparently been even more extensive.[37]

ARSENOPYRITE

The more abundant occurrence of arsenopyrite in the neighborhood of the serpentine and in the high-grade shoots of the veins that cut gabbro suggests direct influence of the wall rock, either as a precipitant or perhaps through accession of iron from these iron-rich rocks. The pyrite and other sulphides, however, show no close dependence on the type of wall rock.

MARIPOSITE

Evidence that the serpentine has yielded mineral matter to the ore deposits is found in the association of the green mariposite with the serpentine, an association which has also been noted in the mines of the Mother Lode [38] and Nevada City [39] as well as at Alleghany. It seems reasonable to suppose that the chromium content of the green mariposite has been derived from the serpentine, which is known to carry chromite. Although at Alleghany the association of mariposite with serpentine is not quite as close as descriptions indicate it to be in the veins of the Mother Lode, and mariposite occurs within the quartz veins as well as in the altered country rock, yet the green mariposite is abundant only close to serpentine and is comparatively rare in veins that do not cut serpentine masses.

CHROMITE

No detailed study was made of the relation of the chromite and serpentine. During the World War small chromite masses were mined from the eastern serpentine belt in the valley of Oregon Creek. At the only one of these where the exposures are good the chromite is present in a narrow streak, a few inches wide and several feet long, in a zone of sheared serpentine. Study of thin sections from this deposit shows that the chromite grains, though idiomorphic, are veined by the serpentine (pl. 20, B), which contains small crystals of the chrome-bearing garnet uvarovite, indicating that the serpentinization was later than

the development of chromite, although the vein-like form of the chromite deposit may imply that the chromite itself was later than the peridotite. Small lines of dustlike specks bordering the antigorite grains of the serpentine are probably chiefly magnetite, but in the vicinity of the larger chromite grains these specks may consist of chromite, though their small size makes the identification doubtful. If they are chromite there may have been a recrystallization of a part of the chromite after serpentinization.

GOLD

The fact that a good case can be made out for the leaching of the serpentine to provide the chromium content of the green mariposite leads to a further speculation as to whether the gold of the high-grade shoots, which is contemporaneous with the mariposite and which, as has been shown above (p. 56), is also distinctly more abundant near the serpentine masses, may not likewise have been derived from the serpentine.

Lindgren [40] notes the association of many gold-bearing veins with the " great serpentine belt " of California, and in his general discussion of the geologic relations of the California auriferous veins says: [41]

It must also be conceded that in many places the evidence points to the gabbros and peridotites (from which the serpentine was derived) and to the numerous albite aplite dikes which accompanied the basic intrusions as a source of at least a part of the gold.

Certain features of regional distribution and of the deposits themselves offer rather scant evidence in favor of this hypothesis. As with the ankerite, the regional distribution of the ore deposits suggests the possibility of a source for the gold unconnected with the granite magma. The great Sierra Nevada batholith is flanked on both sides by numerous subordinate granite batholiths. Around these on the California side are clustered the gold-bearing veins that have yielded such an enormous production. On the Nevada side, however, the satellitic batholiths of western Nevada have given rise to sulphide-bearing quartz veins that closely resemble the auriferous veins of California but, though probably containing more abundant primary sulphide minerals, are deficient in gold. Owing to favorable climatic conditions these veins have been productive of silver in the zone of supergene enrichment. A study of the general areal geology, as yet far from complete on the Nevada side, indicates that the great masses of basic intrusives which preceded the granites and allied rocks of the batholiths are widespread in California but nearly lacking in Nevada.

---

[37] Knopf, Adolph, op. cit., pp. 33–34.

[38] Ransome, F. L., U. S. Geol. Survey Geol. Atlas, Mother Lode folio (No. 63), 1900.

[39] Lindgren, Waldemar, The gold quartz veins of Nevada City and Grass Valley districts, Calif.: U. S. Geol. Survey Seventeenth Ann. Rept., pt. 2, p. 115, 1896.

[40] Lindgren, Waldemar, Mineral deposits, 3d ed., p. 618, 1928.

[41] Idem, p. 627.

As serpentinization took place prior to and concomitant with vein formation, the serpentines must have been permeated with emanations derived from the magmatic source and therefore might have yielded their original gold content to the veins, just as they have yielded chromium as coloring matter to the mariposite; though the actual intrusion of the basic rocks was earlier than that of the granite, which in turn preceded the formation of the veins.

It may be that the difference between the very rich but small ore shoots of Alleghany and the low-grade but vastly larger ore bodies characteristic of other California districts is due to the intimate relation at Alleghany between the serpentine and the veins. This would involve the supposition that in the other California districts generally, where there is no close connection between serpentine and ore, the widely distributed gold in the large ore shoots was either not derived from the basic rocks or came from a distance, whereas at Alleghany, where there were small intrusions of peridotite originally rich in gold, leaching during or after serpentinization provided unusual local concentration. After allowing for the extension of the serpentine masses beneath the lavas, it is estimated that serpentine occupies an area of at least 85,000,000 square feet within the limits of the region covered by Plate 1. To a depth of 800 feet, at $10\frac{2}{3}$ cubic feet to the ton, the mass of serpentine present would amount to about 6,400,000,000 tons. It was estimated on page 68 that the total gold within this 800-foot zone, including gold eroded, mined, and still undiscovered, might amount to as much as $42,000,000. Therefore, if the serpentine was entirely leached of its gold content and if all the gold of the veins was derived from the serpentine, a contribution of less than a cent's worth of gold, or about $\frac{1}{3200}$ ounce from each ton of serpentine would be required to provide the high-grade gold of the Alleghany district. Assays of specimens of serpentine, however, taken at a distance from the veins showed no gold or only a doubtful trace. Knopf [42] finds that the Mother Lode belt is particularly poor in gold where intrusions of gabbro, peridotite, and albite aplite are most abundant. In the Nevada City-Grass Valley district [43] there seems to be no direct connection between the major ore shoots and any particular kind of country rock, though it is said that "specimen rock" is particularly abundant in the Idaho-Maryland vein, which has a serpentine footwall.

It would be necessary, therefore, to explain the lack of notable enrichment near serpentine in other California districts by assuming that at Alleghany the basic intrusives were originally peculiarly rich in gold. Although it may be said that this is inherently unlikely, it is no more unlikely than the facts of gold

distribution, namely, that at Alleghany the ore shoots almost without exception are small but of exceedingly high gold content, whereas in the neighboring districts, such as Grass Valley, large bodies of low-grade ore are the rule, although both at Grass Valley and on the Mother Lode small "bunches" of ore of astonishing richness are sometimes found.

Another objection to this hypothesis is the presence of a variety of sulphide minerals, such as tetrahedrite, chalcopyrite, jamesonite, sphalerite, and galena, which are also later than the quartz and whose introduction was about contemporaneous with that of the gold, although they are not everywhere closely associated with the gold and may be abundant where high-grade ore is absent. Such sulphides, being common to all veins of this type in the Sierra Nevada region, both in California and in Nevada, can not be genetically connected with serpentine, and therefore at about the time of the introduction of the gold there must have been an accession of minerals of hypogene origin from some source outside the serpentine.

The observed association of auriferous veins with the "great serpentine belt" may also be explained in another way. It has already been suggested that the peridotites from which the serpentines were derived and the other basic rocks along this belt were intruded along an older fault. There was renewed movement along the edge of this belt of basic rocks, after the intrusion of both the basic and silicic rocks, in the section between Downieville and Washington and in the Mother Lode region, and it is thought possible that a continuous fault extends the entire length of the Sierra Nevada gold belt. Most of the veins, both at Alleghany and along the Mother Lode, occur in fractures auxiliary to this fault, and veins also follow the major fault itself. Therefore the explanation for the association of gold and serpentine may be that the serpentines mark the position of a major fracture which, when later reopened, formed the channel both for the solutions that deposited the quartz and for the slightly later gold-bearing solutions. The comparative lack of gold in the veins associated with the subordinate batholiths on the east side of the Sierra Nevada is not explained by this hypothesis.

GRAPHITE

The graphite that is found in the dark bands in the quartz associated with sericite or mariposite and the later sulphides is an unusual mineral in veins of this type. The carbonaceous slates of the older sedimentary formations might be thought to provide a possible source, but the bands containing graphite occur most abundantly in veins in the eastern gabbro belt, such as the Eldorado and Plumbago; and in the Oriental mine graphite occurs in altered wall rock, supposed to have been originally gabbro. Unless it

[42] Knopf, Adolph, op. cit., p. 48.
[43] Lindgren, Waldemar, op. cit. (Seventeenth Ann. Rept.), p. 124.

can be supposed that the graphite in the Eldorado and Plumbago mines was derived from the east side of the great fault along the eastern boundary or has come from slates present at depth, no wall-rock source for it is apparent. Nor are bands containing graphite especially abundant in the Rainbow vein, which cuts the carbonaceous slates of the Kanaka formation.

PLAGIOCLASE

The variable composition of the secondary plagioclase found both in the wall rocks and in the veins (p. 44) suggests that wall rock may have furnished the lime necessary to cause its deviation from pure albite. In the study of the alteration of the feldspars of the granite it was observed that wherever orthoclase is attacked a nearly pure albite is formed, but the alteration of oligoclase yields a new feldspar of the same composition as the old. The relations of the veinlets of the new feldspar to the twinning planes of the old (pl. 31, B, C) show, however, that the process of alteration is not everywhere simple recrystallization, but in places involves an introduction of material. The feldspar formed in the gabbro is as a rule nearly pure albite, and this may be due to the fact that the saussuritization of the original calcic feldspar of the gabbro yields albite, which, together with the epidote and other minerals formed by this process, is later replaced by clear albite.

BARITE

The fact that barite seems to be confined to the Evans, Oriental, and Kenton veins, which lie close to the belt of granite outcrops, suggests that the barium may have been derived from the granite. No chemical analysis of the granite of the Alleghany district has been made, but the presence of barium oxide is shown in both the analyses of the Nevada City granodiorite given by Lindgren,[44] to the extent of 0.07 and 0.06 per cent.

METHOD OF VEIN FORMATION

Study of many thin sections of the vein minerals did not, as had been hoped, result in definite conclusions on the much-discussed subject of the method of vein formation, and more questions were raised than could be answered. In the following discussion the same general order as that of the section on mineralogy will be followed—first the chlorite stage; then the quartz stage, in which the data bearing on the origin of the quartz of the veins will be considered in some detail; and last the carbonate stage, with the sulphides and gold. It should be kept in mind, however, that the separation between the stages is not as sharp as this method of presentation implies and that there is un-

doubtedly a certain amount of overlap. Moreover, it will be necessary, in places, to use the data furnished by minerals of one stage to explain the textural features of those of another.

The writer's observations offer no evidence as to the force or forces that caused the solution from which the vein minerals were deposited to rise along the fissures. The simultaneous filling of the fissures along both sets of faults may indicate that the quartz was injected during periods of relaxation of compression, but during the time in which quartz was being deposited, the region as a whole may have been subject to compression, or compression may have been effective at a lower level. If so, a mechanism for the forcing up of the solutions is suggested. The great depth of cover suggests that the force of injection exceeded the hydrostatic pressure, or at least that there may have been a local relief of pressure with compensating greater pressure outside the area. Pressure of expanding gas contained in the solutions may have been a competent lifting force,[45] but if the vacuoles within the quartz, as suggested above, were completely filled with liquid at the time of consolidation, positive evidence for the presence of gas is lacking.

CHLORITE STAGE

The earliest stage of mineralization is marked by the development of chlorite from the hornblende and of epidote and a micaceous mineral from the plagioclase feldspar and by at least partial serpentinization of the most basic rocks, the peridotites and pyroxenites. None of these changes involve any addition of material other than water and, perhaps, carbon dioxide. The solutions that effected the changes must have been sufficiently thin or under sufficient pressure to allow them to penetrate the wall rocks for a considerable distance from the fissures. The formation of the serpentine involved an increase in volume, which prevented the development within the serpentine of channels that could be followed by the solutions of the succeeding stage and also caused the development of subordinate branching fissures that were later occupied by quartz and give the observed bending and feathering of the quartz veins near the serpentine. Minor effects of such swelling are shown in the splitting of the chromite grains (pl. 20, B) and their veining by antigorite and uvarovite.

QUARTZ STAGE

POSSIBLE RECRYSTALLIZATION

The texture of the quartz veins as revealed by the microscope was studied in considerable detail, in the hope that some light would be thrown on the method

---

[44] Lindgren, Waldemar, op. cit. (Seventeenth Ann. Rept.), p. 38.

[45] Ross, C. S., Physico-chemical factors controlling magmatic differentiation and vein formation: Econ. Geology, vol. 23, p. 881, 1928.

of vein formation. As the study progressed, however, the writer became doubtful as to how far this texture was original. Certainly if the present grain is entirely the result of recrystallization, the solution of the problem for this district can not be reached by this method. Adams [46] has shown that in the process here called microbrecciation and in strain twinning there is recrystallization of the quartz, and additional evidence of such recrystallization is given in the description of vein quartz on page 44. If there were no further recrystallization it would be allowable to regard the texture as on the whole original. There are, however, certain features which suggest that the present texture

gold. It is possible in places to trace gradations from ribbon quartz with straight banding to quartz showing crinkly banding, and, as shown above, the ribbon quartz has been formed by shearing parallel to a thrust or developed by shearing of narrow septa of country rock within the vein. (See pl. 45, B, C.) The crinkly banding seems to be best developed where original irregularities of the veins, such as the bend of the Eldorado vein, have favored the crossing of the vein by minor thrusts. Where there was opportunity for detailed study of the relations of quartz showing crinkly banding to the straight ribbon quartz and to the structure of the vein (fig. 19) it was found that the crinkly banding occurs in such relation to the minor thrusts, particularly in the wedges of quartz between two such thrusts, as to suggest that the crenulations represent deformation in the quartz after crystallization.

Although the crinkly banding has a superficial resemblance to the stylolites of limestone, a similar origin is not possible. Stylolites [47] are formed by solution proceeding irregularly outward from bedding planes or fissures, with simultaneous closing in of the walls as ma-

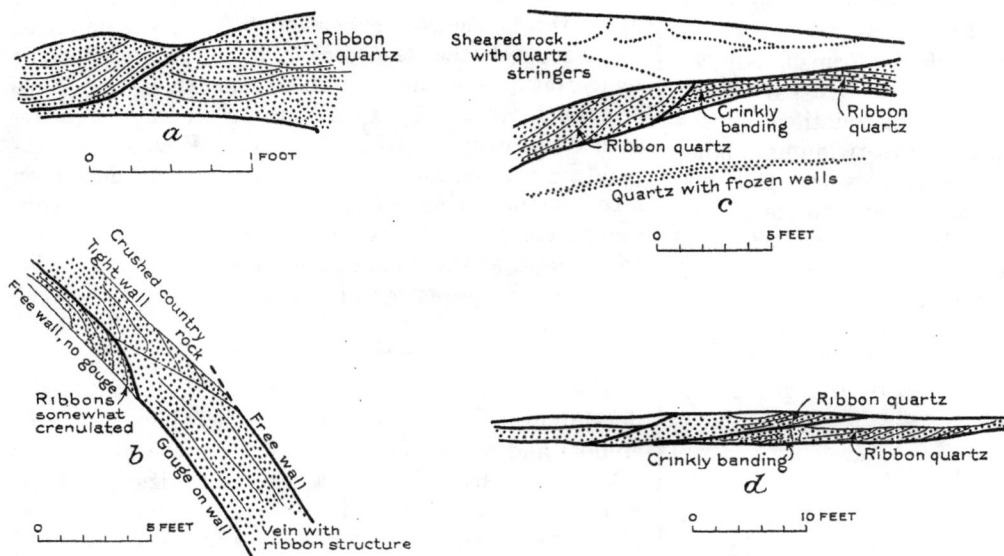

FIGURE 19.—Sketches showing relation of crinkly banding to ribbon quartz: a, Two series of ribbons, the later set following a shear which truncates the earlier, German bar mine, drain level; b, ribbon quartz, bent along shear planes and grading into crinkly banding below lower fault, Gold Canyon mine, stope above drain level; c, gradation between crinkly banding and ribbon quartz, Yellow Jacket mine, adit level; d, crinkly banding in wedge between shear planes, Yellow Jacket mine, adit level

of the quartz may be secondary to the original texture, that the recrystallization involved in the zones of microbrecciation was a local change which was superposed on an earlier and more widespread recrystallization, and that the original texture of the vein quartz may be largely obliterated.

The presence here and there in the veins of the peculiar crinkly banding noted on page 36 has a possible bearing on this phase of the problem. The crinkly banding (pls. 19, B; 40, D; 44) consists of intricately crenulated and curving bands crossing the quartz, in general roughly parallel to the walls but very irregular in detail. The crenulation is generally prominent only in one direction, and surfaces parallel to the axes of the crenulations show only straight or gently curving ribbons. These dark bands contain carbonate, graphite, sericite, sulphides, and rarely

terial is removed, and the dark lines represent, for the most part, the residuum of insoluble material. But the crenulations of the crinkly banding are most pronounced in one plane and lacking in the plane at right angles to that of their best development. Moreover, the crenulations of the crinkly banding are all curved, not composed of a series of plane surfaces, and show no abrupt transection of an earlier texture.

Except for those portions of the veins which exhibit crinkly banding there is no good evidence for deformation or complete recrystallization of the quartz, but certain other features are suggestive. Quartz veinlets, such as those shown in Plate 4, A, may owe their form either to introduction along curved partings in the already crumpled schist or to later folding, involving both schist and quartz. But in Figure 20, a, the quartz stringers between the

[46] Adams, S. F., A microscopic study of vein quartz: Econ. Geology, vol. 15, pp. 647–655, 1920.

[47] Stockdale, P. B., Stylolites—their nature and origin: Indiana Univ. Studies, vol. 9, No. 55, 1922.

two hanging-wall thrust planes are curved as if folded, while those adjoining the frozen footwall are irregularly branching. The relations sketched in Figure 20, b, also suggest folding of the vein, indicated by the sharp bend in the hanging-wall fault (g) while the later faulting (h, h) is straight. Dragged and broken arsenopyrite crystals (pl. 46, E) might also be taken to indicate movement in the quartz after the crystallization of the arsenopyrite.

If there has been considerable deformation in the vein the quartz in the folded portions must have been so far recrystallized that the original texture is lost, and yet the texture of the quartz in the areas of crinkly banding is not as a rule different from that of other parts of the quartz vein. In one thin section of quartz with crinkly banding, however, the quartz crystals show a parallel elongation (pl. 44, D), suggestive of the "shredded texture" described by Adams,[48] as if the recrystallization had been effected under sufficient pressure to cause the formation of a schistose texture. Strain twinning is common in some sections of quartz with crinkly banding (pl. 46, A) but absent or sporadic in others. Where zones of microbrecciation are present in such quartz they tend to end against the dark bands and show no definite relation to the crests of the folds. It is therefore thought probable that the pressure which caused the partial recrystallization shown by the strain twinning and microbrecciation was subsequent to that which caused the folding and complete recrystallization. This recrystallization and folding may have been aided by the presence of solutions in the fissures, and such solutions, by soaking the quartz and dissolving a part of it under pressure, may have produced a condition similar to that under which ptygmatic folding[49] of small granite and aplite dikes takes place. The presence of the sericite, graphite, and sulphides in the dark bands (pls. 40, D; 44, A, B), as well as lack of any definite relation of the microbrecciated zones or areas of strain twinning to the crenulations (pls. 44; 45, B, C; 46, A, B), implies that the folding was completed before the introduction of minerals of the carbonate stage and that the dark bands at the time of folding were either merely fractures or consisted of films of wall rock or gouge which were later replaced by minerals of the carbonate stage. There are, however, in places faint slickensides along the dark bands parallel to the axes of the folds, indicating that there has been slight movement in this direction after the introduction of the later minerals.

As the texture in other parts of the veins is, in general, similar to that in the areas of crinkly banding it is possible that the quartz of the veins as a whole has

undergone recrystallization, and certain features, such as the veining of the arsenopyrite and pyrite crystals by quartz (pls. 21, A, B; 22, A, B; 46, F), the apparent dragging of the arsenopyrite, and the outward growth of quartz crystals from arsenopyrite (pls. 23, D; 46, D) and from unsupported inclusions (pl. 23, A, B, C), might be considered suggestive of such recrystallization, although other explanations (see pp. 81-84) may be preferable.

On the other hand, certain features suggest that the original texture may remain in areas in which there

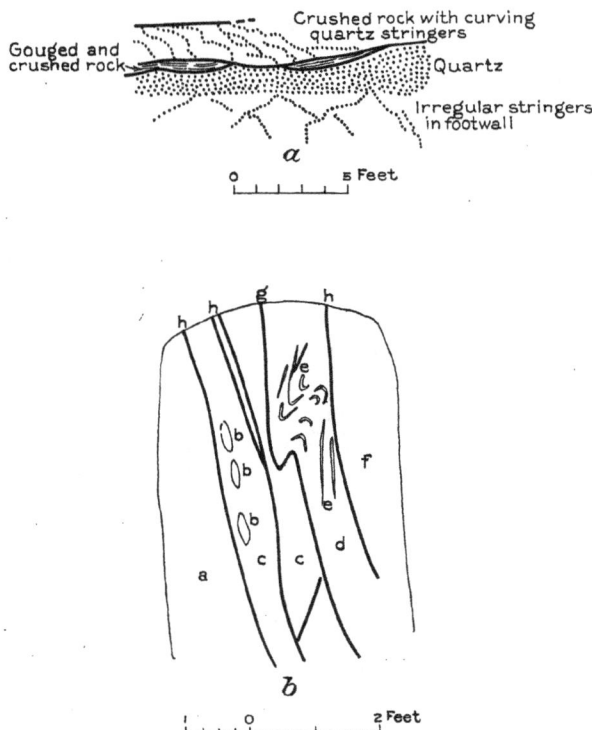

FIGURE 20.—Sketches showing possible folding of vein quartz: a, Yellow Jacket tunnel; b, Rainbow mine, drift on Clinton vein (a, Green schist, in part replaced by carbonate; b, inclusions altered to carbonate and mariposite; c, vein quartz; d, crumpled black slate; e, quartz seams; f, unaltered black slate; g, gougy slip carrying gold; gold extends into quartz for about half an inch, in narrow streaks; h, slips carrying a little gouge but no gold)

has been no folding. In the areas of crinkly banding the vacuoles are all in linear or random arrangement, and the quartz crystals that show a zonal arrangement of vacuoles were found only outside such areas. Vugs were not found in the folded areas. The veinlets that cross the country rock sharply (pls. 12, A, C; 13, A, D) show a comby texture at their borders (pl. 24, A), but in the central portion this grades into the same allotriomorphic texture which is characteristic of the larger quartz strands.

It is concluded, therefore, that although there is good evidence that the quartz in the areas showing crinkly banding has undergone recrystallization, the

[48] Adams, S. F., op. cit., p. 649, pl. 27, A.
[49] Sederholm, J. J., Über ptygmatische Faltungen: Neues Jahrb., Beilage-Band 26, pp. 491–512, 1913.

original texture may persist in other parts of the veins, though the possibility of widespread recrystallization throughout the veins must be admitted. On this assumption the structural and textural features of the vein quartz will next be considered in their bearing on several hypotheses as to the origin of vein quartz.

REPLACEMENT

The evidence for replacement of solid rock by quartz is chiefly to be obtained from observation of the veins in place and inspection of hand specimens. (See pls. 19, A; 20, C, D.) In many places fragments of schist included in the quartz show accordance of structure, but the outlines are frayed and blurred against the quartz. (See pl. 19, B.) A single strand of the vein may consist of what appears to be a breccia of wall-rock fragments so completely replaced by quartz that only a faint difference in color shows that they were originally foreign to the vein. Thin sections of such material show, however, little to confirm the impression given by the rock in place or the hand specimen. The wall-rock inclusions that in the hand specimens are distinct in color and structure are seen in thin section to be completely replaced by minerals of the carbonate stage, which also to a slight extent replace the quartz, thus blurring the original outlines. The grain of the quartz is the same throughout, and where apparent replacement has been nearly complete the only difference between replaced fragment and surrounding vein quartz is in the presence of numerous small shreds of carbonate and sericite which are crystalline against the quartz and in places specks of sulphide minerals and minute indeterminable specks of opaque material.

Knopf [50] has described similar features in the veins of the Mother Lode region and concludes that the lack of evidence from the microscope is opposed to the hypothesis of replacement; Lindgren [51] considers that "aluminous rocks (shales, schists, and igneous rocks) can not be replaced by coarse-grained quartz without leaving abundant traces by structure, texture, and relics."

On the other hand, inasmuch as the carbonate and sericite are later than the quartz the lack of identifiable relics of country rock is explained by the widespread later replacement by these minerals. The presence of parallel plates of mariposite within the vein quartz (pl. 35, A) may be interpreted as preserving the structure of shreds of schistose country rock which are residuals of quartz replacement, later completely replaced by mariposite. In places also inclusions of wall rock within the veins, now altered to later minerals, particularly carbonate and sericite, show irregular embayments of quartz which are suggestive of replacement of the original wall rock by quartz. (See pls. 30, B; 45, B.)

The grain of the quartz as a whole and even the outward growth of quartz crystals from isolated inclusions (pl. 23, A, B, C) and from arsenopyrite crystals (pls. 23, D; 46, C, D) are compatible with the assumption of origin by replacement, and, as suggested above, it is possible that there has been widespread recrystallization of the quartz, which would have erased any prior wall-rock texture that had been preserved by the quartz.

As the veins follow preexisting faults, it is possible that replacement occurred largely in the gouge along the fault planes and only to a minor extent in the solid rock. The presence of isolated quartz crystals embedded in beidellite (p. 45) suggests the possibility of such a process. It may be supposed that along the planes of the preexisting faults the rock was altered to a claylike gouge, which later crystallized to beidellite. The next step would then be the development of the early pyrite and arsenopyrite in this gouge, which would support both these newly formed minerals and such larger rock fragments as might be present in the gouge between the closely spaced parallel fault planes. The entire fissured zone would be soaked by the silica-bearing liquid, and after the crystallization of the arsenopyrite the smaller particles would be completely replaced, while the larger rock fragments, of greater volume in proportion to the exposed surface, would be replaced to a less extent.

Growth of the quartz crystals might then take place freely and more or less at random within the gouge, with a tendency for unreplaceable nuclei and the already formed arsenopyrite crystals to control the positions of early crystal growth. Zoned crystals might develop through slight changes in the character of the solutions rising through the gouge or variation of the speed at which crystallization took place. The growth of zoned replacement crystals is illustrated by the carbonate crystals shown in Plate 33, A.

Breccias, such as that illustrated in Plate 20, D, in which the rock fragments are not in orientation and are largely replaced by carbonate and mica with apparent minor attack by the earlier quartz, suggest that some such process of gouge replacement may have been effective locally.

FISSURE FILLING

Although it is clear that there has been some replacement of both solid wall rock and gouge, there is also positive evidence of fissure filling. Thin sections cut from the rather rare specimens that show contacts of vein quartz and wall rock not obscured by later movement, principally the small stringers (pl. 13, D)

[50] Knopf, Adolph, The Mother Lode system of California: U. S. Geol. Survey Prof. Paper 157, p. 40, 1929.

[51] Lindgren, Waldemar, Magmas, dikes, and veins: Am. Inst. Min. and Met. Eng. Trans., vol. 74, p. 83, 1926.

rather than the major veins, reveal a distinct reduction in the size of the grain of the quartz close to the wall rock and a tendency for the elongation of the crystals to be normal to the contact (pl. 24, A). In a few small veinlets (pl. 24, B, C) the quartz is in cross-fiber structure across the veinlet. Both of these textures are suggestive of fissure filling rather than replacement. Positive evidence of fissure filling by the later quartz is furnished by the veinlet that crosses the twin lamellae of an earlier quartz crystal. (See pl. 37, A, B.)

The fact that on the whole the quartz is thickest in the flatter portions of veins that follow reverse faults is suggestive of the filling of open spaces. But this may indicate not actual open spaces but merely areas in which compression was less severe. The inability of the quartz to penetrate the larger serpentine masses (p. 31), although these are cut to some extent by the fissures followed by the veins, likewise suggests that deposition of the quartz was favored by either actual open spaces or regions of less than average pressure.

It is concluded, therefore, that a portion of the quartz was deposited as a filling of fissures, though the relative amounts of quartz formed by fissure filling and replacement are indeterminable. On the assumption that fissure filling was an important if not the dominant method of vein formation, several hypotheses which have been advanced to explain the method of deposition will be briefly reviewed in connection with the data furnished by the texture and structure of the Alleghany veins. These hypotheses are:

1. The veins were formed by gradual deposition on the walls of open spaces.

2. Fissure walls were forced apart as the quartz was deposited by pressure exerted by the accretion of new material added to the quartz already formed.

3. The veins are " vein dikes " resulting from crystallization of a relatively " dry " siliceous magma.

4. The quartz was deposited from colloidal solution as silica gel with later change to crystalline quartz.

5. The quartz crystals accumulated and grew from a tenuous liquid.

1. The deposition of the quartz was not a single event; the relation of the different strands with their intervening septa of altered country rock indicates that the filling of the veins was a pulsatory process and that movement in and along the veins took place concomitantly with the deposition of quartz. Hulin[52] has called such a process " deposition by accretion." But, although it is not required that open spaces of the size of the present quartz veins should have existed, single strands several feet thick and extending for considerable distances in the plane of the vein seem to have been formed as a result of a single pulse of the quartz-depositing solution; or else the original boundary septa have been replaced and all trace of their existence destroyed by the recrystallization of the quartz. Although it may be supposed that veinlets in solid rock of the type shown in Plates 12, C, and 13, D, could have been formed by the filling of open spaces, even at great depth, it is difficult to imagine that such open fissures could have existed or long continued in sheared rock such as that shown in Plate 13, B, C.

Another objection to this hypothesis lies in the texture of the quartz, on the assumption that this is original, and the relations of the quartz to the early formed arsenopyrite crystals and to certain of the wall-rock inclusions. Although there is in places a tendency for the growth of quartz crystals outward from the wall, particularly in the smaller veinlets (pl. 24, A), a parallel texture is generally lacking, and the widespread allotriomorphic texture suggests that, if the grain is even in part original, crystallization must have taken place almost simultaneously throughout each strand of quartz, so that the crystals have hindered each other's growth. Also the clusters of outward-pointing crystals (pl. 23, A, B, C) and individual doubly terminated crystals (pls. 27, D; 28, A, B) suggest free growth within the vein.

The sulphides belonging to this stage of mineralization are crystalline against the quartz but are also veined by the same early quartz, and quartz crystals show growth outward from the faces of the sulphide crystals. (See pls. 23, D; 46, D.) The major relations shown in Plate 21, B, can be interpreted as growth of the sulphide outward from the wall of the vein, followed by injection of additional quartz between the arsenopyrite and the wall. But the small veinlets within the sulphide may be formed of the same crystal of bubbly quartz against which the sulphide is crystalline (pl. 46, D), and isolated sulphide crystals in the veins are veined by the quartz that incloses them (pl. 22, A, B). Therefore, if the texture of the quartz is original, not only must the early arsenopyrite and pyrite have crystallized prior to the early quartz, but also, if this portion of the vein was formed by some process of fissure filling, the medium from which the quartz crystallized must have been temporarily capable of sustaining the arsenopyrite. The clusters of bladed arsenopyrite crystals that show at most only minor fracturing imply that in parts of the vein in which they are found recrystallization, if it has taken place, was effected without substantial movement of the surrounding quartz.

Inclusions of country rock within the vein may in most places represent material caught by successive reopening of the fissures, and elsewhere they are sufficiently close together to suggest mutual support, but isolated inclusions surrounded by outward-pointing quartz crystals (pl. 23, C) are found which can not be

[52] Hulin, C. D., Structural control of ore deposition: Econ. Geology, vol. 24, p. 23, 1929.

so explained and which, like the arsenopyrite crystals, suggest the necessity for support at the time the surrounding quartz crystallized.

2. The method of vein filling suggested by Taber [53] requires spreading of the walls of the fissures by growth of the quartz, involving the increase in volume resulting from minute additions to the crystal from a film of supersaturated solution between the crystal and wall and to a less degree the "linear force of growing crystals." The growing crystal is thought to make room for itself either by expediting the solution of the wall rock, which would give replacement without change in grain of the quartz, or by mechanical movement of the wall.

The experimental work of Becker and Day [54] indicates that the force exerted by a growing crystal is of "the same order of magnitude as the ascertained resistance which the crystals offered to crushing stress." If this order of magnitude holds for crystalline quartz, whose crushing strength is "two or three times the crushing strength of high-carbon steel and a hundred times that of most building stone or concrete," [55] then the force of crystallization alone is presumably competent to wedge the walls apart, even against greater pressure than may be supposed to have been present at the time of vein formation. Taber [56] considers that

It is probable that the pressure effects observed during crystallization are due chiefly to the separation of solid matter from solution rather than to the growth of crystals, and, under favorable conditions, the pressure developed in this way may greatly exceed the crushing strength of the substance.

If, as suggested above (p. 77), vein formation took place in intervals of actual or relative tension between periods of compression the force required need not be great. The growth of the quartz crystals may be considered as having taken up the slack in the times of tension or less active compression.

It would seem, however, that veins formed in this manner would, if the texture is original, necessarily show a tendency to cross-fiber structure, inasmuch as the crystals should tend to be elongate in the direction of easiest growth. If the force exerted is proportional to the force of crystallization and this to the crushing strength of the crystal, a cross-fiber structure would seem to be required, for the crushing strength of quartz, according to Sosman, [57] "when compressed in the direction of the principal axis is about one-tenth greater than perpendicular to the axis." Parallel growth normal to the wall is indeed seen in many of the small stringers in the wall rocks (pl. 24, B, C), and to this extent the process may have been operative, but the granular interlocking texture of the larger strands of the veins seems inconsistent with such a method of vein growth.

The observed relations of the zoning of crystals within the vein and the presence of cores of bubbly quartz are also difficult of explanation under a hypothesis which presupposes very slow growth of the quartz from nourishment supplied by a subcapillary film of solution. It is difficult to see how a phenocrystlike crystal of zoned quartz such as that shown in Plate 27, D, could have grown entirely surrounded by other quartz crystals, nor indeed how the gradual building of the quartz crystal postulated by this hypothesis could have permitted the entrapping of the liquid and gas now present in the vacuoles. Taber, however, mentions the presence of fluid and gas filled cavities in the quartz of veins which he considers to have been formed by this method. [58]

3. In his first study of the ores of the Alleghany district [59] the writer considered that to account for the relations of the early arsenopyrite and quartz it was necessary to suppose that the veins were formed by the sudden crystallization of a solution, which, by implication, consisted essentially of silica in liquid form. Such an origin has also been suggested by Spurr [60] for similar veins. Under such a hypothesis it is assumed that wherever preexisting fissures afforded routes of easiest access, they were followed, but where these were lacking the liquid tended to follow the schistosity of the country rock. To some extent the liquid must have been capable of dissolving the adjoining rock and fragments of wall rock within the fissures in the wall rocks. A change in character to greater viscosity may be supposed to have taken place through the escape of volatile constituents, new conditions of temperature and pressure, and change in character of the liquid by the accession of material dissolved out of the wall rocks. This viscous stage need not have been of long duration, but its existence is essential to the argument, for it can not be supposed that the liquid could have had a specific gravity close to

[53] Taber, Stephen, Geology of the gold belt in the James River Basin, Va.: Virginia Geol. Survey Bull. 7, pp. 222–231, 1913; Pressure accompanying the growth of crystals: Nat. Acad. Sci. Proc., vol. 3, pp. 297–302, 1917; Origin of Bendigo quartz veins: Econ. Geology, vol. 13, pp. 538–546, 1918; Mechanics of vein formation: Am. Inst. Min. Eng. Trans., vol. 61, pp. 3–36, 1919; Metasomatism and the pressure of growing crystals: Econ. Geology, vol. 21, pp. 717–727, 1926; The linear force of growing crystals: Econ. Geology, vol. 23, pp. 335–336, 1928.

[54] Becker, G. F., and Day, A. L., The linear force of growing crystals: Washington Acad. Sci. Proc., vol. 7, pp. 282–288, 1905.

[55] Sosman, R. B., The properties of silica, p. 480, 1927. (On page 482 are quoted data showing an average strength under compression, parallel to the axis, of 24,470 kilograms per square centimeter, equivalent to 348,037 pounds per square inch.)

[56] Taber, Stephen, Pressure phenomena accompanying the growth of crystals: Nat. Acad. Sci. Proc., vol. 3, p. 302, 1917.

[57] Sosman, R. B. op. cit., p. 480.

[58] Taber, Stephen, Geology of the gold belt in the James River Basin, Va.: Virginia Geol. Survey Bull. 7, p. 213, 1913.

[59] Ferguson, H. G., Lode deposits of the Alleghany district: U. S. Geol. Survey Bull. 580, pp. 167–168, 1914.

[60] Spurr, J. E., The origin of metallic concentrations by magmatism: Econ. Geology, vol. 18, pp. 631–633, 1923; Ore magmas, vol. 1, pp. 142–143, 1923; The Camp Bird compound vein dikes: Econ. Geology, vol. 20, p. 148, 1925.

that of arsenopyrite. Some differentiation must have taken place within the liquid, for when crystallization began the first minerals to crystallize out—the sulphides, feldspar, and barite—tended to form along the walls, but nevertheless they must have been developed in a medium capable of supporting them, for they occur also within the quartz. Crystallization of the quartz took place at first inward from the walls and around whatever favorable nuclei, such as sulphide crystals and wall-rock fragments, were present in the vein, but after the first start had been made the simultaneous growth of crystals in all parts of the vein caused mutual interference. But if so, motion could not have continued during the viscous stage, for the texture of the veins is nowhere indicative of viscous flow. Druses might well be formed under such conditions, for a loss in volume at or shortly preceding consolidation is to be assumed.

Knopf [61] has doubted the validity of the " vein dike " hypothesis for the Mother Lode veins, because of the lack of evidence of a chilled margin close to the walls. But it may be questioned whether under deep-seated conditions and in rocks above the temperature required by the normal temperature gradient the difference in temperature of " ore magma " and wall rock was sufficient to require a sharply marked chilled border; certainly such a border is not generally characteristic of the small granite and aplite dikes at Alleghany. Moreover, some of the thin sections from the rather rare places in the Alleghany veins where the original wall is present show a distinct increase in the size of the primary grain (distinct from change in grain size due to later fracturing) inward from the wall. (See pl. 24, A.)

Such an explanation assumes the possibility of alteration of the wall rock by a liquid consisting essentially of silica, with relatively little water. Under such a hypothesis the replacement of the wall rock by quartz could not have been replacement in the sense ordinarily understood but must have involved a certain amount of solution of the wall rock by the liquid or " magma," with the place of the dissolved solid filled not immediately by a solid but by the same liquid which filled the fissure, the whole crystallizing simultaneously. This conclusion offers an explanation of the identity of grain of fissure filling and replaced wall rock.

On the other hand, growth by accretion, together with changing character of the surrounding solution, suggested by the banded structure of individual quartz crystals (pls. 27, D; 28, A, B, C), seems inconsistent with the assumption of rapid consolidation of the minerals of the vein required by this hypothesis. Another difficulty with this hypothesis seems to lie not in its failure to fit the observed conditions but in the question whether the existence of a magma composed essentially of silica and capable of passing from a highly mobile through a viscous to a crystalline state and also capable of corroding the wall rocks is possible.

4. Boydell [62] has argued that the quartz crystals of veins were not deposited originally from a molecular solution but that the original colloidal solution of silica became a jelly, which in time crystallized to quartz. The loss of volume in change from gel to crystal is considered to have been taken up by successive injections of new material. The gelatinous silica may have changed first to opal and then with further dehydration to chalcedony, or crystalline quartz, [63] or the quartz crystals may have formed directly from the gel. [64]

The present texture of the crystalline quartz may be considered to have been a later development after the vein was filled with solid colloform silica. The veining of the arsenopyrite crystals may be supposed to have occurred in the gel state or during readjustments accompanying the change from gel to crystalline quartz; the puzzling features of both veining by quartz and outward-pointing quartz crystals from the faces of the arsenopyrite are explained as the result of the later development of the crystalline form of quartz; the support required by the isolated arsenopyrite crystals and the small isolated inclusions is provided on the supposition that they were held in place in a stiff gel. [65] These would form nuclei for the crystallization of the quartz from the surrounding gel and crystal framework.

The change from gel to crystalline quartz involves a considerable loss of volume, but as there is evidence of successive deposition of the vein quartz it might be supposed that as space developed from the transformation of the gel to crystalline quartz it was taken up by successive injections of gelatinous silica. Therefore the evidence of a long-continued process in the formation of the early quartz may be considered favorable to this hypothesis.

Certain other features are also suggestive. The later veins of clear quartz that cut the cloudy quartz contain both chalcedony and opal, in places in the same vein (pls. 36, A, B; 37, D), showing that deposition as a gel did take place to some extent in this stage; and if this is true for the later quartz, such a process may have been carried to completion in the

[61] Knopf, Adolph, The Mother Lode system of California : U. S. Geol. Survey Prof. Paper 157, p. 47, 1929.

[62] Boydell, H. C., The rôle of colloidal solutions in the formation of mineral deposits : Inst. Min. and Met. (London) Trans., vol. 34, pt. 1, pp. 145–337, 1925.
[63] Scott, A., Application of colloid chemistry to mineralogy and petrology : British Assoc. Adv. Sci. Fourth report on colloid chemistry, pp. 219–223, 1922.
[64] Storz, Max, Die Sekundäre authogene Kieselsäure in ihrer petrogenetisch-geologischen Bedeutung, Teil 1 : Mon. Geol. und Pal., ser. 2, Heft 4, pp. 28–32, 1928.
[65] Merrit, C. A., Angular inclusions in ore deposits : Econ. Geology, vol. 20, p. 603, 1925.

earlier stage. Also a spherule of chalcedony (pl. 32, D) was found entirely inclosed in a single large crystal of the older quartz and apparently of contemporaneous development. On the other hand, the persistence of the opal shows that it has, in part at least, been stable since the formation of the veins. It might therefore be argued that if the change had taken place in part there is no reason why it should not have continued to completion and that the presence of opal, instead of indicating original deposition of all the quartz as a gel, merely shows that under certain conditions opal and quartz could form simultaneously.

The presence of lines of vacuoles which are considered to have developed later than the quartz has already been noted (p. 42). Tronquoy [66] has explained a similar feature in the French tin-bearing veins as due to the entrapping of rising trains of gas bubbles in the gelatinous silica and preservation in the later crystalline state. The objections to this argument, so far as it applies to Alleghany, are, first, that it assumes that the crystalline quartz occupied the same space as the colloid, for contraction would distort the line, and, second, that at Alleghany these lines of vacuoles appear to be dependent on minerals which are later than the development of the present crystalline texture in the quartz and do not show parallel orientation.

The presence of the zonally arranged vacuoles containing liquid and vapor is hard to reconcile with this hypothesis if it is supposed that the silica once existed as chalcedony, but if the change to crystalline quartz took place directly from the gel it might be supposed that the entrapped liquid was originally present in the gel and that the zoning was due to different rates of crystal growth or to the formation of the quartz, first, as skeleton crystals which entrapped portions of the still hydrous gel. Storz [67] has found that zoned quartz crystals form in this way from a gel under supergene conditions and regards the formation of such skeleton crystals of quartz as evidence of crystallization from a gel.

The folding that resulted in the development of the crinkly banding might be supposed to have taken place while the silica was in the condition of a stiff gel. This would avoid the requirement that folding should have taken place in the crystalline quartz soaked with liquid and that recrystallization and folding should have taken place simultaneously. But it does not seem possible that the resultant loss of volume would permit the preservation of these delicate crenulations.

5. Another possible method of vein formation involves crystallization from a tenuous solution moving in the fissure and presumably exerting some pressure on the walls. It may be considered that at some stage

the degree of saturation necessary for the beginning of crystallization, possibly dependent on change of temperature, would be reached. Crystallization would begin for the most part against the walls. In so far as the crystals formed against the walls they would tend to be stable, but crystals forming within the vein, either free or about nuclei of foreign matter, might descend to portions of the fissure where the solution was still unsaturated and thus be dissolved. There would, however, conceivably be a stage at which the crystals formed more rapidly than they were dissolved. Then a mixture of solid crystals loosely packed in an environment of liquid would fulfill the conditions necessary for the support of the newly formed arsenopyrite crystals and spalled-off fragments of wall rock, which wherever favorably situated would serve as nuclei for the growth of new crystals. The already formed quartz crystals of the vein, receiving accretions from the surrounding silica-bearing liquid, would exhibit the observed phenomena of growth by accretion. Such a system, if confined, would be capable of yielding by shearing.[68] Therefore, the fracturing of the arsenopyrite might have taken place at this stage and the fractures have been filled by quartz deposited from the surrounding liquid and crystallizing in optical continuity with the older crystals bordering the arsenopyrite. Similarly it is possible that the shearing that produced the banded quartz might have taken place while this condition of mixed solid and liquid prevailed.

Such a process would imply that on the whole crystallization proceeded down the vein. Escape of the residual solutions deficient in silica must have been effected along the walls or through fractures in the quartz. Possibly this affords an explanation of the deposition of the minerals of the carbonate stage, if microbrecciation and strain twinning followed closely upon the crystallization of the quartz, for carbonate replaces microbrecciated quartz. An objection to this hypothesis is that there was rarely observed any tendency to parallel alinement of the quartz crystals. Also, if the arsenopyrite were fractured and veined by quartz there should be similar fracturing and veining of the quartz crystals themselves, but unless the relations shown in Plate 28, A, B, can be so interpreted, no evidence for this was found, though the shearing resulting in the banded quartz and the microbrecciation—features developed prior to the carbonate stage—show that the quartz was fractured soon after it attained its present crystalline form.

CONCLUSIONS

In the foregoing paragraphs the writer has endeavored to determine how far different hypotheses of

[66] Tronquoy, R., Contribution à l'étude des gîtes d'étain: Soc. française minéralogie Bull., vol. 35, pp. 384–389, 1912.
[67] Storz, Max, op. cit., p. 33.

[68] Mead, W. J., The geologic rôle of dilatancy: Jour. Geology, vol. 33, p. 697, 1925.

vein formation will meet certain conditions imposed by phenomena, chiefly of a textural nature, observed in the veins. The object has been to test each hypothesis, as far as possible, against actual observation, without reference to conditions imposed by postulates of physics and chemistry.[69] Objections are found to each, and none seems to fit the observed facts completely.

Evidence is believed to have been found that there has been complete recrystallization in places and that the present grain of the quartz is therefore not everywhere an original feature. On the other hand, there are indications that for the veins in general the original texture has been essentially preserved. If recrystallization has been widespread the lack of textural evidence for replacement is explained, and even if the texture is original it is thought that replacement may have been an important or even the dominant process in the formation of the quartz veins. It seems certain, however, that fissure filling played a part in the vein formation, and various hypotheses are considered. If the present texture of the quartz and the relations of quartz and early arsenopyrite are even in part original features, the formation of such portions of the veins as are due to fissure filling is best explained by a hypothesis which involves either a viscous, gelatinous, or slushy state prior to crystallization. If, however, it can be considered that the evidence for recrystallization where the quartz is folded can be applied to the veins as a whole, then the texture of the quartz throws no light on the mode of fissure filling.

CARBONATE STAGE

Although the presence of carbonate and sericite (pl. 32, A) inclosed in quartz crystals shows that there was some overlap in the deposition of the minerals of these two stages, there is abundant evidence that essentially all of these two minerals was deposited not only after the quartz had crystallized in its present form but after the quartz had been fractured and faulted.

In contrast to the type of deposition prevailing during the filling of the quartz veins, the minerals of the carbonate stage formed most abundantly in the wall rocks and in fragments of wall rock in the veins, evidently by replacement, for on a large scale, though not in detail, they preserve the schistose structure of the original wall rock. To some extent they have replaced the quartz close to the walls but without preservation of the preexisting texture. This is particularly true of the carbonate.

The dark bands of the ribbon quartz and crinkly banding, which now consist of minerals of this stage, were probably originally for the most part shear

planes in the quartz, possibly marked by films of gouge. The later minerals have completely replaced this gouge and even narrow septa of country rock, and to some extent the adjoining quartz as well.

Minor filling of open spaces by minerals of this stage is shown by the breccias consisting of fragments of early quartz cemented by carbonate and later quartz (pls. 34, A; 36, C) and by the veinlets of carbonate (pl. 33, D) and later quartz (pl. 36, A, B), particularly in the relation of the quartz veinlet to the lamellae of earlier strain-twinned quartz (pl. 37, A, B, and p. 49).

The writer can offer no adequate explanation for the unusual concentration of gold found in the Alleghany district. In other California districts "high-grade" ore of this sort is rare and large low-grade bodies the general rule. In the description of the ore shoots (pp. 56–58) it is shown that the high-grade ore tends to occur where structural conditions have favored the reopening of the vein, during this stage of mineralization. But in other California districts the gold is also deposited after consolidation and fracture of the quartz.[70] The derivation of the gold from serpentine has been suggested (p. 75), but the evidence for this is weak, and the association of gold and serpentine may be explained on other grounds. It is possible that temperature conditions were so delicately adjusted as to favor the deposition of more abundant gold in a roughly horizontal zone, which has been preserved at Alleghany because the veins here crop out at higher altitudes than in most other California districts and which has been eroded elsewhere. But as it has been shown (p. 69) that gold was deposited in the California veins over a total range of many thousand feet, it seems unreasonable to assume that conditions were so delicately adjusted as to favor this unusual concentration in a single zone.

Another feature of the deposition of certain of the minerals of the carbonate stage is the close association of the gold with the earlier quartz, while the bulk of the associated minerals replace the wall rock. The sulphides belonging to this stage, except pyrite and arsenopyrite, are also confined to the quartz. Free gold has been reported to occur in the altered wall rock, particularly the carbonate-mariposite mixture ("blue jay"), but in nearly all such occurrences seen by the writer the gold was actually within quartz veinlets which had cut the rock prior to carbonatization and had been partly replaced by the carbonate and mariposite. The only specimens seen that seemed to present exceptions to this position were one from the Brush Creek mine showing gold in talc but close to the quartz and suggesting replacement of quartz by

[69] For a summary of our present knowledge of physical and chemical conditions governing vein formation, see Ross, C. S., Physico-chemical factors controlling magmatic differentiation and vein formation: Econ. Geology, vol. 23, pp. 864–886, 1928.

[70] Howe, Ernest, The gold ores of Grass Valley, Calif.: Econ. Geology, vol. 19, p. 619, 1924. Hulin, C. D., Structural control of ore deposition: Econ. Geology, vol. 24, p. 31, 1929.

the talc, and one of high-grade ore from the small Clinton vein of the Rainbow mine, where a small thread of gold penetrated the dark slate for a distance of about a centimeter. On the other hand, the carbonatized country rock adjoining the quartz of the high-grade shoots is essentially barren.

This close association of at least the coarser gold with the quartz has also been noted in other localities.

Lindgren [71] has suggested that the walls may be considered

as forming a septum permeable only for a part of the solution, according to osmotic laws, especially for substances which act chemically upon the minerals of the rocks. The latter in general are shown to be permeable for the carbon dioxide and alkaline carbonates; also for carbonate of calcium; further for hydrogen sulphide or sodic sulphide and for arsenic sulphide. On the other hand, they are less permeable for silica and gold and almost entirely impermeable for the other metallic sulphides.

But this explanation seems hardly adequate to account for the constant association of the gold and quartz, for, as the gold was deposited in the quartz after fracturing of the quartz, there must also have been some actual fissuring of the adjoining country rock, with consequent opportunity for passage of the gold-bearing solutions. Moreover, in places arsenopyrite is abundant in the country rock and would presumably be as effective a precipitant as the arsenopyrite within the quartz veins. Spurr [72] considers that in quartz veins formed at depth, in which the gold is apparently later than the quartz, the slow freezing of the quartz permitted a fractional crystallization, so that the gold crystallized distinctly later than the bulk of the quartz, although belonging to the same period of ore injection. Boydell [73] considers that gold and sulphides of certain quartz veins were originally contained in the silica gel, peptized by the silica and localized in shrinkage cracks as the gel contracted with loss of water.

All three of the foregoing explanations assume that the gold and quartz are closely connected in origin, although the crystallization of the gold is slightly later than that of the quartz. In the Alleghany veins, however, although the gold is spatially closely related to the quartz, a long series of events, including movements in the vein, folding and recrystallization of the quartz, microbrecciation, and introduction of at least a part of the minerals of the carbonate stage, intervened between the deposition of the gold and that of the quartz. Therefore, unless it can be supposed that

the gold is in its present position as the result of re-solution and redeposition, it seems as if a longer period intervened between the deposition of the quartz and that of the gold than is implied in the foregoing explanations. As gold has replaced quartz, it may be supposed that the change of composition of the gold-bearing solution upon dissolving the quartz compelled the deposition of the gold.

SUMMARY

The veins of the Alleghany district are considered to have been formed at about the end of Upper Jurassic time, and various lines of evidence lead to the conclusion that the depth of formation was at least 10,000 feet and may have been much greater. The reverse faults of two systems, which are the sites of most of the veins, are considered to be auxiliary to the major reverse fault that borders the district on the east. The later reverse movement along the veins, which continued during the whole period of ore deposition, may have been in part caused by the progressive serpentinization of the basic intrusives. As thrust faulting was active during the entire period in which the veins were being deposited, there must have been compression in addition to the pressure due to weight of overlying rock, but it is possible that the actual deposition of minerals took place during intervals of relaxation. The low vapor content of the vacuoles in the quartz suggests a relatively low temperature at the time of final crystallization of the quartz.

The minerals of the earliest stage of mineralization were formed by the alteration of wall-rock minerals without the addition of material except water and carbon dioxide. Derivation of the chromium content of the mariposite from the serpentine seems also to be well established. For the other minerals, both of the altered wall rock and of the veins themselves, it is considered that evidence of derivation of any of their components from the wall rock is not conclusive.

The vein quartz in places shows evidence of recrystallization after solidification but prior to the introduction of minerals of the carbonate stage. This recrystallization may possibly have been widespread throughout the veins, but it is thought more likely to have been confined to certain localities that were subject to greater than average pressure. There is evidence that both the solid wall rock and the fault gouge have been replaced by the vein quartz, but the extent is uncertain, and there is also evidence of fissure filling. If the grain of the quartz is original, the textural conditions are best met on the supposition that the quartz which filled fissures passed through a viscous, gelatinous, or slushy state prior to crystallization. If, on the other hand, the present grain of the quartz is the result of recrystallization, no conclusions can be drawn from the texture.

[71] Lindgren, Waldemar, The gold quartz veins of Nevada City and Grass Valley, Calif.: U. S. Geol. Survey Seventeenth Ann. Rept., pt. 2, p. 183, 1896.

[72] Spurr, J. E., The Camp Bird compound vein dike: Econ. Geology, vol. 20, p. 135, footnote, 1925.

[73] Boydell, H. C., The rôle of colloidal solutions in the formation of mineral deposits: Inst. Min. and Met. (London) Trans., vol. 34, pt. 1, pp. 216–217, 1925.

The minerals of the carbonate stage were deposited after the final crystallization of the quartz and largely by replacement of earlier-formed minerals, both of wall rock and of veins, but also to a minor extent in open fissures. No acceptable explanations are found for the peculiar concentration of the gold in high-grade shoots or for the fact that the gold, though associated with minerals that have replaced the wall rock, is always found in association with the earlier quartz.

MINE DESCRIPTIONS

The principal lode mines and prospects of the Alleghany district are described briefly in the following pages. Reference to the drift mines, even those that are accessible, is confined to those whose tunnels have cut or developed veins. No description is therefore given of the following drift mines: Omega, Lucky Dog, Brown Bear, Bald Mountain, Bald Mountain Extension, Mammoth Springs Placer, Gold Star, Bonanza.

Except for a few of the mines that were again visited in the summer of 1928, the information here presented refers to the state of development in the summer of 1925. (The mine map of the Sixteen to One mine shows developments to the summer of 1930.) For several of the mines where up-to-date mine maps were not available some development work prior to 1925 has not been included, or such work has been roughly surveyed and added to the mine maps. For many of the smaller mines and several of the old mines no maps were available, and sketch maps were made by Mr. Gannett which are believed to be essentially accurate on the scale here shown. As far as possible in the following descriptions emphasis is placed on the geologic features that have controlled the distribution of the high-grade shoots.

Although the writer visited all the mines here described and studied parts of several in detail, the greater part of the detailed study of the mines was the work of Mr. Gannett, and his field notes are the principal source of the following descriptions, except for the additional data gathered in 1928.

The mines are described by drainage areas, and in each section a general order of west to east is followed.

MINES OF OREGON CREEK DRAINAGE BASIN

BRUSH CREEK MINE

The Brush Creek mine, near the Mountain House saddle, in the extreme northwestern part of the district, is an old mine that has been recently reopened.

The old production of which there is record is said to have amounted to $890,000, and an old estimate [74]

gives the production to 1875 as $1,000,000. According to Raymond [75] the output for the year ending June 1, 1870, was $95,000.

The mine was reopened in 1923 by the French Gulch Mining Co. and sold in 1927 to the Kate Hardy Mining Co. Recent production has amounted to only a few thousand dollars. The old workings, which are now inaccessible, consisted of an inclined shaft 520 feet in length a short distance south of the Mountain House saddle. From this short levels were run and a lenticular body of quartz, which was in places extremely rich, was mined. It is said that all of the old output was obtained within 30 feet to the south of the shaft.

Recent work consists of several tunnels driven from the gulch north of the saddle, north of the area shown on Plate 1. Only the two nearest the mapped area, which are also the most extensive, were visited. From the upper tunnel a winze has been sunk and drifts run at two levels.

The vein follows the contact between the Cape Horn formation and the serpentine. For the most part it lies within the slate, though closer to the contact than at the Kate Hardy mine, but in places serpentine forms one wall.

The American tunnel (fig. 21), whose portal is about three-quarters of a mile north of the Mountain House saddle, follows the vein for a distance of 430 feet. The footwall is serpentine, here intensely altered to the usual carbonate-mariposite mixture. Along the hanging wall is an andesitic dike of post-mineral age. In one place this dike crosses the vein and apparently intrudes also the serpentine footwall. The andesite is rarely in actual contact with the quartz of the vein; in most places a strip of crushed slate a couple of inches wide intervenes between dike and vein. The vein here strikes about north and dips between 43° and 65° W. Near the tunnel portal it is well defined and has a width of nearly 20 feet, with a well-marked footwall slip separating the vein quartz from the altered serpentine. Beyond the point where it is crossed by the andesite dike the vein is greatly contracted, and in places definite quartz filling is lacking. Elsewhere the quartz rests without sharp demarcation on the altered serpentine, and the "wall" passes to the hanging wall between the quartz and the narrow band of crushed slate between the vein and the dike. At about 300 feet from the portal the vein has been stoped in two places, but the output is unknown.

The crosscut near the south end of the tunnel passes through 50 feet of altered serpentine, cut by well-marked gouge bearing slips, and enters the unaltered serpentine. There is no gradation between the intensely altered material and the fresh serpentine at

[74] Min. and Sci. Press, vol. 31, p. 378, 1875.

[75] Raymond, R. W., Mineral resources of the States and Territories west of the Rocky Mountains for 1870, p. 45, 1871.

the end of the crosscut, but a westward-dipping slip forms the boundary between altered and fresh rock.

The portal of the Brush Creek tunnel (fig. 22) is higher up the ridge to the south, only a short distance north of the Mountain House saddle and just north of the area included in the detailed map. The vein here also follows closely the contact of serpentine and

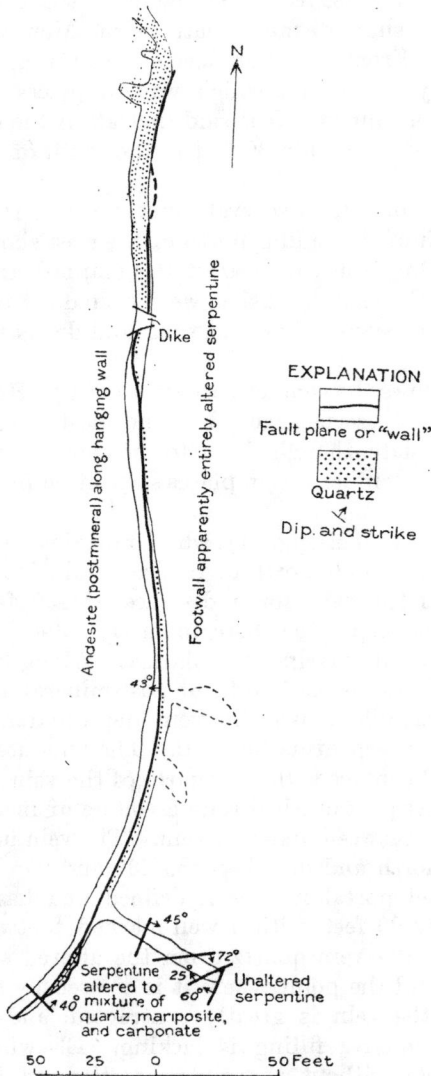

FIGURE 21.—American tunnel, Brush Creek mine

slate, but where strongest lies within the slate, though close to the serpentine. Small "prospects" were obtained here and there, but no production was made from the tunnel level. The serpentine, even where in contact with the vein, is fairly fresh and nowhere shows the intense alteration observed in the American tunnel. The Brush Creek tunnel reaches the old workings, and much difficulty has been caused by explosions of gas generated by the decaying timber of the old works. A winze has been sunk in the hanging-

wall slate to a vertical depth of 290 feet on a 74° incline, and from this a portion of the vein not explored by the old workings is being developed on two levels (fig. 22), at a depth of 230 feet and at the bottom of the winze.

The vein consists of an irregular body of quartz about 200 feet in length. The dip between the two levels is much steeper than in the tunnels, from 80° W. to vertical. A zone about 35 feet wide of intensely altered Cape Horn slate close to the serpentine contact is cut irregularly by quartz, which has a maximum thickness of 15 feet, but the quartz in these lower levels

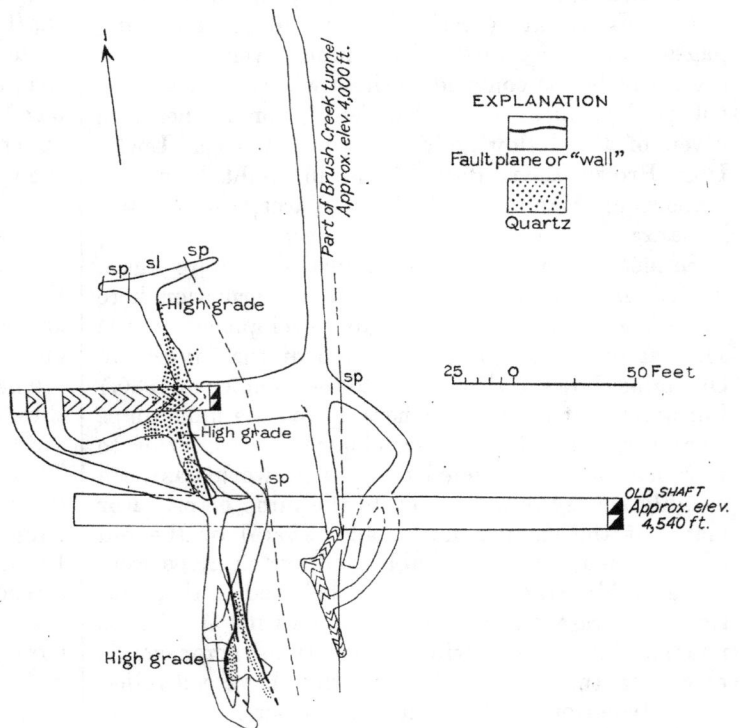

FIGURE 22.—Drifts from winze, Brush Creek mine. sp, Serpentine; sl, slate

is not in contact with the serpentine. The slate between the vein and the serpentine is intensely altered, principally to a mixture of carbonate and mariposite, with irregular stringers of quartz, but is here and there replaced by talc, which also replaces the vein quartz. The serpentine, however, shows relatively little alteration. On the west side of the vein the slate is less altered, and in the winze there are large numbers of horizontal veinlets cutting unaltered slate. (See pl. 13, D.) At both ends of the lens the quartz fingers out irregularly into the slate.

Postmineral "walls" adjoin the quartz in the southern part of the 230-foot level, but elsewhere later movement seems to have been much less extensive than in other mines, and the quartz tends to finger out into the slate, particularly along the hanging wall. No definite geologic controls for the three small bunches of high-grade ore found on the first level could be observed. One occurred in the northern part, where the vein lies between two serpentine masses and begins to widen southward; another in the widest part of the vein but near the footwall; and the third near the south end of the drift, close under the hanging wall. A small shoot of high-grade ore yielding a few thousand dollars was found on the lower level. Here there is a decided change in both dip and strike of the vein. The gold was for the most part not associated with coarse arsenopyrite, though a little was present in one of the bunches on the first level, but tended to follow rather indistinct dark streaks in banded quartz. At one point talc, apparently replacing quartz (pl. 35, C, D), carried gold, and in the shoot on the second level a veinlet of very dense bluish quartz crossing the white coarse quartz was reported to have carried much coarse gold.

Besides the high-grade bunches, much of the quartz, particularly where mixed with the carbonate and mariposite that have replaced the slate, yielded $8.50 a ton in the mill. Although this ore consists principally of the carbonate-mariposite mixture, a little quartz is everywhere present and probably contains the gold.

The gold has an average fineness of 811 parts per 1,000.

The available data from the tunnels and recent work from the winze, as well as the reported mode of occurrence in the old workings, suggest that in this mine the quartz occurs in a series of southward-pitching lenses, within the slate but close to the serpentine contact.

FINAN PROSPECT

The Finan prospect, about half a mile north of the Kate Hardy mine, is in the same zone along the contact of the serpentine with the Cape Horn slate in which the veins of the Kate Hardy and Brush Creek mines have been developed. The workings (fig. 23) consist of two small tunnels on opposite sides of one of the little gulches tributary to Brush Creek. No production has been made.

The contact of the serpentine and slate has been exposed in both tunnels and shows unusually gentle southwesterly dips. In places the serpentine cuts the slate sharply; elsewhere there is a crushed zone as much as 3 feet in width along the contact. It is thought probable that the slate yielded unevenly to the pressure exerted during serpentinization, so that in one place the contact of the two rocks is clear and sharp, while a few feet away there may be intense crushing in the slate.

Quartz is found in the slate close to the contact but only in small discontinuous stringers. Some of these stringers follow the slaty cleavage; others cut across it with contacts as sharp as those of the stringers of the Brush Creek mine. The quartz must have been introduced later than the crushing of the slate, for the quartz stringers in the crushed slate are massive and unbroken.

KATE HARDY MINE

The Kate Hardy mine is in the valley of Oregon Creek about 2 miles west-southwest of Forest. In 1925 it was operated by the Kate Hardy Mining Co. The mine was idle in 1928. The property comprises two patented claims, the Derelict and Kate Hardy, and several unpatented.

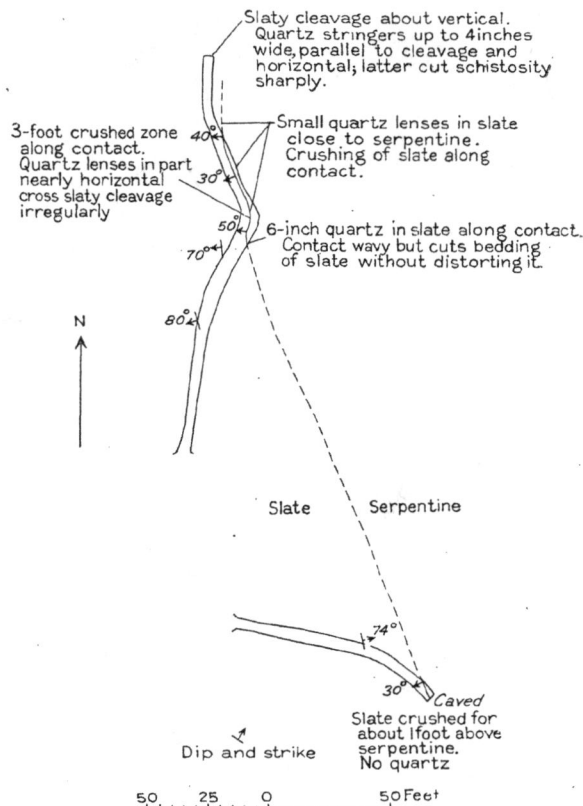

FIGURE 23.—Sketch map of Finan prospect

As might be supposed from the prominent outcrop the mine was one of the first discovered in the district. It has been worked at intervals since about 1860, and the total production is said to have been about $300,000. As the upper tunnels are in part caved it is not possible to estimate closely the ratio of the production to development work. Probably it is close to $100 a foot.

The principal workings are south of the creek and consist of several tunnels (pl. 47), the lowest a few feet above creek level and the highest 250 feet above the creek. A lower level about 100 feet below the drain tunnel is connected with the tunnel by a 75° inclined shaft. There is also a small tunnel north of the creek.

Except for the vein outcrop surface exposures are poor. The surface geology as far as it could be worked out is shown on Figure 24. The vein lies within the

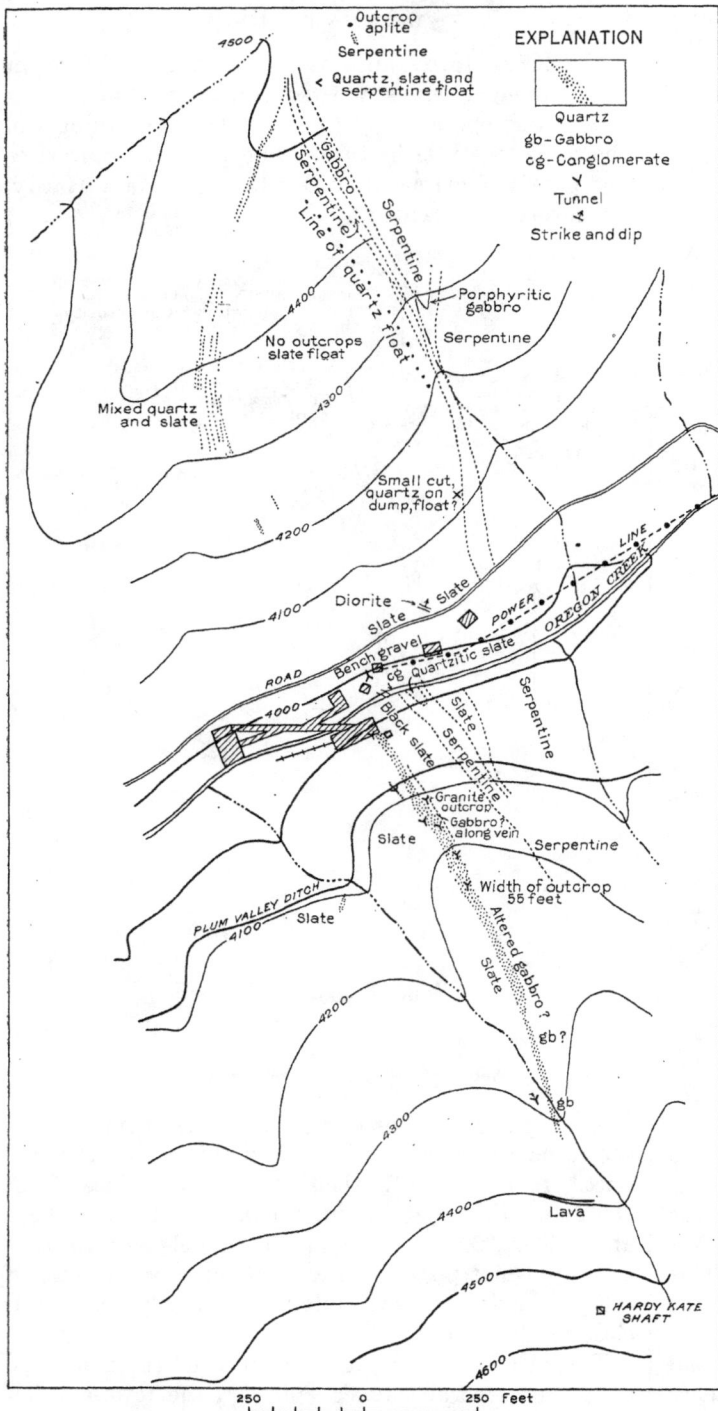

FIGURE 24.—Surface geology, Kate Hardy mine

usually found in this formation, but at one point there is a lens of conglomeratic black slate somewhat similar to that occurring in the Kanaka formation. There is also close to the vein a massive green rock that suggests a tuff but may possibly be a fine-grained gabbroic dike. East of the vein lenses of quartzite may represent gradation from the Relief formation.

The great serpentine mass that crosses the entire district lies 100 to 200 feet to the east, and small dikes of serpentine extend from it nearly to the vein.

Several dikes of gabbro and related rocks cut both the slates and the serpentine. The largest one shows a wide variation in composition, for it grades within a short distance from an olivine gabbro to diorite and in places shows a porphyritic phase containing orthoclase.

Three outcrops of acidic intrusive rocks were found. An aplite dike from 4 to 8 feet wide crosses Oregon Creek a few feet upstream from the Kate Hardy tunnel but could not be followed on either bank. Another outcrop of aplite was noted on the ridge 1,200 feet to the north. A small outcrop of pyritized granite consisting essentially of quartz and orthoclase was encountered at the portal of the first tunnel above the drain tunnel on the south bank of the creek.

The principal vein is straight and well defined, with an average strike of N. 23° W. and a dip of 75°–80° W. It cuts the slate at a small angle on both strike and dip. The outcrop, which is traceable uphill to the south for a distance of over 1,000 feet, nearly to the lava contact, is the most prominent in the district. North of the creek quartz float and scattered outcrops indicate a probable northward extension for about 1,200 feet, though for about 400 feet immediately north of the creek quartz is either lacking or present in so small an amount that no outcrops are visible.

Both along the outcrop and in the drifts south of the creek the vein is unusually thick. The maximum thickness is 55 feet, and it is rarely so thin that both walls are exposed in a drift. On the ridge to the north the outcrop is less prominent, but a zone of quartz stringers and silicified slate exceeds 100 feet in width. On the crest of the ridge the strike changes to a few degrees east of north. The quartz of the vein is in places mixed with slate, and there are splits in the vein giving elongate horses of slate which are cut by flat crossovers of quartz. (See pl. 12, B.) Along most of the vein there is no marked development of later walls, and the contact of quartz and

slates of the Cape Horn formation, which is here cut irregularly by dikes of gabbro and serpentine. The slate is chiefly of the dark fissile fine-grained type altered slate is tight and without gouge. The crumpling of the slate adjoining the quartz shows that the early movement was in the reverse direction, though of unknown amount. The grooving along such contacts pitches a few degrees to the south of the dip. Ribbon structure is common and is in general due to successive fissuring, leaving narrow streaks of slaty matter, but is in places due to fracturing with filling by later minerals. Rarely ribboning of the second type is superposed on the first with opposite dip. In a few places a horizontal sheeting (pl. 18, C) was observed and also veining by later quartz crossing the earlier structure.

The slate is more or less altered close to the vein, but the alteration is less intense than in the Brush Creek mine. For the most part the slate is replaced by carbonate close to and within the vein. Carbonate within the vein, replacing either the quartz or included rock fragments, seems to be most abundant in the zone from which the high-grade ore was obtained. What appears to be talc is developed in the slate in places along the footwall for 130 feet south of the portal. Mariposite is erratic in distribution. It is most abundant in the portions of the two lower levels north of the shaft and seems to be more abundant on the footwall side, nearest the serpentine, but is not confined to the vicinity of the serpentine. Arsenopyrite, pyrite, and galena were the only sulphides noted.

There has also been later movement along and across the vein after the completion of mineralization. In the drain tunnel north of the shaft the vein is cut obliquely by a vertical fault along which the horizontal component of offset may be as much as 25 or 30 feet. There is evidence of post-mineral movement along the hanging wall in the crosscut near the shaft on the drain level and at the north end of the lower level. In the southern part of the drain level the postmineral fault along the hanging wall is at a small angle to the strike of the vein, and farther south it may completely cut off the vein, for the outcrop is not traceable as far as the lava capping.

The distribution of high-grade ore in this mine is erratic. Localities from which such ore is known to have been obtained are indicated by "Au" on Plate 47. A large number of small bunches, most of which yielded under $1,500 each and some less than $100, have been taken out from a zone along the footwall extending about 700 feet to the south of the portal of the drain tunnel. Almost all the output has been obtained above the level of the drain tunnel. The hanging wall has not proved as productive but has not been as thoroughly prospected. The largest pocket, said to have yielded $40,000, was mined from the surface on the hanging-wall side of the vein, near the creek. The high-grade shoots are described as being in general nearly vertical and in places ending against prominent sheeting zones. In the neighborhood of the high-grade shoots the sheeting generally dips less than 10° N., except in the southern part of the mine, where the dip is 30°–40° S. This sheeting is later than the ore, but where it is prominent there is also a noticeable amount of later quartz in small veinlets cutting the older vein quartz, in the same direction as the sheeting. The gold of the high-grade shoots is commonly associated with coarsely crystalline arsenopyrite. In the vicinity of these shoots sulphides are abundant, giving a concentrate of about 4 per cent with a tenor of $60 to $100 a ton.

As far as could be determined there is no serpentine in contact with the developed portion of the vein, but serpentine occurs on the footwall side (fig. 24) close to the vein, near the zone from which most of the high-grade ore was obtained. The vein and serpentine diverge uphill to the south, in spite of the westward dip of the vein, but in the productive zone upward projections of vein and serpentine would meet at about 50 to 100 feet above the present surface. It is possible, therefore, that here as in other mines the serpentine influenced the localization of the high-grade ore.

Other veins on the property have not been developed. There is a prominent quartz outcrop just above the Plum Valley ditch, about 400 feet south of the main tunnel portal and 200 feet west of the vein. The apparent strike is north and the dip vertical. The vein can not, however, be traced on the hillside. On the north side of the creek an outcrop of quartz and a line of quartz float indicate the presence of a vein that follows closely the contact of the slate and a narrow serpentine dike. Still another quartz vein crops out on the summit of the knoll north of the creek and east of the main vein.

The lack of success in the development of the lowest level caused the closing down of the mine in 1927. If, as suggested above, the high-grade ore owes its position to the proximity of the serpentine, the promising region for future prospecting would seem to be north of the creek, for on the north ridge the vein again approaches the serpentine. The small vein that seems to follow the serpentine contact closely also merits investigation.

HARDY KATE PROSPECT

The Hardy Kate shaft (fig. 24) was sunk through the lava capping south of the Kate Hardy mine to prospect the continuation of the Kate Hardy vein. According to report a crosscut from the shaft cut a small vein. The shaft was inaccessible at the time of visit.

TOMBOY PROSPECT

The Tomboy claim is in the valley of Lucky Dog Creek, north of the Eureka prospect and about a mile west of Forest. As far as known there has been no noteworthy production, though a few small "bunches" are said to have been found. Two small tunnels, partly caved, and a shaft develop a small vein, which here strikes about northeast and dips 60°–75° SE. Surface cuts near the Lucky Dog gravel tunnel, a short distance to the east, suggest, however, that the average strike is more easterly. The intersection of the vein with the Eureka vein and with the serpentine to the west has not been explored. The vein lies within the gabbro. In the lower tunnel near the shaft a dike of granite not over 3 feet wide cuts the gabbro and is cut by the shear zones accompanying the vein.

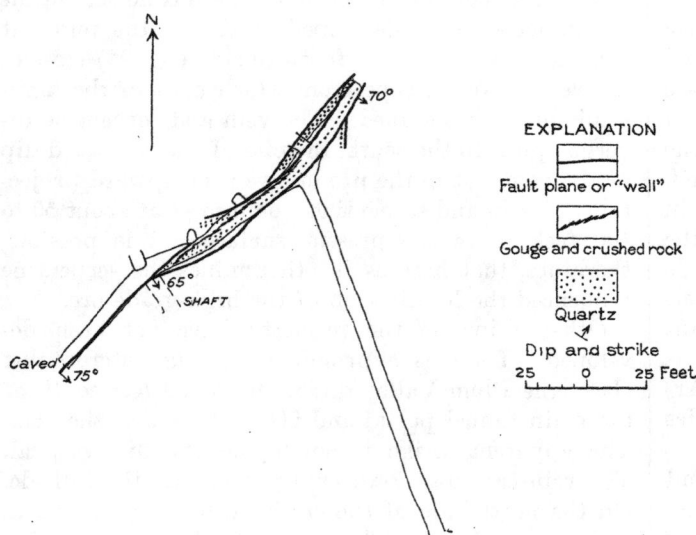

FIGURE 25.—Map of lower tunnel, Tomboy prospect

The vein (fig. 25) shows intense postmineral shearing, and the quartz is much crushed and irregular, in places lacking entirely. Here and there minute crystals of pyrite and arsenopyrite were noted. Wall-rock alteration to carbonate is not prominent except close to the vein.

EUREKA MINE

The Eureka property, consisting of the Eureka and Eureka Extension claims, lies north of Oregon Creek in the valley of Lucky Dog Creek about 1¼ miles west of Forest. It has yielded a small production in recent years, probably not exceeding $2,000.

The vein, which has a northwesterly strike and dips rather gently to the northeast, has been prospected in a number of short tunnels and shallow shafts from the ridge above the Oregon Creek road nearly to the Mountain House road. The vein lies within gabbro and closely parallels the serpentine contact, about 200 feet distant at the southernmost tunnel and 40 feet at the last outcrop below the lava to the north. At the south end of the property a few fragments of granite were found on the dump of one of the small shafts.

These may represent the northward extension of the zone of small granite dikes that cut the gabbro at Oregon Creek.

The vein is in a well-developed shear zone and in most places shows well-marked gouge walls. It is irregular in both strike and dip and intersected though apparently not faulted by vertical shear zones which are themselves mineralized. There are several minor branches at small angles to the strike and dip. Near the south end of the Eureka claim at the shaft north of the Oregon Creek road, the vein forks to the southeast. The eastern branch may be the same as that prospected on the neighboring Oak prospect. The western branch can be followed only a short distance to the south of the shaft.

The gold produced was obtained from the shaft close to this junction. The quartz showed no visible gold but contained abundant small pyramidal crystals of arsenopyrite and was said to have yielded from $300 to $500 a ton. In many places the vein contains abundant pyrite, largely oxidized. The wall rock is locally altered to carbonates and heavily pyritized, and the pyrite here also is largely oxidized. Both vein and altered country rock are said to have a tenor of $3 to $5 in gold to the ton over long distances.

The facts that the vein verges toward the serpentine contact to the north and that the intersection of the Tomboy vein should also be encountered a short distance north of the most northerly prospect suggest that exploration in this direction is warranted.

OAK PROSPECT

The Oak or Evans prospect (fig. 26) is in the valley of Oregon Creek east of the southern workings of the Eureka and about a mile west-southwest of Forest. There are several small tunnels on both sides of the creek.

The major vein seems to strike northwest and dip northeast and is presumably continuous with the eastern branch of the Eureka vein. There are also numerous small veins of irregular attitude but with dominantly northerly strike and westerly dip. The gabbro is here cut by several small granite dikes which have a general northerly strike. In several of the smaller of these dikes are irregular veinlike segregated deposits of quartz that contain also a little chalcedony.

MUGWUMP MINE

The Mugwump mine, formerly known as the Diamond Peak and also as the Young America, was first developed as a gravel mine, and has had a considerable placer production from the Young America and Gold Star placer claims on a gravel channel that is presumably the same as that at Forest. Gravel was first mined here in 1852, and at that time some gold is said to have been taken from the vein as well. A vein cut in the bedrock tunnel is now being developed, but the lode production has been small, probably not over $5,000, from about 1,300 feet of drifts on the vein.

The portal of the tunnel is on the south bank of Oregon Creek about 1,500 feet southwest of Forest. The tunnel (pl. 48), which is about 3,000 feet long, has a general south-southwesterly direction and is entirely within the Kanaka formation; the serpentine mass that crops out near the South Fork mine is not found in the workings. Different horizons of the Kanaka formation are encountered. In the upper level the tuffaceous or effusive members are for the most part deeply altered to a whitish clay, but the dark slates are easily recognizable. Near the gravel workings at the south end there is much conglomerate. This was of interest as containing the largest boulders found in this formation; one with a maximum diameter of 14 feet and one of 9 feet were noted. At the end of the accessible workings, east of the conglomerate, the gravel rests on a gray clay which retains its schistose structure and may possibly belong to the Tightner formation.

There was still a little gravel being mined when the mine was first visited, in 1924, from a small tributary channel. This was very coarse, with boulders as much as 6 feet in diameter, but was rich, averaging over $4 a yard. The gold, which was unusually coarse, was in part sunk into the clayey bedrock, making it necessary to mill the clay. A notable feature of the gravel was the large number of boulders of coarsely crystalline marble similar to the marble band in the Tightner formation cut by the North Fork crosscut.

A vein was cut about 100 feet from the portal and followed for some distance in a drift, now caved. Where cut by the main tunnel the vein has both walls in tuffaceous slate, strikes a few degrees west of north, and has a steep westerly dip. A westward-dipping slip carrying a little quartz, cut at about 1,000 feet from the portal, may be the same vein. At about 1,400 feet from the portal a northerly fault with nearly vertical westerly dip carrying a little quartz has been explored for a short distance. About 200 feet farther on a southwestward-dipping vein is encountered. On the assumption that this is the same vein cut in the tunnel north of the fault, this segment must be displaced about 100 feet to the south by the northerly fault. It is possible, however, that the caved branch of the tunnel follows the vein and that the slip just

mentioned is not significant. If so, the displacement of this segment is to the north and the fault a reverse fault. The vein itself occupies a fault plane, for where it is developed in the southern part of the tunnel level the footwall is black conglomeratic slate and the hanging wall a dense green rock supposed to be altered tuff. The vein on this level has a maximum width of 3 feet at the winze, where the dip is flattest, and narrows to an inch or two of quartz with well-defined gougy walls at the end of the drift. Wall-rock alteration is not conspicuous, and mariposite is

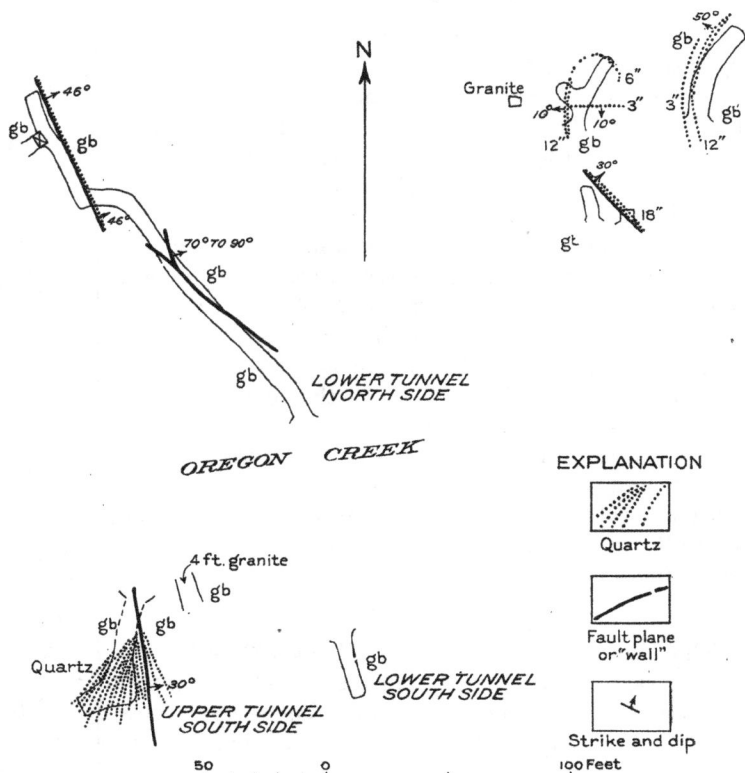

FIGURE 26.—Sketch map of workings, Oak prospect, Bob Evans claim. gb, Gabbro

lacking. At the widest part gold associated with coarse arsenopyrite was found and followed for a short distance down the winze. A second level about 125 feet vertically below the first extends for 850 feet southeast from the winze. In the winze at about 25 feet below the upper tunnel the black conglomeratic slate in the footwall gives place to greenstone, and on this level both walls of the vein are greenstone with streaks of black slate. The vein is generally well defined for about 600 feet from the winze but pinches beyond, and although a "wall" continues to the end of the drift there are only here and there small lenses of quartz.

At the south end of the quartz there is a sharp change in strike and a flattening in dip. The vein is here wider than elsewhere, having a maximum width of about 4 feet. Rather spotty high-grade ore was being stoped at this point in 1928. The gold is associated with quartz showing crinkly banding and carrying mixed sulphides. A little coarse arsenopyrite is also present.

Both places where high-grade ore has been found are where the dip is flatter than usual, and, as would be expected, the quartz is thicker than the average, because it is to be presumed that this vein follows one of the steeply dipping reverse faults.

GOLD BUG PROSPECT

The Gold Bug prospect (pl. 48), owned in 1925 by Father O'Reilly, of Nevada City, is in the valley of Oregon Creek about a quarter of a mile southwest of Forest. The development work consists of three small tunnels, the longest about 100 feet long, on the north bank of the creek. The vein, which is within the tuffaceous facies of the Kanaka formation, nowhere exceeds a foot in thickness, and in places quartz is entirely lacking, though distinct walls are present. The strike varies from north to N. 40° W., and the dip is to the west, averaging about 60°. At the north end of the upper tunnel the vein is cut off by a vertical fault striking N. 5° W. As the tunnel is caved beyond this point, it is not known whether the vein was recovered. The Gold Bug vein is the same as that explored in the northern part of the Mugwump tunnel and probably the same as that which has yielded a small production from the southern part of the Mugwump.

FEDERAL MINE

The Federal tunnel (pl. 48), about 700 feet south of Forest, was run many years for gravel and is now inaccessible. It is reported to have cut several veins, one of which may be the same as that explored in the South Fork mine.

SOUTH FORK MINE

The portal of the South Fork tunnel (pl. 48) is on the south bank of Oregon Creek just south of the town of Forest. The property consists of several lode and placer claims owned in 1925 by the Sierra Alleghany Gold Mines Co. The tunnel runs in a general easterly direction for about 5,500 feet and develops a gravel channel from which a large production is said to have been obtained. Several veins were cut by the tunnel, one of which has yielded a little high-grade ore (said to have amounted to about $1,000 prior to 1925) and has been developed to some extent. The mine was idle at the time of visit. Some work was done between 1925 and 1927, but the mine was closed when the district was revisited in 1928, and data on the recent operations were not obtained.

The tunnel, as far as it was accessible, is in the schists of the Tightner formation except for a narrow belt of serpentine, probably the southward extension of that cropping out in Oregon Creek at the eastern edge of the town of Forest. Another serpentine belt is just west of the portal. The Tightner formation is in part represented by its more slaty facies, in places suggesting the dark slates of the Kanaka formation.

The first vein encountered in the tunnel lies between the two serpentine belts and has a northerly strike and easterly dip. It may be the same which has yielded gold in the Amethyst workings, a short distance to the north, and its projection (pl. 10) falls in approximate line with the system of eastward-dipping veins in the valley of Kanaka Creek. The drift from the tunnel level is longer than is shown on the map, but at the time of visit was largely inaccessible on account of bad air. Recent developments have carried work on this vein much farther.

The vein where seen consists of lenses of quartz which have a maximum thickness of 8 feet but which pinch to a shear zone at about 300 feet south of the adit. The hanging wall is a well-defined slip, which cuts off the quartz, but the footwall is indefinite. A zone of quartz stringers on the footwall side grades off into the country rock. This zone is widest, about 50 feet, where crossed by the tunnel and narrows to the south. Gold associated with coarse arsenopyrite was found in the hanging-wall side of the vein, close against the hanging-wall slip. The vein has been further developed by a winze and two short levels, the deeper one 75 feet below the tunnel level and not accessible at the time of visit. If the dip averages 40° the serpentine should be encountered on the hanging-wall side of the vein within a vertical depth of 200 feet.

Other veins that have been cut by the tunnel but not considered worthy of development are shown on Plate 48. At about 400 feet east of the vein just described a small vein with northerly strike and easterly dip has been cut but not developed. Another vein, scarcely more than a shear zone, which dips steeply to the west, was crossed 1,000 feet farther in, nearly 5,000 feet from the portal. A westward-dipping vein cut in the portion of the tunnel which is not now accessible is indicated on the old mine maps.

AMETHYST MINE

The workings of the Amethyst mine (pl. 48) in the southern part of the town of Forest, close to Oregon Creek, are no longer accessible. The property is part of the South Fork group. The vein strikes a few degrees west of north, and the old shaft indicates a rather steep easterly dip. Several small shoots of high-grade ore, according to the most conservative estimates, yielded a total of about $6,000. These were reported to have been found at the junctions of the main vein with minor branches. The vein is in the Tightner formation between two belts of serpentine and may be the same as that developed in the South Fork mine.

At 300 feet west of the Amethyst shaft a small vein, essentially a shear zone carrying a little quartz, crops out near the forks of the creek. It has a northerly strike and dips 60° W.

DIADEM MINE

The workings of the Diadem mine (pl. 48), on the west bank of the North Fork of Oregon Creek in the western part of the town of Forest, are now caved. The mine is said to have produced about $20,000. In 1899 [76] 5 tons of high-grade ore, which yielded $5,000, was mined. The vein lies at the contact of serpentine with the slaty schist of the Tightner formation and has a northerly strike and steep westerly dip. It may be the same vein as that developed in the North Fork mine.

A shear zone with some quartz which is exposed in the creek bed strikes nearly north and dips steeply to the west. It is probably the same as that which crops out west of the Amethyst shaft.

NORTH FORK MINE

The North Fork mine, on the west side of the North Fork of Oregon Creek about half a mile north of Forest, was formerly a gravel mine, but of late years attention has been devoted entirely to developing a vein discovered in the old placer workings. This vein was discovered in 1875, when it became necessary on account of swelling in the serpentine bedrock to change the course of the tunnel then being run to reach a gravel channel. The vein was followed by a drift for a distance of 800 feet, and a very rich shoot 75 to 100 feet in drift length was discovered. The gold is reported to have occurred " in arsenical and iron sulphurets and some galena." The ore was extremely rich, and it is reported that on May 31, 1875, $5,000 was taken out in eight hours.[77] Apparently mining of this rich shoot continued to 1877, and it is said that the total yield was $100,000,[78] of which $40,000 was extracted with a hand mortar.[79] Some sinking seems to have been done on the ledge,[80] but no attempt at systematic development of the vein was made for many years after the rich pocket was extracted.

The present management began work about 1910, with the object of exploring the vein beneath the old

workings. In recent years there has been a production of not more than $25,000 from a small high-grade shoot on the vein. The drifts on the principal vein amount to slightly over 1,000 feet, which on the assumption of a production of $25,000 gives a ratio of $25 per foot of drifting. The footage of the winze, which crosses the vein from footwall to hanging wall, is not included.

The shaft is in andesite breccia, and bedrock is reached at a horizontal distance of 1,000 feet from the shaft by a 13° incline having a course of N. 53° W. (See fig. 27.) Here a tunnel follows the same course for about 250 feet and thence runs S. 50° W. for 550 feet to the winze on the vein. From the third level on the vein, about 200 feet below the tunnel level, a crosscut has been run eastward for 900 feet. These two crosscuts give a section of bedrock at a considerable distance north of the nearest outcrop.

At the foot of the incline the crosscut tunnel begins in much-weathered schist of the Tightner formation, here cut by a small dike of serpentine. At the bend the tunnel enters Tertiary gravel. Bedrock is exposed for a short distance and shows black slate overlain by the more characteristic schist of the Tightner formation along a fault with a dip of about 35° E. On the basis of the relations in the lower crosscut, where exposures are better, it is thought that this black slate is a slaty facies of the Tightner formation rather than a part of the Kanaka formation.

Beyond this point for 180 feet the tunnel is entirely in gravel. From the end of the gravel nearly to the winze the prevailing rock is the black-slate conglomerate of the Kanaka formation with a single band of chert. Close to the winze this is overlain by the green dense rock considered to be altered tuff.

From the third level, 200 feet below the tunnel, there is a long crosscut to the east, at an average distance of 200 feet south of the upper tunnel. The face is in serpentine, possibly the same dike as that cut at the east end of the upper tunnel. To the west is typical amphibolite schist of the Tightner formation. At a point 130 feet west of the face a small westward-dipping quartz vein with accompanying postmineral shear has been followed by a drift for a short distance. It is likely that this vein follows the same fault that forms the boundary between the black slate and schist in the level above. Westward from this vein the tunnel is within the Tightner formation for a further distance of 350 feet. Here the formation consists for the most part of the same amphibolite schist but contains a belt of coarsely crystalline white marble, about 70 feet wide, that is similar to that found within the Tightner formation at the Eldorado mine and on the bank of the Middle Fork of the Yuba River. Although about 200 feet below the Tertiary surface, it contains large solution channels filled with mud and

[76] Min. and Sci. Press, vol. 79, p. 694, Dec. 16, 1899.
[77] Idem, vol. 31, p. 378, Dec. 11, 1875.
[78] Idem, vol. 34, p. 319, May 19, 1877.
[79] Idem, vol. 35, p. 86, Aug. 11, 1877.
[80] Idem, vol. 35, p. 197, Sept. 29, 1877.

containing boulders of Tertiary gravel with rare fragments of wood.

There is a small dike of gabbro, faulted on the east between the Tightner and Kanaka formations, on this level. The dip of both formations wherever observable generally exceeds 45°.

gabbro; others are of uncertain composition and probably represent originally more basic rocks. A small dike of aplite was found at one point on the hanging-wall side of the vein. The interpretation of the position of the vein in the different levels assumes that there are two branches with parallel strike, a few de-

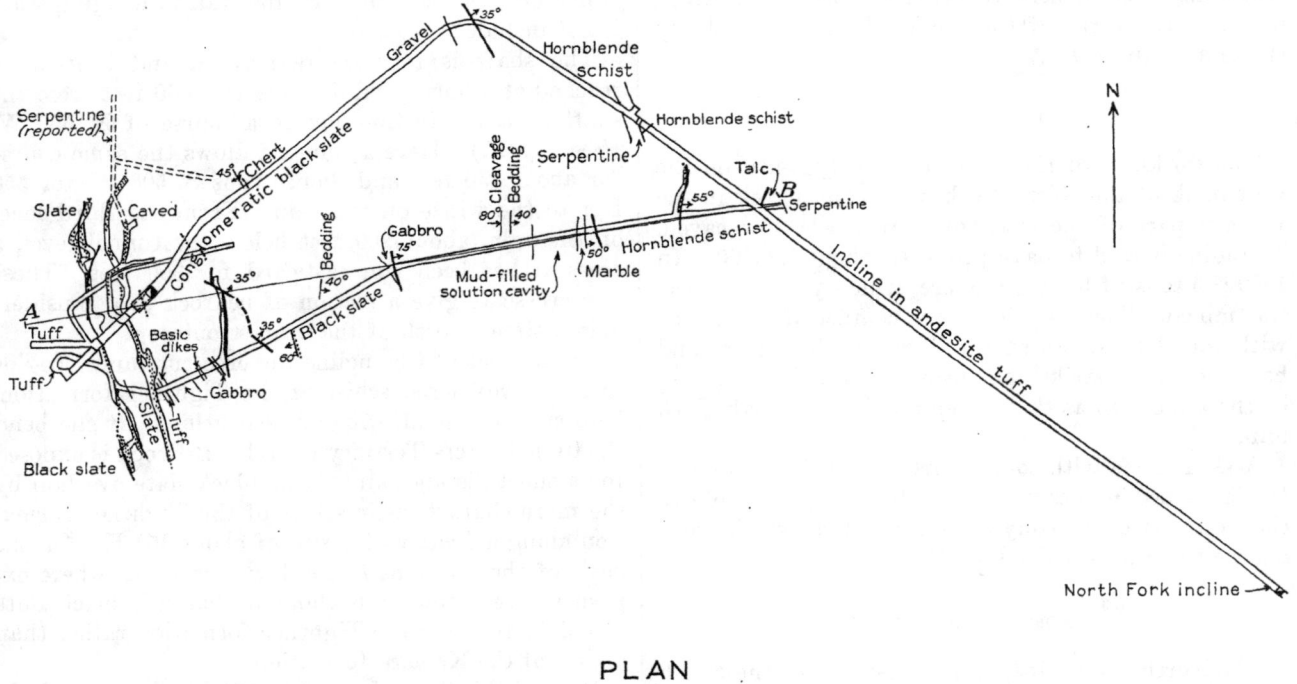

PLAN

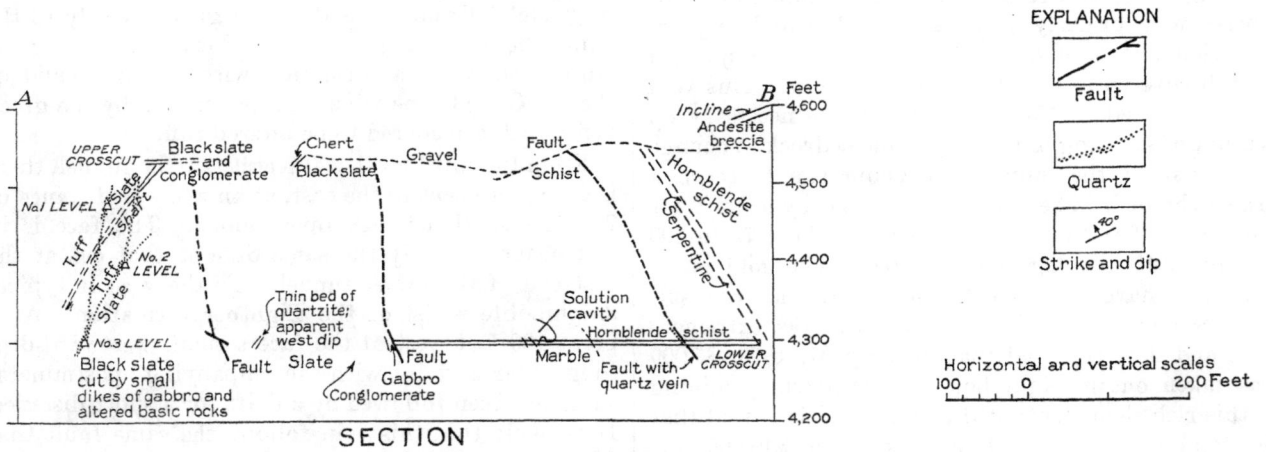

SECTION

FIGURE 27.—Plan of workings of North Fork mine and section along line A–B

West of this fault the tunnel is in the Kanaka formation to the vein; the rock is chiefly black conglomeratic slate, which near the vein is overlain by altered tuff. The small chert band crossed in the upper tunnel is not found in the lower one.

Near the vein on both levels the Kanaka formation is cut by a number of small dikes, everywhere considerably altered. A few of these are recognizable

grees west of north, which unite above the No. 3 level, but the hanging-wall branch was not identified in the winze. On all the levels the dip is steep to the west. Movement along the vein is shown not only by the gougy walls but by the fact that greenstone or tuff is in places present in the hanging wall and the black conglomeratic slate in the footwall. These relations indicate reverse faulting, but the throw need not be large. On the projection (pl. 8) the vein falls in approximate alinement with the Diadem vein, to the south.

The vein has been explored on three levels for a maximum distance of 450 feet, but the northern part of the upper drift is now caved. There is a distinct splitting observable to the north, and this is one of the few veins in the district on which feathering in this direction is clearly evident. The vein carries from a few inches to a maximum of 10 feet of quartz. In places where there is a frozen wall, particularly the hanging wall, and in small stringers near the vein there is abundant albite close to the walls. Later quartz veins cross the earlier quartz (pl. 11, A), and with these is associated pyrite, commonly not actually within the later vein but close to it, and carbonate. A few small veinlets of later quartz were observed to grade off into carbonate stringers.

There is also abundant evidence of postmineral opening of the fissures. Nearly everywhere the footwall of the vein is well defined, with a little gouge, but the hanging wall is more commonly frozen and grades off into irregular curving stringers in the crushed slate. In places, however, fault planes diverge from the footwall, cross the vein diagonally, faulting it for short distances, and disappear in the hanging wall.

The effects of wall-rock alteration, consisting of replacement by carbonate, are relatively inconspicuous, though observable in places close to the vein, but no mariposite was seen.

High-grade ore, said to have amounted to about $25,000, in which the gold was associated with arsenopyrite, was found on the second level near the point where the crosscut from the shaft reaches the vein. At this point there is a rather sharp change in the dip. Splitting of the vein is very pronounced, and the footwall, as shown by the crosscut to the west, contains small dikes of a flaky talcose schist, possibly derived from serpentine. The geologic conditions accompanying the larger high-grade shoot found in the old placer workings above are unknown other than the fact that serpentine was reported.

A steep eastward-dipping shear zone explored on the lowest level about 200 feet east of the vein is probably to be correlated with small quartz stringers cut in the crosscut on the first level and in the tunnel and with a small vein cut in the crosscut on the second level. On the third level it is faulted slightly by a reverse fault with gentle northeasterly dip. This is an exception to the general rule that the steeply dipping fissures fault those of gentle dip.

The eastward-dipping vein cut at the end of the third-level crosscut is of interest because of its alinement in projection (pl. 8) with the veins developed in the Amethyst, South Fork, and Red Star mines.

MINES OF KANAKA CREEK DRAINAGE BASIN

KENTON MINE

The Kenton property includes the Kenton and King Solomon lode claims and the King Solomon mill site, on the south side of Kanaka Creek south of the Oriental mine and about 1¼ miles southwest of Alleghany. The mine is one of the oldest in this district and is said to have been discovered in 1860 [81] and to have had a stamp mill installed as early as 1864. There has been no noteworthy production since 1892, although the mine has been reopened several times for short periods since that date. According to some reports the production was $175,000, but the most conservative estimates give a total of about $90,000, practically all from the upper levels. On the assumption that the two caved levels extended as far to the north as the lowest level, this is equivalent to $46 per foot of drifting, but the ratio may be nearly double this figure if the higher estimate for total production is correct.

Development work (fig. 28) consists of tunnels at approximate altitudes of 3,720, 3,810, and 3,890 feet; the two upper tunnels are now partly caved.

The vein is entirely within gabbro, but at its south end it approaches the serpentine contact. In the northern part the strike is about N. 25° E. To the south it swings nearly parallel to the serpentine contact, or about N. 30° W. The dip is easterly, averaging about 45°, but varying in different places from 25° to 80°. The amount of quartz present is variable; a maximum thickness of 10 feet was observed, but in places quartz is lacking entirely or present only as small irregular stringers along the sheared walls.

The fissure has been the site of repeated shearing. In places, particularly along the footwall, the post-quartz shear wall is lacking, and the original wall, with abundant quartz stringers running from the vein into the country rock, is present. Shearing later than the vein filling has produced in places a crushed sugary quartz but more commonly a quartz breccia cemented by quartz and in one place a breccia consisting of sheared and partly silicified gabbro fragments cemented by quartz. The later shearing is irregular both in distribution and in intensity. In places there may be little or no gouge, though a few feet away the vein is mashed and sheared extensively. Most of the later shears give evidence of movement of the hanging wall up the dip and also to the north over the footwall. Piling up of one quartz strand above another was noted in the main tunnel just south of the adit, where there is an unusual thickness of quartz. To the south, near the serpentine, the postmineral shears are steeper and cut the vein irregularly.

[81] MacBoyle, Errol, Mines and mineral resources of Sierra County. p. 97, California State Min. Bur., 1920.

The old stopes indicate two localities from which ore was obtained—one in the southern part of the vein, near the serpentine, and the other 400 feet farther north, where the evidence of repeated shearing is most pronounced. The northern stopes seem to have been the more extensive, but it is said that the southern one produced the more gold. The high-grade ore is said to have been associated with coarse arsenopyrite.

Mariposite is not prominent and is found chiefly in the altered wall rock at the south end of the workings. Carbonate also seems to be most abundant toward the south, and the gabbro wall rocks in the northern part of the main tunnel show little alteration.

Since 1892, when important production ceased, the mine has been reopened several times with uniform lack of success. If, however, further development work is done, attention might be directed to exploration down the dip of the thick lens of quartz that has been stoped above the lower tunnel. Here there is abundant evidence of shearing later than the first quartz—in many places a favorable indication. (See fig. 10.) The vein seems to split in the south end of the lower tunnel, and the relations here are confused by steep postmineral slips at a slight angle to the vein. Exploration of this portion does not, however, seem to have been exhaustive in view of the productivity of the Oriental vein in the vicinity of the serpentine contact.

ROYE-SUM PROSPECT

The Roye-Sum prospect, on the ridge southeast of the Kenton mine, is owned by W. J. Seymour, of Alleghany, who has, unaided, run a tunnel that in 1924 was 2,300 feet in length. The tunnel has a general southeasterly direction and follows a shear zone in schistose gabbro rather than a single definite vein. Along this zone several lenses of quartz, the largest of which is 100 feet in length and 3 feet in maximum thickness, have been developed. Parallel lenses and stringers within the same broad shear zone have been exposed in short crosscuts. The strike is N. 30°–50° W., and the dip averages 50° SW.

Shearing is much more pronounced than usual in westward-dipping veins, possibly owing to the influence of the serpentine. The quartz carries sulphides that follow parallel planes, the result of shearing after the first filling, and a little free gold is found in places close to these bands. At one point veinlets of

chalcedonic quartz carrying pyrite cut the older quartz.

The suggestion from a hasty inspection of this prospect is that an original definite vein has been sheared by postmineral movement into a series of discontinuous lenses.

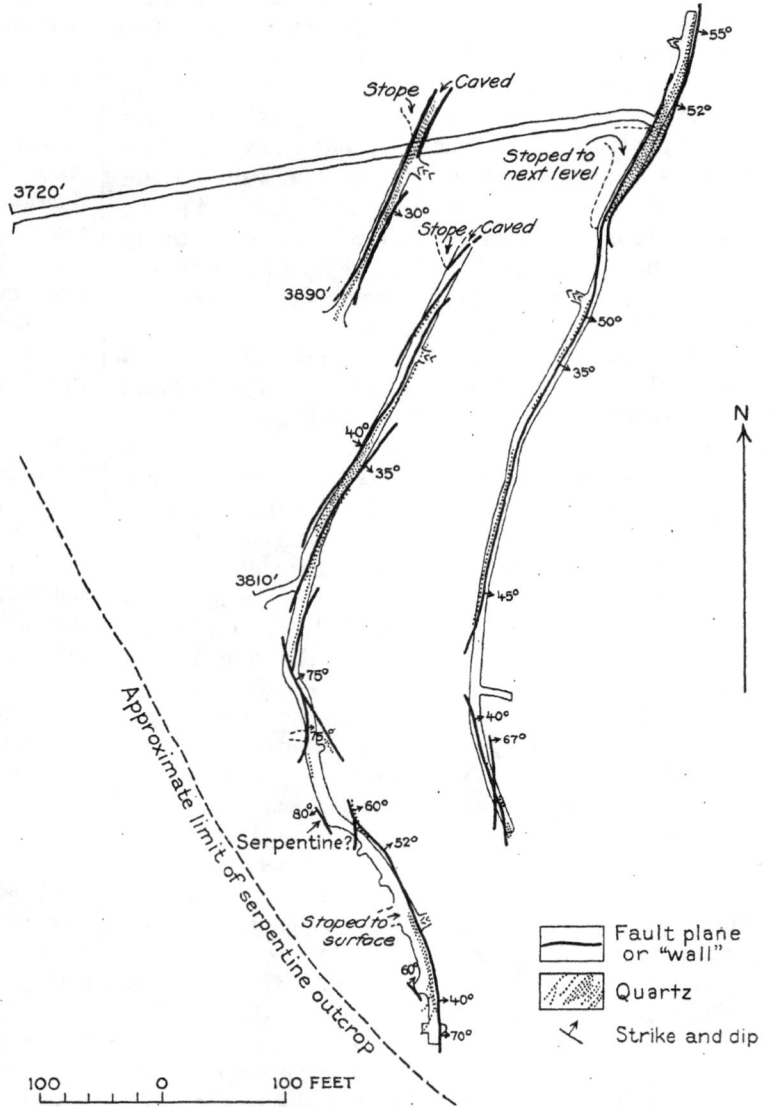

FIGURE 28.—Sketch map of accessible workings of Kenton mine. (Relative position of tunnels taken from topographic map and therefore not accurate to this scale)

ORIENTAL MINE

The Oriental mine is in the valley of Wet Ravine north of Kanaka Creek, about a mile southwest of Alleghany. The property includes nine patented claims covering an area of 153.4 acres. The old Gold Star placer mine, a short distance to the west, is under the same ownership. The mine was operated under lease from 1923 to 1925 and was purchased in 1925 by

Austin, Gamble & Wilson from its former owner, the Oriental Gold Mining Co. It has produced a total of over $2,000,000, mostly prior to 1890 and from the upper levels. It is reported that $734,000 was obtained from an area on the vein measuring only 14 by 22 feet. Exclusive of the very productive workings above the third level, whose production and extent are uncertain, and crosscuts and exploratory work that are not on the vein, the workings have yielded about $1,260,000 from about 5,000 feet of drifts and shaft, or about $250 to the foot of development work on the vein. The total of drifts on the vein as shown on the map, together with the shaft, compared with the total estimated production, gives a ratio of nearly $400 a foot, indicative of the great concentration of high-grade ore in the upper workings. For the adit level alone the ratio of production to the drifting on the vein is about $40 a foot.

The mine is developed by an inclined shaft on the vein, with levels at short intervals, to a vertical depth of 630 feet. At a depth of 780 feet below the outcrop an adit 4,150 feet long intersects the vein and is connected with the shaft by a raise through country rock. All the recent work has been done from the adit level. The workings on the Alta vein, about 500 feet southeast of the Oriental shaft are caved.

The surface geology (pl. 49) in the neighborhood of the veins is complex and greatly obscured by brush, talus, and mine dumps. The oldest rock is the Relief quartzite, which is present in small irregular areas surrounded by gabbro. This formation is here largely schistose quartzite, which grades in places into a fine-grained mica schist with thin quartzitic bands. Here and there small bands of darker schist similar to that of the Kanaka formation were encountered. Quartzite was not found in the underground workings.

Gabbro is the principal rock in the neighborhood of the mine and forms the wall rock of most of the veins. It is of the usual type, with the feldspar almost unrecognizable and the original augite almost completely altered to hornblende.

A small mass of serpentine, possibly connected beneath the lava with the larger serpentine area on the west, crops out a short distance west of the shaft and forms the footwall of the vein at the west end. The boundary between gabbro and serpentine is here probably in part a fault contact, but other contacts of serpentine and gabbro in the vicinity are sharp and without the gradational phases found in other places.

The Oriental mine is the only one of the Alleghany mines in which granite forms a wall rock. Several small masses, the largest about 800 feet in length, crop out and cut the older rocks. A large mass, probably connected with that outcropping north of the shaft, is encountered in the underground workings. Although

89277—32——8

the gabbro is distinctly schistose, the granite shows no trace of parallel structure. The grain of the larger masses is coarse, averaging greater than a centimeter, but the smaller dikes have a much finer texture. The large mass south of the shaft, although outcropping directly above the adit tunnel, was not cut underground.

The productive portion of the Oriental vein lies in gabbro between serpentine on the west and granite on the east. (See pl. 50.) The granite is 50 feet wide in the upper workings and widens to about 350 feet on the adit level. The vein has been followed through the granite and into the gabbro beyond but, though well defined in the granite, especially on the adit level, has never proved productive outside of the narrow band of gabbro between the granite and the serpentine. The granite on the adit level is cut by a large number of small quartz veins, none over an inch or two in thickness, but closely spaced.

The vein occupies the plane of a reverse fault, whose displacement measured by the displacement of the granite-gabbro contact along the vein in the lower part of the shaft is about 200 feet (pl. 51), but owing to the small angle between the dips of contact and vein and the obscuring by alteration this figure may be considerably in error. In the workings at the west end of the adit level there is an apparent displacement of about 120 feet, here divided among the three branches of the vein. As in other veins of the district, minor thrust faulting within the vein has also accompanied and followed mineralization. (See pl. 17, B.)

The vein as a whole is irregular in strike and dip; the strike ranges from west to N. 65° W., and the dip from 28° to 50° N. The northwesterly strike is most marked at the east end, near the serpentine, and flatter dips prevail in this part. The change in strike near the serpentine contact is particularly noticeable in the adit level. (See fig. 7.) Here the various stringers into which the vein divides, which 50 feet away strike at a 45° angle to the contact, turn sharply and for short distances follow the contact before dying out in the serpentine. The western part of the 400-foot level, on the other hand, follows a well-marked shear zone in the serpentine for a distance of 350 feet, but no quartz is found more than 50 feet from the contact.

Splits and branches are numerous. On the adit level, as shown in Figure 7 and Plate 50, the vein near the serpentine breaks up into a complex of smaller veins and stringers. On the other hand, in the upper workings, particularly above the sixth level, the vein splits into several branches, which diverge in strike away from the serpentine contact. Only one of these has been developed. On the sixth level there is a peculiar sharp turn and branching of the vein at the granite contact. The sharp change in dip between the fifth and sixth levels is correlated with this splitting of the vein. Owing to the fact that the vein has been followed down from the outcrop only one of several branches has been explored. Had a footwall branch been followed the apparent change in dip would have been much less marked.

The quartz present ranges from a mass about 20 feet wide to mere stringers. The greatest width is found on the seventh and eighth levels but does not continue for more than about 100 feet. On the seventh level the change from a width of 20 feet to almost nothing takes place within a distance of 50 feet.

The country rock is much altered, chiefly to carbonate, but the results of an earlier stage of albitization are also prominent, particularly in the granite. In places fine-grained arsenopyrite and pyrite occur in the altered wall rock, the former most abundant in the gabbro and the latter in the granite. Mariposite is not abundant except as an alteration product of the serpentine. On the adit level the serpentine of the footwall is in places altered to the carbonate-mariposite mixture, and fragments, probably originally gabbro, included within the veins are similarly altered, but although the gabbro on the hanging wall contains carbonate, mariposite is there lacking. On the 400-foot level, where the fissure has been followed into the serpentine, mariposite is present in the carbonate rock on both walls.

As far as known the high-grade ore was everywhere associated with coarse arsenopyrite, and the arsenopyrite is said to have been most abundant in the stopes along the serpentine contact. No shoots of mill rock were encountered, but assays of $2 to $5 to the ton have been obtained from the pyritized granite and little quartz stringers which cut it in the neighborhood of the main vein on the tunnel level.

The upper workings from the 700-foot level to the surface have yielded the principal production. So far as could be determined from an inspection of the old stopes, two major features controlled the occurrence of the high-grade ore. The vein seems to have been everywhere productive at the contact of the gabbro and serpentine, where for a short distance the quartz rests on serpentine. Stopes that follow this contact from the 900-foot level to above the 300-foot level are reported to have produced $150,000. The richest shoot, however, including the small area that yielded $734,000, was west of the serpentine contact, and appears to coincide with the change of dip of the main vein at the point where it splits into two or more branches. The principal junction is well marked on the 400-foot and succeeding higher levels, and the rich shoot occurred on the 400-foot level just below the shaft. The stopes that followed this junction from the 400-foot level to the surface are said to have yielded over $1,500,000. Similar junctions on the 400, 500, and 600 foot levels have been followed by stopes, which are reported to have produced about $300,000. Possibly this is the junction of the Alta and Oriental veins. (See pl. 19, A.)

The work on the adit level has so far produced only about $60,000. For much of the ore the yield was from about $1,000 to $2,000 a ton, much less than the usual return from the high-grade ore of the district, and the distribution was erratic. In this part of the mine the productive zone is confined to the serpentine contact, and here the feathering of the vein has made exploration more difficult. No ore has been found at the junction of the minor branches within the gabbro. As in the upper levels, the gold was in association with coarse arsenopyrite, and even arsenopyrite that showed no visible gold gave assay returns of over $1,000 a ton. Adjoining the productive portions of the vein the gabbro, both in wall rock and in inclusions, contains much fine-grained arsenopyrite in minute octahedronlike crystals. Other sulphides, except pyrite, are not abundant, nor is the ribbon quartz and crinkly banding as prominent as in some other mines.

The Oriental mine, in spite of its large production, exceeded only by the Sixteen to One and Plumbago, contains a larger proportion of undeveloped ground than most other Alleghany mines. The vein has been productive wherever the serpentine contact has been explored. It is therefore probable that the portion of the contact between the adit level and the tenth level is worthy of development. In the upper part of the mine, particularly above the fifth level, where the dip changes, there are a number of branches in the vein, only one of which has been developed. It is possible that some of the others may be productive.

Besides the Oriental vein and its branches there are several other veins on the property. (See pl. 49.) The shaft of the Alta mine, now caved, is about 500 feet south-southeast of the Oriental shaft. The outcrops and caved workings indicate a vein with northwesterly strike and northeasterly dip. The dump of the Alta shaft shows quartz with coarsely crystalline arsenopyrite and pyrite. The northwestward extension is uncertain, but it may be the same as that developed by shallow workings about 150 feet south and southwest of the Oriental shaft and may unite with the Oriental in the broken ground above the fifth level, or its junction with the Oriental may be the same as the split in the Oriental which controlled in part the localization of the high-grade ore. (See pl. 19, A.) The vein at its west end shows the same sort of bending and feathering toward the serpentine contact as the Oriental vein, but with the serpentine on the hanging wall.

The production of the old Alta workings is unknown, but a few thousand dollars' worth of high-grade ore is known to have been obtained from a small incline which followed the vein below the serpentine. In 1928 exploration of the vein in this incline was being continued on a small scale, and a little coarse arsenopyrite with free gold had been found.

A vein with strike a few degrees west of north and westerly dip is traceable on the surface for about 250 feet along the ridge west of Wet Ravine and is probably the same that crops out at the forks of the road. It has been cut by a tunnel a few feet below the outcrop.

Several other small veins south of the shaft (pl. 49) have been cut by tunnels but have not been developed.

FRANCES D PROSPECT

Uphill and south from the Spoohn workings a 150-foot tunnel and several open cuts have prospected a small vein that strikes about N. 30° W. and dips about 50° SW. No stoping has been done on the vein.

SPOOHN MINE

The Spoohn mine, on the south side of Kanaka Creek nearly a mile southwest of Alleghany, is one of the older mines in the district, but nothing definite is known of its history or past production. The workings, which appear to have been extensive, are now inaccessible. The vein is entirely within the gabbro and, as far as can be determined from the position of the workings, strikes N. 15°–50° W. and dips about 45° SW. in the lower workings, steepening to 70° or more higher up the hill. Granite cuts the gabbro in the creek near the caved lower tunnel.

GENERAL SHERMAN PROSPECT

The General Sherman claim lies between the Wyoming group on the north and the Spoohn on the south. The tunnel portal is close to the Foote road about three-quarters of a mile southeast of Alleghany. Thomas Bradbury, of Alleghany, was the owner in 1925. There has been no production.

The country rock is coarse to fine grained gabbro, here less schistose than usual. At the road 300 feet east of the tunnel portal are small irregular stringers of granite, and 400 feet to the east a narrow belt of quartzitic slate of the Relief formation intervenes between the gabbro and the serpentine belt that lies between the great gabbro mass and the Kanaka formation. The vein, however, as far as developed, is entirely within the gabbro.

The vein is irregular, as shown in Figure 29. The average strike is N. 5° W. and the dip 35°–65° W., with an average of about 55°. The outcrops on the road show three veins of quartz with 70 feet between the uppermost and lowest. Several branching strands, particularly on the footwall side, were noted in the

tunnel. These are irregular, and the walls show no evidence of movement. Minor faults nearly parallel to the vein, which themselves carry a little quartz, were observed north of the winze. There has also been postquartz movement in the plane of the vein. Where observable the grooving was found to strike N. 57° E.

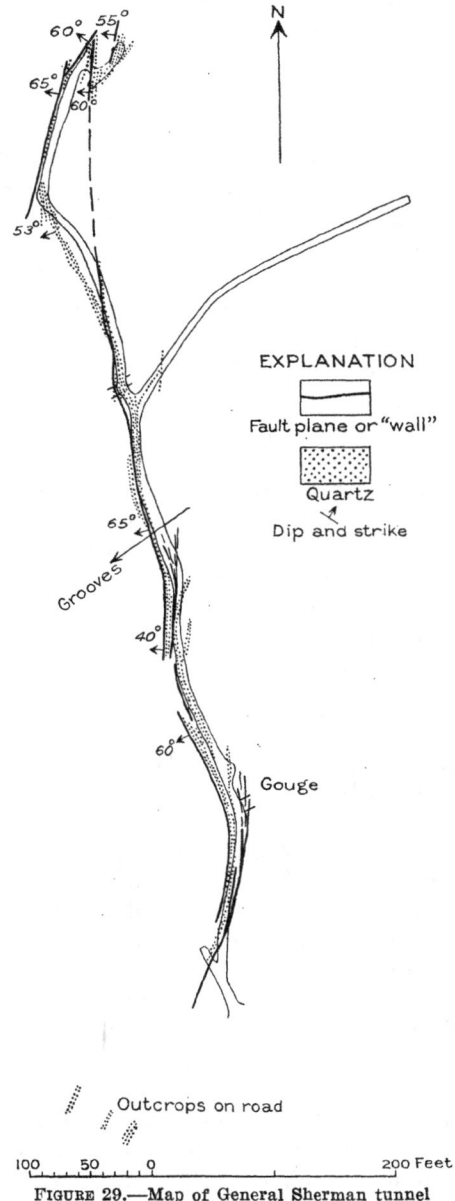

FIGURE 29.—Map of General Sherman tunnel

The quartz is commonly "watery" to white or rusty with rather abundant vugs. It is seamed with limonite, from pyrite or arsenopyrite originally present. At one point minute strands and veinlets of later quartz perpendicular to the major vein were noted. The wall rock is locally replaced by vein quartz, but evidence of carbonate alteration was found only at one point. The mine water carried abundant iron oxide.

A continuation of the drift should reach the serpentine contact within 200 or 300 feet, if the serpentine holds the same course as on its outcrop south of the lava. The vein follows the same course as that of the Dead River vein, to the north, and also lines up well in projection with the Spoohn vein, to the south, but there is a gap of several hundred feet in which there are no known outcrops.

WYOMING GROUP (DEAD RIVER)

The Wyoming and Wyoming Extension lode claims, on the lava ridge north of the General Sherman claim, cover a vein discovered in the old Dead River placer tunnel. A shoot of high-grade ore was found at the point of discovery, but the placer tunnel was allowed to cave before the vein was developed. The vein is said to have the same strike and dip as the General Sherman, but from the relations observable south of the lava it may be that a belt of serpentine intervenes or is faulted by the vein. The position of the vein indicates that it forms part of the system of northerly veins with westerly dips extending south from the North Star at least as far as the Spoohn and including the Mugwump. The high-grade ore is said to have been obtained at the serpentine contact. Crosscuts to develop the vein in depth have been started from the adit level of the Oriental mine and from the Tightner tunnel (250-foot level of the Sixteen to One mine) but not completed.

WONDER PROSPECT

The Wonder group of two claims adjoins the Sixteen to One on the west and is owned by the Wonder Gold Mines Co.

The vein is entirely within the Kanaka formation, which here consists of green chloritic schist, probably derived from tuff and interbedded black slate. The upper tunnel, on the flat at the old hydraulic workings south of Alleghany, develops a small vein which strikes N. 20° W. and dips 45°–60° W. It is said that a little high-grade ore was found on this vein in an old bedrock gravel tunnel just above the present one. The lower tunnel is 400 feet lower, about 200 feet above Kanaka Creek, north of the Rainbow mine. The vein here developed, which is probably not the same as that of the upper tunnel, strikes north to N. 22° W. and dips 50° W. This vein is joined by a small nearly horizontal vein. Both veins contain an unusually large amount of calcite and siderite.

RAINBOW MINE

The Rainbow property extends from the south side of Kanaka Creek southeastward to the top of the ridge, where it adjoins the Oriflamme group. The Central and Rainbow Extension groups bound it on the north. From 1921 to 1925 the mine was worked under bond by the North Star Mining Co., but it was closed down in the fall of 1925 and was idle in 1928.

The mine is estimated to have produced about $2,200,000. The vein was discovered some time prior to 1860 in a gravel tunnel near Chips Flat, about 2,000 feet from the portal. A stamp mill was in operation on the property as early as 1858. For some time the vein was worked by winzes from the gravel tunnel, and some very rich shoots extracted. According to Burchard,[82] as much as $60,000 was taken out in a single day and over $100,000 in a month. It is said that a shipment of 1,953 pounds yielded $116,337, and a single specimen placed on exhibition in San Francisco was calculated to contain gold to the value of $20,468.[83] As early as 1884 the mine is credited with a production of $350,000.[84] About 1883 work was begun on a long crosscut tunnel known as the Groves tunnel to develop the vein at greater depth, and from stopes, now caved, above this tunnel, together with the old workings from the gravel tunnel, it is estimated that about $1,200,000 was obtained. At a later date the vein was developed to a greater depth by another tunnel 420 feet lower than the Groves, and from this there is an inclined shaft on the vein for an additional vertical distance of 170 feet. Allowing 200 feet for the difference in altitude between the Groves tunnel and the apex of the vein beneath the gravel gives a vertical distance of 800 feet through which the principal vein has been developed, equivalent to 1,500 feet on the dip of the vein. Exclusive of the work above the Groves tunnel, the extent of which is unknown, the ratio of production to development work on all the veins is about $145 a foot. Divided between the two productive veins the ratio is about $190 a foot for the Rainbow vein and $22 a foot for the Clinton.

The workings (pl. 52) except for the crosscut tunnels, lie beneath the lava and owing to shearing along and near the veins and extensive alteration, it is difficult to identify the country rock. Apparently the workings are entirely within the Kanaka formation, but the Tightner formation, if its boundary holds the course indicated in the Spiritualist tunnel and on the surface north of the lava, should form the hanging wall of the vein in the southern part of the mine. It may be that the green schist found in that part of the mine and referred to the tuffaceous members of the Kanaka formation is actually chloritized amphibolite schist of the Tightner formation. On the other hand, the boundary between the Tightner and Kanaka for-

[82] Burchard, H. C., Report of the Director of the Mint for 1883, p. 216, 1884.

[83] Hanks, H. S., Second report of the State mineralogist, p. 149, Sacramento, 1882.

[84] Burchard, H. C., Report of the Director of the Mint for 1884, p. 153, 1885.

mations in the head of Jackass Ravine, on the south side of the ridge, lies farther to the east than on the Kanaka Creek side, and a reasonable connecting line could be drawn that would leave the Rainbow workings entirely within the Kanaka formation. It is certain that black pebbly slate which is characteristic of the Kanaka formation forms the wall rock of the veins in many places.

A dike of serpentine is cut by a crosscut in the footwall on the main level, and serpentine is also present on the footwall of the vein in the Groves tunnel.

Three veins have been developed—the Rainbow vein, which has yielded nearly all the production, and the barren Parallel vein, both having northwesterly strikes and northeasterly dips, and a northerly vein, the Clinton, which dips steeply to the west and limits the other two, although the age of fissure filling is the same in all three. (See fig. 12.) The direction of drag of the schists implies reverse faulting on the fissures followed by all three veins. No certain measure of the displacement is possible.

The Rainbow vein has yielded all but about $50,000 of the production of the mine, a reported amount of $1,200,000 from the old stopes above the Groves tunnel and $990,000 below that level, almost entirely above the main tunnel. The average strike in the accessible upper workings is N. 75°–80° W. and on the main tunnel and below about N. 45°–55° W. The dips in different parts of the vein range from 20° to 70° NE. and are on the average steeper below than above the tunnel level. The average thickness does not exceed 2 feet, quartz is lacking in places, and the vein is nowhere over 6 feet thick. A well-marked split occurs in the southern part of the mine, and the junction of the two branches, which rakes to the north down the dip, can be identified on the different levels from a point just below the Groves tunnel to the lowest level. Just north of the junction the vein attains its greatest width and has proved most productive. South of the split neither of the two branches appears to be persistent, at least in the lower workings. The Groves tunnel is now caved at the old stopes, but Mr. Duggleby, the former manager of the mine, informed the writer that he had penetrated it for a distance of 1,400 feet and that throughout this distance there was a continuous vein of quartz from 6 inches to a foot in thickness.

The Parallel vein lies in the hanging wall of the Rainbow vein and about 100 feet to the east. It has been explored only by a drift on the main tunnel level. It is a small vein with a fairly well marked hanging-wall slip but no distinct footwall and carries a continuous though narrow quartz vein for the entire 150 feet of drift. No production has been made from this vein, but the work has not been continued into the region that was productive on the Rainbow vein.

The Clinton or North vein, or group of veins, for the different quartz strands are in places separated by several feet of country rock, has an average strike of about N. 8° W. and dips steeply to the west. The fissuring faults that of the Rainbow system, yet the vein filling is contemporaneous. The vein is well defined only near the junction with the Rainbow vein on the main and lowest levels, but on and above the main level later vertical northerly faulting and faulting parallel to the Rainbow vein has given a confused area with irregular segments of quartz. (See fig. 12.) On the lowest level, where the vein was followed southward, it is represented only by rather indefinite slips. No fissure that can be certainly correlated with this vein is found in the Groves tunnel. Possibly the fissuring faded out against the serpentine dike to the south. The projection northward of this vein intersects the workings of the Rainbow Extension, but no effect is seen on the two veins there developed, unless the sharp bend of the eastern vein at the end of the drift is due to fissuring parallel to the Clinton, or the westward-dipping fault that cuts the Rainbow Extension veins represents the same structure as that followed by the Clinton vein.

Faults of different periods are easily recognizable in the two major veins, and the fact that movement in different directions prevailed at different times makes the region of the junction of the two veins extremely complicated. The drag of the schists on all three veins indicates that they are reverse faults. Fissuring and faulting later than the vein but prior to the introduction of the gold is clearly shown at several places, particularly in the Clinton vein. (See fig. 20.) The effect of post-mineral movement along and in the veins is everywhere visible. In places near the junction, particularly along the main level, there is evidence that the postmineral movement along the Rainbow system was in part later than that parallel to the Clinton vein; hence although the Clinton vein faults the Rainbow and Parallel veins, later faults forming the walls of these veins displace the Clinton. The amount of this later displacement probably nowhere exceeds a few feet.

The displacement on the three faults followed by the veins is indeterminate but probably not large. Black slate forms the hanging wall and green schist, probably an altered tuff, the footwall of the Rainbow and Clinton veins near their intersection on the lower level, but the wall rocks are not well enough exposed to determine the amount of movement involved. A small vein parallel in strike and dip to the Rainbow is found on the hanging wall of the Clinton near the north end of the drift on the lowest level. If this can be correlated with the Rainbow vein it implies a total reverse motion along the Clinton of less than 100 feet. This would accord with the supposition that the fault which cuts the Rainbow Extension veins is the same as the Clinton fault, but this vein does not crop out to the north beyond the lava. The lack of outcrop of anything that can be correlated with the Rainbow vein west of the Clinton, on the hillside north of the ridge, suggests the possibility that the displacement on the Clinton vein may be much greater and that the two veins developed in the Rainbow Extension may be the same as the Rainbow and Parallel veins. This is rendered unlikely, however, by the fact that a vein outcrop near the lava north of Chips Flat checks with the strike of the eastern vein of the Rainbow Extension, east of the point where the Clinton vein should displace it. It is thought more likely that the faulted segment of the Rainbow fissure does not carry quartz, at least at the level to which erosion has reached, and is therefore undiscovered.

Altered wall rock of the carbonate-mariposite type adjoins Rainbow and Clinton veins and is most prominent where the wall is of green schist rather than black slate. The altered rock differs from the usual type in showing a rough banding with alternate green streaks composed of ankerite and mariposite and brown streaks consisting largely of siderite or ankerite with a higher iron content than usual. Possibly this is an inheritance from original differences in composition of the rock. The black slate is commonly less altered, but here and there mariposite-carbonate alteration and replacement by quartz have occurred. Mariposite is rather widely distributed in the altered country rock, even at a distance from the known serpentine. Its abundance in the lower level may indicate that a serpentine mass, possibly the same as that outcropping near the Rainbow Extension, is present on the hanging-wall side.

In the region above the Groves tunnel, which was the richest in the mine, the vein has a footwall of serpentine. Down the dip, between the Groves tunnel and the adit level, the principal stopes follow a rather definite swelling in the vein at the point of intersection of several branches that can not be far from the serpentine. Some gold was also obtained from the same position on the first level below the adit. The two lower levels, where there is little quartz and much irregular feathering of the vein, were barren.

None of the high-grade ore from the stopes on the Rainbow vein was seen by the writer. According to report coarse arsenopyrite was abundant in connection with the high-grade ore obtained above the Grove tunnel and less prominent in the lower stopes. This is in accord with the association of abundant arsenopyrite with serpentine observed elsewhere.

The Clinton or North vein has been far less productive than the Rainbow vein, but several small shoots of exceedingly rich ore have been obtained. Practically all the production came from the neighborhood of the junction with the Rainbow veins. The quartz of the Clinton vein shows much microbrecciation, and the gold in the fractured quartz is associated with a little carbonate and sericite. (See pl. 42, *A*, *B*.) The gold of the Clinton vein, as far as seen, was nowhere in direct association with coarse arsenopyrite or with later sulphides, though pyrite is abundantly and arsenopyrite sparingly present in small crystals in the wall rock. Much of the gold was found in flakes, many of them large, in a gougy slip of presumed intermineral age which cuts preexisting vein structures and appears to have been folded. From this slip the gold extends into the quartz of the vein, but in the small shoot of high-grade ore seen in place it did not enter the quartz for more than an inch from the slip. Later minor movement on the slip has resulted in places in polishing and striating the gold. Thin films of pyrite were found coating fractures in the quartz and plating slickensided surfaces in the gougy slips. This pyrite appears to be of later age than the gold and may be of supergene origin, though definite evidence is lacking.

There are several areas in the Rainbow mine which might be worthy of further exploration should new work be undertaken. The vein was not productive on the two lowest levels, but this barren zone is not as wide as that passed through in the Sixteen to One mine to extremely productive ground at greater depth. It is also possible that the pinching of the quartz of the Rainbow vein on the lowest level is compensated by increased strength of the Parallel vein a short distance to the north, which has been explored only on the main level.

As the Clinton vein was productive near its junction with the Rainbow it might be desirable to explore the Rainbow vein further near this junction.

At the east end the Rainbow vein appears to fray out and be lost in stringers, which steepen in dip and in part follow the cleavage of the inclosing black slate. Here, however, the vein is nearer the serpentine, and it is possible that close to the contact conditions similar to those on the adit level of the Oriental will be found. Therefore it is believed that in further exploration of the eastern portion of the vein in the adit level and the sublevel between the adit level and the Groves tunnel, attention should be paid to the southern branches of the vein.

The ground between the east end of the Rainbow workings and the workings of the Oriflamme, Irelan, and Arcade mines is unexplored. The workings of these mines all show well-defined veins at distances of 2,000 to 2,500 feet southeast of the end of the Rainbow workings.

The Clinton vein has so far proved unsatisfactory, but perhaps a possibility of production still exists in its southern extension on the adit level toward the serpentine found in the crosscut east of the shaft. On the third level this serpentine apparently does not extend far enough west to be cut by the drift.

The large outcrop of gossan, consisting of limonite, carbonate, and mariposite, near Chips Flat, above the northernmost workings on the Clinton vein, suggests contact of a vein with a serpentine mass whose outcrop is beneath the lava.

RAINBOW EXTENSION MINE

The Rainbow Extension mine, on the south side of Kanaka Creek south of the Twenty-one claim, has been idle for some years and was last operated by Doctor Hardie, of Alleghany, and associates. Two veins, the southward extension of the two branches of the Sixteen to One vein, are developed by four tunnels for a total distance of about 800 feet. (See pl. 52.) A small amount of high-grade ore, estimated to have yielded not more than $5,000, has been produced. This gives a ratio of about $3 to the foot of drifting on both veins, or $4 for the eastern vein alone. A fifth tunnel, known as the Spiritualist tunnel, was run many years ago by an old miner who is said to have depended entirely on extramundane advice. The results seem to have been even less satisfactory than those obtained where reliance has been placed on human expert assistance.

The veins lie within the amphibolite schist of the Tightner formation, close to the western limit of the formation. The black slaty conglomerate that forms the base of the Kanaka formation is found in the Spiritualist tunnel about 250 feet west of the westernmost vein. A small mass of serpentine lies along the footwall of the southern part of the eastern vein in the upper tunnel and on the surface.

The western vein, known as the Twenty-one vein, is the better defined of the two in the underground workings, though its outcrop, except close to the creek, is less prominent than that of the other vein. It is possible that the mass of gossan at the edge of the old Chips Flat placer diggings represents the edge of a mass of serpentine altered by this vein, although as suggested above (p. 104), this gossan may lie on the northward extension of the Clinton vein of the Rainbow mine. The vein in the tunnel near the creek is well marked and shows continuous quartz nearly to the face. At the face of this tunnel and where it crosses the Spiritualist tunnel the vein fades to a small and poorly defined slip. It is also weak in the small tunnel above the creek tunnel. A crosscut from the main tunnel recovers the vein about 250 feet southeast of the face of the creek tunnel. Here it is cut off on the north at an acute angle by a westward-dipping fault, presumably the same as that which faults the eastern vein in the middle tunnel. From this point southward for 300 feet to the face the vein is well defined. The average strike (N. 30° W.) is more northerly than in the northward continuation of the vein in the Twenty-one tunnel, north of the creek, and the dip (50°–75° NE.) is much steeper. The fault that cuts the vein has a westerly dip of 45° in the Spiritualist tunnel and of 35° in the main tunnel and appears to be a reverse fault of small displacement. It follows approximately the strike of the Clinton vein and may represent the same fissure, though the observed dip is much flatter. What may be the same fault displaces slightly the eastern vein in the middle and upper tunnels. Here the strike is more nearly north, and the dip is 48° W. in the middle tunnel and 40° W. in the upper one.

The eastern vein in the lower tunnel is represented merely by a wall with here and there small lenses of quartz, from the portal southward to the point where it intersects the Spiritualist tunnel. From this point southward for 300 feet it is well defined in both the lower and middle tunnels, though less regular than the western vein. A sharp bend in the vein in the middle tunnel is not repeated in the lower tunnel and seems to be due to a curve in the original fissure rather than a local diversion along a fissure belonging to another system. A marked split in the vein occurs at this point, and the high-grade ore was obtained from the main vein just north of this split, principally, it is said, from ground above the middle tunnel, though the ground above the lower tunnel has also been stoped.

South of the split the veins are less definite. The eastern branch has been followed for over 300 feet in the middle tunnel, but the quartz decreases toward the south until at the face there are only small quartz stringers with a rather indefinite hanging wall, which here strikes east of north. In the lower tunnel this branch fades out a short distance from the split. The western branch has been followed only in the lower tunnel and likewise fades out a short distance to the south. The upper tunnel prospects what is probably the eastern branch of the eastern vein. Near the portal serpentine forms the footwall. To the south the vein, here dipping about 40° NE., leaves the serpentine contact, which dips steeply to the east and is found in the schist a few feet to the east. The same northerly fault that cuts the vein in the middle tunnel here displaces the vein but apparently not the serpentine contact.

The position of the small high-grade shoot in the eastern vein appears to have been dependent on the junction of the two branches, and it is possible that exploration of this junction at greater depth might be profitable.

No production has been made from the western vein, and, so far as known, no favorable prospects were encountered. Apparently its regularity and lack of bends and splits is an unfavorable feature. The vein is well defined where developed in the lower tunnel and may be worthy of further development. The small serpentine mass cut in the upper tunnel lies almost directly above this part of the vein, though, as far as seen, the wall rocks are schist.

SIXTEEN TO ONE MINE

The Sixteen to One mine (pl. 53 and secs. A–A' to G–G', pl. 10), owned by the Original Sixteen to One Mining Co., now includes also the mines formerly known as the Twenty-one and Tightner, and as they develop the same vein and are now worked as a single mine they will be described together.

The Twenty-one mine formerly comprised the portion of the present Sixteen to One mine south of about line G–G' on Plate 53. It was the first of the three mines to be developed and was in operation prior to 1868 [85] but never made any notable production, except from ground afterward determined to be in trespass on that of the Sixteen to One. In 1919 apex litigation with the Sixteen to One resulted in a judgment for $93,000 against the Twenty-one, and the mine was purchased by the Sixteen to One for $60,000. The Twenty-one tunnel now forms part of the 800-foot level of the Sixteen to One mine.

The Tightner mine included the portion of the vein lying within the Contract and Contract Extension claims, north of the Sixteen to One claim. The vein had been found during placer operations in the early days of the camp, but no attention was then paid to it. In 1902, in the course of clean-up operations from the old Knickerbocker tunnel, some drifting was done to the north of the tunnel without much success, and the property was bonded to H. L. Johnson, who drifted to the south and in 1907 obtained the first high-grade ore. The old workings of the Tightner, in the vicinity of the Knickerbocker tunnel, yielded the largest high-grade shoot which this portion of the mine has yet produced. At least $375,000 in coarse gold is said to have been obtained from picked ore, besides lower-grade ore that was left on the dump and later milled at a profit. Later the portion of the Eclipse property lying north of the creek was purchased and the present Tightner tunnel was run. This tunnel cuts the vein at about 400 feet vertical depth below the old Knickerbocker tunnel. The Tightner tunnel now connects with the 250-foot level of the Sixteen to One mine. The mine was sold in 1909 to the Tightner Mines Co. A. D. Foote, of the North Star mine, at Grass Valley, became general manager, and Capt. A. B. Hall superintendent. This company operated the mine until 1918 and is said to have produced $3,000,000, of which about $500,000 was paid out in dividends. The mine was sold in 1920 to the Alleghany Mining Co., but the Tightner Mines Co. reserved the Red Star property. This company continued development in depth down to the present 1,000-foot level. When it was discovered that the apex of the vein beneath the lava left the west side line of the Contract claim instead of passing through the end line, as had been supposed, a compromise was effected with the Sixteen to One mine by which the Tightner withdrew its claim on the vein to the line of the Compromise raise, which was run jointly by the two mines. The ground ceded to the Sixteen to One mine proved exceedingly rich, but though a little high-grade ore was obtained north of the raise the richest ore did not extend into the restricted Tightner territory. As developments in depth were not satisfactory, the mine was sold in 1924 to the Sixteen to One.

The ground covering the present Sixteen to One claim was formerly owned by the Rainbow mine but was never deemed worthy of development. The present claim was located by Thomas Bradbury, the late general manager, in 1908. Mr. Bradbury, assisted only by his brother, drove the present No. 2 tunnel of the Sixteen to One mine. In this portion of the mine the vein is not well defined, and it was some time before his persistence was rewarded by the discovery of small bunches of high-grade ore. The mine was later bonded to some promoters. A little high-grade ore was obtained, but quarrels and lawsuits between the promoters prevented development, and the property reverted to the original owners, who organized the present company. The capital stock consists of 200,000 shares, par value $1, of which 164,030 are outstanding. Dividends of $2,706,495 had been paid to February, 1930.[86] The mine did not become a noteworthy producer until, largely under the stimulus of the conflict with the Twenty-one, the ground between the 250 and 800 foot levels was explored. Since the acquisition of the Tightner property the main development work has been carried on from the Tightner shaft, but in 1925 the Sixteen to One shaft had been deepened to the 1,300-foot level.

The mine has been the most productive in the district, and the total output of the combined properties to 1928 has been about $9,000,000, of which the greater part has come from the area between the Sixteen to One shaft and the Compromise raise. For each foot of development work on the vein, including shafts and principal raises but excluding development raises, there has been a production of about $230. The average yield per ton of material mined during the entire history of the mine has been over $20.

The mine (pl. 53) is developed by two principal inclined shafts and levels spaced at intervals of approximately 100 feet on the dip of the vein in the upper

[85] Browne, J. R., Mineral resources of the States and Territories west of the Rocky Mountains for 1868, p. 139, 1869.

[86] Mines Handbook for 1930, p. 636, 1931.

workings and 200 feet in the lower workings. The lowest level in 1928, the 2,100-foot level from the Tightner shaft, is 1,100 feet lower than the No. 2 tunnel and 1,360 feet lower than the apex of the vein beneath the lava. (In the summer of 1930 exploration had been carried to the 3,000-foot level, 1,700 feet below the No. 2 tunnel.) At the time of visit (1928) no stoping had been done below the 1,300-foot level. The 1,500-foot level, however, had been carried for some distance from the shaft and showed excellent prospects.

The Tightner formation, here consisting of hornblende schist, in places much altered, forms the hanging wall of the vein. The footwall rocks include both the Kanaka formation and the Tightner formation. Although the strike of the vein is at an angle to that of the inclosing rocks, the Kanaka formation reaches its greatest depth, about the 250-foot level, on the footwall in the southern part of the mine. (See sec. G–G', pl. 10.) From this point northward, owing in part to change in the amount of displacement along the fault followed by the vein, but probably in the main to change in strike of the inclosing rocks, the Kanaka formation does not extend to as great a depth, and north of the line of section D–D', Plate 10, and in the lower workings the vein seems to have the Tightner formation on both walls. However, as the wall rocks are much altered the distinction between the two formations is not everywhere certain. Two dikes of serpentine cut the Tightner formation and are faulted by the vein.

The vein follows a reverse fault whose displacement, measured along the dip, ranges from about 900 feet along the line of the Tightner shaft (sec. C–C') to less than 500 feet in the northern part of the mine (sec. A–A') and less than 300 feet at the south end (sec. G–G'). The only data available for determining the displacement are the positions of the two serpentine dikes faulted by the vein, and as these dikes are nearly parallel to the vein and dip in the same direction, though more steeply, these figures are only approximate. The eastward-dipping vein is cut by two groups of faults, which for the most part dip steeply to the west but in a few places have vertical or even, for short distances, steep easterly dips.

The upper group of faults cuts the fault followed by the vein between the No. 2 tunnel and the 300-foot level in the area between the two shafts. As the average strike of the steep faults, N. 24° W., is at a slight angle to that of the vein, the intersection is at a higher altitude northward. The number of faults present apparently varies in different parts of the mine. This is in part due to incomplete exploration in certain areas, but it is believed that the major part of the displacement in some areas is along a single fault, whereas in others it is divided among a group of small faults.

Like that of the fault followed by the Sixteen to One vein, the displacement along the dip appears to be at a maximum, about 135 feet, in the region of the Tightner shaft (secs. B–B' and C–C', pl. 10) and decreases both to the north and south, to about 50 feet on the lines of sections A–A' and G–G'. As in the upper workings the principal vein tends to branch upward, it is not everywhere certain whether the segment followed on one side of a fault is the same as that on the other side, so these measurements may be in error.

The lower group of westward-dipping faults has been encountered in the shaft between the 1,800-foot and 2,100-foot levels, with a maximum displacement of about 250 feet. At least three faults are present where cut by the shaft (sec. C–C', pl. 10), but owing to lack of data the group is represented on the other sections by a single fault. The fault seems to be a part of the same group that is present in the Morning Glory and Extension of the Minnie D mines.

The faults of both groups contain quartz that grades into that of the main vein (pl. 7, B, C, and figs. 6 and 13), and therefore they must have been formed after the formation of the Sixteen to One fault but prior to the vein filling. The two groups occur at approximately the borders of an area of relatively flat dip on the main vein, and it is possible that this change in dip may have been a factor in determining their position, but both groups of faults seem to be parts of series of westward-dipping fissures that cross a large part of the district.

Throughout most of its developed distance the vein has its apex beneath the lava capping, but it can be traced on the surface from the edge of the gravel-covered area near Alleghany southeastward as far as the Sixteen to One mill, though it nowhere makes a conspicuous outcrop. Southeastward from the mill for 900 feet there is no distinguishable outcrop, owing in part to faulting by the westward-dipping Tightner fault but probably also to pinching out of the quartz. From about the 4,040-foot contour on the point southeast of the mill the outcrop can be followed to the Twenty-one tunnel. A footwall branch, which has not yet been developed, is exposed in the road cut near the mine office. (See pl. 7, B.)

In the southern part of the mine the vein (pl. 53) has an average strike of about N. 30° W.; farther north, in the Tightner workings, the strike is from a few degrees east of north to north as far as the region north of the North shaft, where the northwesterly strike again prevails. The dip also varies greatly, from less than 20° to 60°, but over most of the productive part of the mine the prevailing dip is between 25° and 30° E. To the north, in the vicinity of the North shaft, the dip steepens to 35° and over, and in the workings below the 1,500-foot levels the average dip is about 40°. Steep dips (40° to 60°) also prevail in the workings above the No. 2 tunnel level in the southern part of the mine. Here, however, there is a suggestion that there is a split in the vein and that the steeper hanging-wall branch has been followed.

The vein is well defined throughout the developed portion of the mine, and quartz is essentially continuous, though the thickness varies greatly. There is no apparent relation between the presence of serpentine in the wall rock and the amount of quartz present, though where serpentine forms both walls, owing to superposition by faulting, the thickness of the quartz present is less than the average. For the most part

a crosscut about 6 feet of banded quartz on the walls and an interior portion consisting of white coarsely crystalline quartz with large vugs. In the sublevel above the 1,000-foot level this interior coarse quartz ends sharply on its strike against horses of slightly altered wall rock.

To the north and south of the productive part of the mine the dip steepens considerably and the vein tends to split up into a number of small stringers. At the north end there is also an increase in the amount of quartz parallel to the structure of the inclosing schists. The faces of some of the drifts south of the shaft in the upper part of the mine show a fading and irregular branching of the vein, with pinching out of the quartz. Quartz is again prominent near the south end of the Twenty-one tunnel.

There are several minor branches. In the Twenty-one workings the vein forks southward, but the point of intersection on the level of the Twenty-one tunnel is in an area in which only a small amount of quartz is present. Possibly the irregular fractures observed on the south end of the 600 and 400 foot levels indicate the neighborhood of the junction. As in the Twenty-one workings the walls of both veins are formed by the Tightner formation, it is not known which branch follows the principal fault.

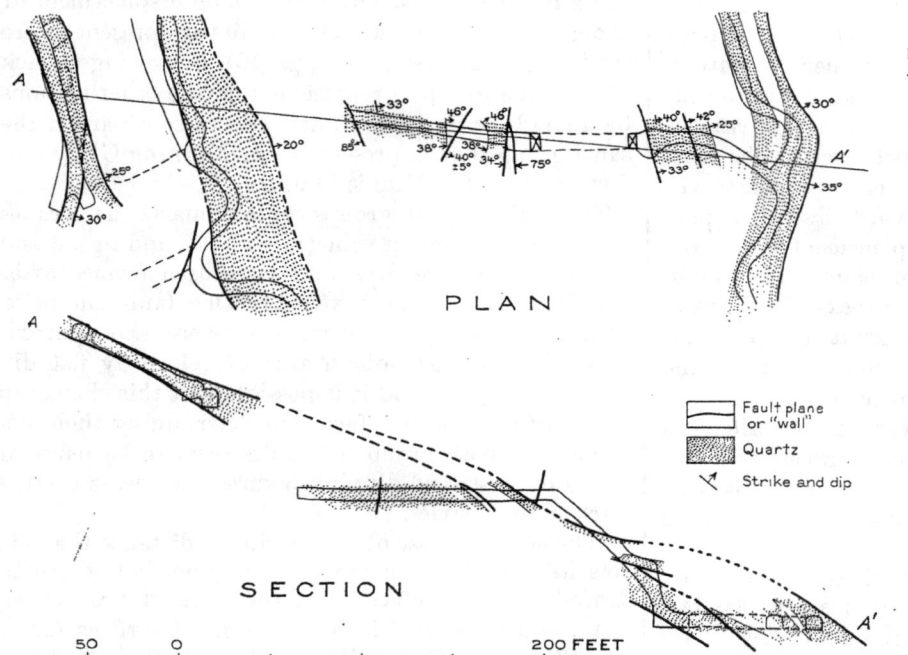

FIGURE 30.—Plan and section of parts of the 900-foot, intermediate, and 1,000-foot levels, Sixteen to One mine

the portions of the vein with flattest dips are those with thickest quartz, but this does not everywhere hold true. There are three lenticular areas where the quartz is several times the average thickness. (See pl. 53.) The largest, which extends southward down the dip from the 600-foot level between the Tightner and North shafts to the 1,500-foot level at the Tightner shaft, has a maximum thickness of 50 feet. Another on the same pitch is intersected by the North shaft and the 250-foot level (Tightner tunnel). The third, which is less regular in outline and pitches about down the dip, lies to the south of the Compromise raise. Each of these enlargements occurs in a region of flat dip and also marks a change in strike of the vein. Neither of them consists exclusively of quartz, but in each quartz greatly predominates over the included horses of country rock. All seem to be associated with minor branching of the vein. The large swelling in the Tightner workings (fig. 30) showed where cut by

The footwall branch that crops out on the road above the mine office has not been certainly identified underground. As far as traced on the surface it seems to be nearly parallel in strike to the principal vein. The steeper average dip in the No. 2 tunnel and workings above suggests that the junction lies at about this level. The drag of the conglomeratic slate at the outcrop of the footwall branch (pl. 7, B) shows that there has been movement in the reverse direction.

The Ophir vein joins the Sixteen to One south of the Sixteen to One shaft. Although of slightly steeper average dip and more westerly strike, the Ophir bends toward the Sixteen to One near the junction, so that the trace of the junction on the Sixteen to One vein, instead of pitching to the south, is irregular and apparently follows the direction of the dip of the major vein. Where connection between the two mines has been established the Ophir vein can be seen grading into the Sixteen to One vein, and as far north as the Sixteen to One shaft it is identifiable as the hanging-wall strand of the Sixteen to One vein.

A small vein, probably a branch of the Eclipse vein, was followed by the Tightner tunnel for a distance of about 500 feet from the portal. This may be the same vein as that which joins the main vein on the 250-foot level, but if so the strike must change close to the main vein even more sharply than that of the Ophir.

The workings of the Morning Glory mine show that the Morning Glory vein, like the Ophir, bends toward the Sixteen to One, but the point of junction is not certainly established. It is thought likely to be at the north end of the large mass of quartz north of the Tightner shaft.

The high-grade shoots are richer than in most of the other mines. One small shoot yielded nearly $1,000,000, and several others have yielded over $200,-000 each. A tenor of $50 and over a pound is not unknown in the richest ore. A lot of 80 pounds from one of the rich shoots mined in 1924 yielded $5,000, and in 1928 a chunk of ore weighing 160 pounds netted $28,000. As a rule the gold is not closely associated with the early coarsely crystalline arsenopyrite but occurs in the quartz unaccompanied by other minerals than a little sericite and carbonate. A little coarsely crystalline arsenopyrite was, however, found in most of the shoots, generally below the richest ore. Sulphides of the second generation, principally galena, sphalerite, tetrahedrite, and jamesonite, are present in the productive portions of the mine but are not constant associates of the highest-grade ore. In several of the stopes "headcheese" breccia was present above the ore, and in a few places (pl. 39, B) gold could be observed in the cement of such a breccia.

High-grade ore has been mined throughout the portion of the mine in which the vein has a relatively flat dip, from the level of the Knickerbocker tunnel to the 1,300-foot level. The ground about the 1,300-foot level immediately south of the Tightner shaft and between the 1,300 and 1,500 foot levels had not been developed at the time of visit, but some high-grade ore was encountered in drifting on the 1,500-foot level.

There seem to be two poorly marked zones, each with a flat pitch to the south, in which the high-grade ore is particularly abundant. The upper zone extends from the Knickerbocker tunnel to the Sixteen to One shaft between the 250 and 800 foot levels, and the lower zone from the region north of the raise above the North shaft to the Sixteen to One shaft below the 1,100-foot level. (See pl. 53.)

Between these two productive zones from a point at least as far north as the Tightner shaft between the 300 and 600 foot levels to the Sixteen to One shaft between the 800 and 1,100 foot levels lies a zone which, though physical conditions are apparently as favorable as elsewhere, is completely barren of high-grade ore. It is possible that a similar barren zone bounds the lower productive zone, but development has not gone far enough to determine this point.

Although the best ore seems to be found where there is a fair thickness of quartz, 5 to 15 feet, little high-grade ore has been obtained from the large lenticular swellings described above. The edges of these enlargements, however, particularly where they begin to fade out along the strike, have been very productive.

Junctions of the vein with minor branches seem to have been effective in localizing the high-grade ore. The ore shoots in the upper part of the mine, just below the line where the dip begins to steepen, probably lie near the junction of the vein with the unexplored footwall branch. The quartz-bearing faults in this portion of the mine may have also been effective in localizing the gold. On the 200-foot level, north of the Sixteen to One shaft, very rich ore extended up to the fault vein, but, although the quartz filling was continuous in the two veins, the gold does not enter the fault vein. It is thought possible that the fracturing of the quartz necessary for the introduction of the gold in the main vein was facilitated by the presence of quartz in the fault vein. (See p. 57.)

The small high-grade shoots at the south end of the 250 and 400 foot levels lie near the probable junction of the two branches that fork to the south and crop out above the Twenty-one tunnel. The junction of the Ophir and Sixteen to One veins may have caused the localization of the high-grade ore obtained just south of the Sixteen to One shaft. The Ophir vein was also rich close to the junction. High-grade ore was found close to the vein junction (supposed Eclipse junction) near the Tightner shaft above the 250-foot level, but although the junction is recognizable as far down as the 600-foot level it does not seem to have been productive below the 250-foot level. Another split in the vein, doubtfully correlated with the Morning Glory vein, occurs in the vicinity of the high-grade shoots near the North shaft, above the 250-foot level.

On the other hand, there are several branches from the main vein with which no high-grade ore is associated, and for some of the richest shoots, especially in the area between the Sixteen to One and Tightner shafts, no connection with branching veins can be established. In this area there appears to be some connection between the high-grade ore and small vertical faults. Such a fault, which displaced the quartz about a foot but which did not cut the hanging-wall slip, was seen in the very rich stope above the 800-foot level near the Compromise raise, and another which displaced both quartz and hanging-wall slip crossed the very rich shoot above the 1,300-foot level.

OPHIR MINE

The Ophir mine (figs. 31, 32; sec. G–G', pl. 10), which is reported to have produced over $60,000, was not being operated at the time of visit to the district. It is now owned by the Original Sixteen to One Mining Co. The vein lies on the hanging-wall side of the Sixteen to One, and has been followed by several tunnels. Two shafts from the lowest tunnel have developed the vein at depth, but the full extent of the feet the vein is marked only by a slip. Near the northern shaft quartz comes in again and continues to the point where the vein blends with the Sixteen to One. The quartz, however, shows greater continuity on the next level above but nowhere exceeds 5 feet in thickness. The apparent thickening in the curve near the Sixteen to One junction is largely due to the flatter dip.

Two branch veins are known; one leaves the footwall of the main vein about 100 feet from the portal of the main level, and the other on the hanging-wall

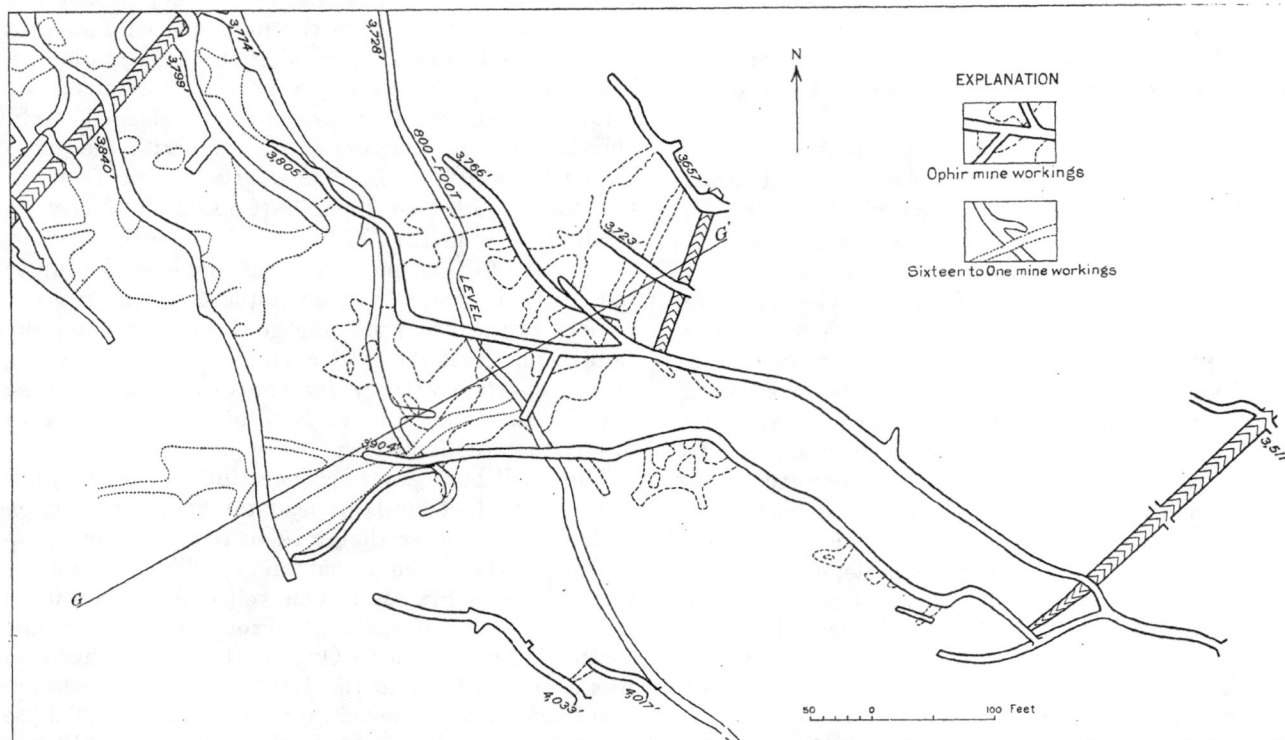

FIGURE 31.—Map of Ophir mine and adjoining workings of Sixteen to One mine.   G–G', Line of section G–G', Plate 10

lower workings is not known. Including only the drifts and shafts shown on the accompanying map, the ratio of production to development is about $20 a foot.

The country rock is schist of the Tightner formation and serpentine. The vein follows a normal fault with a displacement of only a few feet on the hanging-wall side of the Sixteen to One vein. Over most of its course the average strike is N. 62° W. and the dip 50°–70° NE. Near the Sixteen to One vein both strike and dip change sharply toward parallelism with the major vein. (See fig. 31.) Although the Ophir vein follows a normal fault, the portion close to the Sixteen to One vein shows the same type of minor thrusting that characterizes the latter vein.

The quartz is not continuous throughout. On the drain level (fig. 32) there is quartz for about 150 feet from the portal. Thence for a distance of over 300 side near the north shaft. Both have been followed by drifts, but the quartz does not appear to continue far from the main vein.

It is believed that practically all the production came from the region of the bend in the vein. Here serpentine forms the footwall and in part the hanging wall also. There are several small stopes above the upper of the two tunnels, but as far as known these yielded only a small production. The main vein near the two branches has been explored by shafts and drifts, apparently without success. Probably the most promising area is that near the junction with the Sixteen to One vein below the present workings.

ECLIPSE MINE

The Eclipse patented claim lies between the Ophir and the Morning Glory and extends across Kanaka Creek and its North Fork. That part lying north of the North Fork was owned in 1925 by the Original Sixteen to One Mining Co. and the remainder by the Ophir Mining Co. The production from the workings near the North Fork is said to have been about $20,000, or, according to some accounts, $50,000.

The vein is opened by three tunnels on the south side of the North Fork. (See pl. 54.) Two of these are now caved, but a part of the lowest one is accessible from the middle tunnel. A winze from which an unknown amount of drifting has been done is now flooded. A tunnel, now caved, on the Kanaka Creek side is said to have developed two parallel veins. As the total extent of the workings is unknown the ratio

vein. The projection of the fault which displaces the Sixteen to One vein in the lower levels indicates that the Eclipse vein must be cut by the same fault, a short distance below the lowest accessible workings (sec. F–F′. pl. 10).

Quartz is not continuous throughout and does not exceed 5 or 6 feet in thickness. The thickest portion lies along the junction of the two branches, near the winze.

The ore shoot mined occurred at the end of the enlargement of the vein near the junction, apparently following both the junction and a sharp roll in the vein and raking a few degrees east of the dip. Near the junction the quartz is crushed and sheared and the

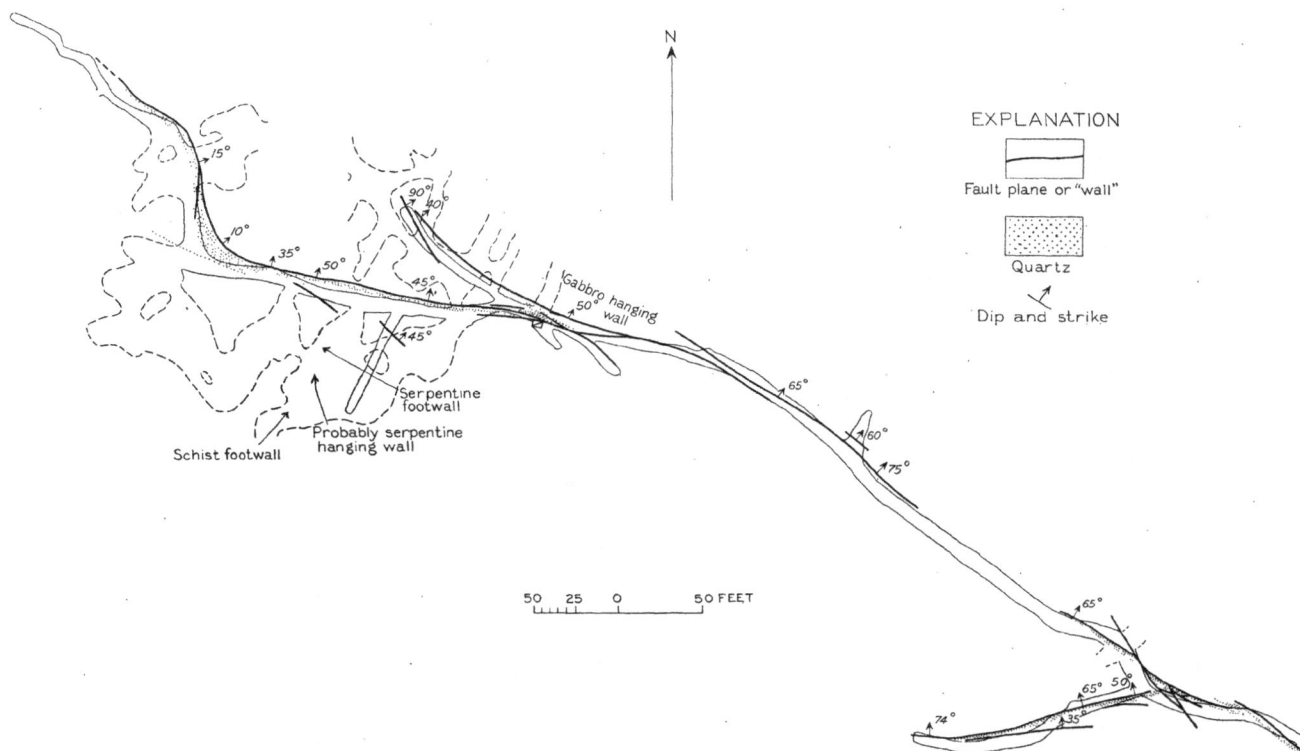

FIGURE 32.—Map of drain level, Ophir mine

of production to development can not be given with any approach to accuracy. It is probably less than $40 a foot.

The vein, which probably follows a normal fault of small displacement, lies wholly within the Tightner formation, which here consists of massive amphibolite schist. The vein in the tunnels strikes N. 45°–60° W. and dips 45°–60° NE. It is traceable to the southeast across the ridge for about 600 feet. It is not known south of Kanaka Creek, unless the Snowdrift represents its continuation. The vein developed in the mine is thought to be the same as that followed by the first part of the Tightner tunnel (p. 109); the vein that branches from the main vein on the hanging-wall side near the raise may connect with the Morning Glory

hanging-wall slip cuts across the quartz strands. The country rock is here crushed and intensely altered, but a short distance away it is quite fresh.

MORNING GLORY MINE

The Morning Glory mine (pl. 54 and secs. C–C′, D–D′, E–E′, pl. 10), adjoining the Tightner, just east of the town of Alleghany, was owned in 1925, together with the Extension of the Minnie D, by Martin Rohrig and was then being worked under lease by the Alleghany Morning Glory Mining Co. It was idle in 1928. The underground workings consist of a drift tunnel and an inclined shaft following the vein to a vertical depth of about 200 feet, or 380 feet on the dip of the vein, from which four levels have been run. The total amount of development work consists of about 2,500 feet of drifts and 600 feet of shaft and raises; the estimated total production is $80,000 to $100,000, which gives a ratio of about $27 to $32 a foot.

The vein lies entirely within rather massive amphibolite schist of the Tightner formation and is probably continuous with one branch of the Eclipse vein. The strike in the productive part of the upper levels is N. 20°–60° W. The dip, which ranges from 30° to 60° NE., is flattest in the north. The vein in the lower levels therefore shows a divergent strike, averaging about N. 32° W. The similarity in position of the Morning Glory vein to the Ophir vein suggests that it may follow a similar small normal fault. On the other hand, there are several small thrust faults which cut the quartz in the productive portion of the mine, near the shaft. None of these has an apparent displacement of more than 10 feet.

The vein is small, and the quartz is not continuous. The maximum thickness is about 7 feet, but over most of the distance where continuous quartz is present the thickness ranges from 1 to 3 feet. The vein is fairly well defined in the workings immediately adjoining the shaft but becomes very irregular toward the north, as it nears the Sixteen to One vein. Development has not gone far enough to the northwest to determine whether there is a bend close to the junction similar to that of the Ophir vein. In the lowest level a sharp change to northerly strike appears to be due to premineral faulting of the northwesterly fissure by a fissure belonging to the northerly system. Beyond this point the vein is lost.

Intermineral faulting is well marked in a few places, particularly near the ore shoots, by brecciated and recemented quartz and shears approximately parallel to the walls.

Postmineral faults of two kinds are prominent throughout the vein. One group consists of small thrusts approximately parallel to the vein but crossing it in places so that the footwall slip crosses to the hanging wall down the dip or along the strike. The other group, which fault the vein fissure, consists of steeply dipping fissures with northerly strike and generally a steep westerly dip. Most of these, as in the northern part of the drain level and on the third level near the shaft, displace the vein only a few feet and in the reverse direction, but at the south end of the fourth level the vein is cut off by a fault belonging to this group. The fault that cuts the vein on the drain level north of the shaft may be the same as that followed by the Lucky Larry vein, but it does not displace the Sixteen to One vein. The fault that cuts off the vein in the lowest level may be the same as that which displaces the Sixteen to One between the 1,800 and 2,100 foot levels. If so, the displacement of the

Morning Glory vein may be as much as 250 feet. (See secs. C–C', D–D', E–E', pl. 10.)

The schist wall rock is heavily pyritized in places, apparently chiefly in the productive part of the vein. A little mariposite was seen, but this mineral is much less abundant than in most other mines. Neither carbonate nor albite was noted in association with the vein quartz.

The principal output was obtained above the drain level, with a minor amount from the third level. The apparent cause for localization is the change in strike of the vein. All the production was derived from high-grade ore, mostly in small bunches. These were in nearly flat streaks with gentle pitch to the south, suggesting that the gold was introduced along minor fault planes. In several of the localities from which high-grade ore was obtained brecciated and recemented quartz was observed along the hanging wall. In places flaky gold was found directly on the hanging-wall slip. It is said that most of the gold was free within the quartz and not commonly associated with arsenopyrite.

A crosscut was being run in 1925 from the fourth level to intersect the vein developed in the workings of the Extension of the Minnie D.

EXTENSION OF THE MINNIE D MINE

The Extension of the Minnie D is an irregular claim on the east bank of the North Fork of Kanaka Creek between the Morning Glory and Colorado Extension patents. The claim is part of the Morning Glory property, but as the workings are independent it is described separately.

Several cuts and three tunnels (pl. 54, fig. 33, and secs. D–D' and E–E', pl. 10) show the presence of three fairly distinct veins and minor branches, together with a complex system of faults of different ages. There are in all about 850 feet of drifts and about 300 feet of crosscuts, exclusive of a crosscut 262 feet long, which was being driven in 1925 beneath the old workings from the lowest level of the Morning Glory. A small production, probably less than $10,000, has been made, giving a ration of about $12 to the foot of drifting.

The rocks in the vicinity of the veins all belong to the Tightner formation. Three varieties are recognizable on the surface and in the crosscuts—a banded chlorite-amphibolite schist of rather slaty aspect, believed to represent original water-laid tuff and shale; an amphibolite schist composed of needlelike hornblende crystals; and a more massive amphibolite, probably originally a basic flow. These distinctions are obscured near the veins.

Three systems of premineral fissures are present. The oldest, represented by a small vein in the southern part of the middle tunnel and a small vein near the face of the upper tunnel, have a northwest strike and dip northeast, conforming to the productive fissures of the immediate vicinity. The two principal veins, which strike N. 20° W. and dip at fairly low angles to the west, constitute a second system, probably in part contemporaneous with the first. In the middle tunnel fissures of this system and the first cross without displacement. In the upper tunnel the vein that dips to the northeast ends against the westward-dipping vein, but here the quartz filling is contemporaneous in the two veins and later than the controlling fault. The westward-dipping veins and faults form part of the same group which cuts the eastward-dipping veins in the Morning Glory and the lower workings of the Sixteen to One.

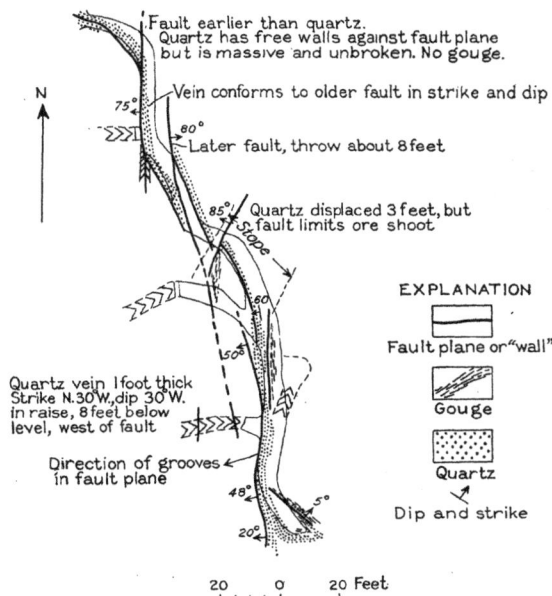

FIGURE 33.—Map of upper tunnel, Extension of Minnie D mine, showing complex faulting of vein

In the upper tunnel near the portal (fig. 33) a fissure striking about N. 2° W. and dipping 75° W. is obviously premineral, for the southwestward-dipping vein turns and follows this course for a short distance and then continues on the other side. It is not certain whether the segment west of the fault is to be correlated with the eastern or the western of the two veins known in the middle tunnel. In either case movement down to the west of the fault—that is, in a normal direction—is indicated. The parallel nearly vertical fault 10 feet to the east may also have originated at the same time, but on this there has been postmineral movement, displacing the vein in the opposite direction about 9 feet. This fissure appears to fault a northeasterly fissure with steep northwesterly dip, which itself displaces the vein about 3 feet. Although there clearly has been postmineral movement

on this last fissure, it is of probable premineral or intermineral origin, for it sharply limits the small ore shoot stoped south of this point.

The effects of intermineral movements are small and ill defined. Evidence of their occurrence is afforded by grooves covered with pyrite on the walls, pyrite along free walls without gouge, and rare sulphide banding in the quartz.

The major postmineral faulting took place along the premineral fissures already mentioned and also parallel or nearly parallel to the veins.

The only production came from a small stope in the upper level. The small shoot seems to have been directly dependent on a fault of intermineral age which displaced the quartz slightly but controlled the position of the ore, which was found only to the south of the fault.

MINNIE D AND LUCKY LARRY MINES

The Minnie D and Lucky Larry claims lie on the west bank of the North Fork of Kanaka Creek between the Morning Glory and Osceola properties. The claims were owned in 1925 by Albert Holmes. Two tunnels over 500 feet in combined length develop the Minnie D vein, and there is also a short tunnel on the Lucky Larry vein. (See pl. 54 and sec. B–B', pl. 10.) There has been no production worthy of note.

The country rock belongs to the Tightner formation, which is here a much crenulated amphibolite schist, on the whole rather massive and with abundant small quartz stringers parallel to the schistosity.

The Minnie D vein is parallel to the Morning Glory, having an average strike of N. 55° W. and dip of 50° NE., with many minor splits. In a raise in the upper tunnel the gougy slip that forms the footwall in the drift cuts across the vein and passes out in the hanging wall, carrying one strand of quartz above another. In the main tunnel near the portal a westward-dipping northwesterly wall definitely cuts off the vein walls; this may be the same fault which cuts the Sixteen to One vein in the lower levels. (See sec. B–B', pl. 10.) A little farther in a parallel slip abuts against the vein wall without displacement.

Although the mine has not yet yielded high-grade ore, good prospects have been found at two points, one in the upper tunnel above the thickest quartz, where the strike changes to the west in the lower tunnel, and the other in the lower tunnel near the face, where the main vein is joined on the footwall side by a minor vein striking a few degrees west of north and dipping 40° E. The quartz is massive and on the whole fairly free from wall-rock inclusions but contains numerous vugs in which crystals of siderite and ankerite occur on the faces of the quartz crystals. Where the best prospects are found the quartz carries dark ribbons containing chiefly pyrite. Gold in minute flakes is found in these seams. A specimen collected in 1913 showed small specks of gold on the faces of pyrite crystals in the vein. Pyrite and rare grains of sphalerite, galena, tetrahedrite, and arsenopyrite are also found in the vein quartz. Crystals of albite are found here and there on the borders of the vein in those parts where later fault walls are absent. The chief wall-rock alteration consisted in silicification, but carbonates were noted on the footwall side of the large quartz lens in the central part of the lower tunnel.

The Lucky Larry vein can be traced by outcrops, small cuts, and float from the ravine containing the Minnie D nearly to the Osceola vein. The strike in the southern part is N. 10° W. and the dip 55° W. About 200 feet south of the Osceola, however, a strike of N. 35° E. with a dip of 60° W. was noted, suggesting a turn of the minor fissure toward the major one. The trace of the outcrop extends across the strike of the Minnie D vein. The Lucky Larry vein may follow the same fissure that cuts the Morning Glory vein on the adit level, but this fissure does not cut the Sixteen to One vein.

OSCEOLA MINE

The Osceola mine (fig. 34) is on the west bank of the North Fork of Kanaka Creek, just north of Alleghany. The property includes the Osceola (patented)

and Osceola Fraction claims. The owner in 1925 was R. W. Gillespie.

The mine has been operated intermittently since the early days of the district, and the total production is uncertain. Local reports give information of a shoot in the middle tunnel that yielded $30,000 and one in the lower tunnel that yielded $60,000. This must have been prior to 1895. The mine was reopened between 1900 and 1902, and about $12,000 was produced. It was again reopened in 1913, and a small amount of development work done, probably without notable production. It was idle at the time of visit.

There are three tunnels on the vein. The uppermost, a short distance beneath the lava, has 120 feet of drifts, with two raises and a winze. The middle tunnel is caved 30 feet from the portal. This is said to have been the most productive and to contain extensive stopes. The lowest or main tunnel close to the North Fork of Kanaka Creek has a total length of 1,550 feet but is now caved 200 feet from the portal. In the portion now accessible the ground is partly stoped above the tunnel level and stoped in places below. Near the portal is a winze said to be 150 feet deep, now flooded. As the total extent of the workings is unknown, the ratio of production to development can not be given closely. It must, however, be less than $50 a foot.

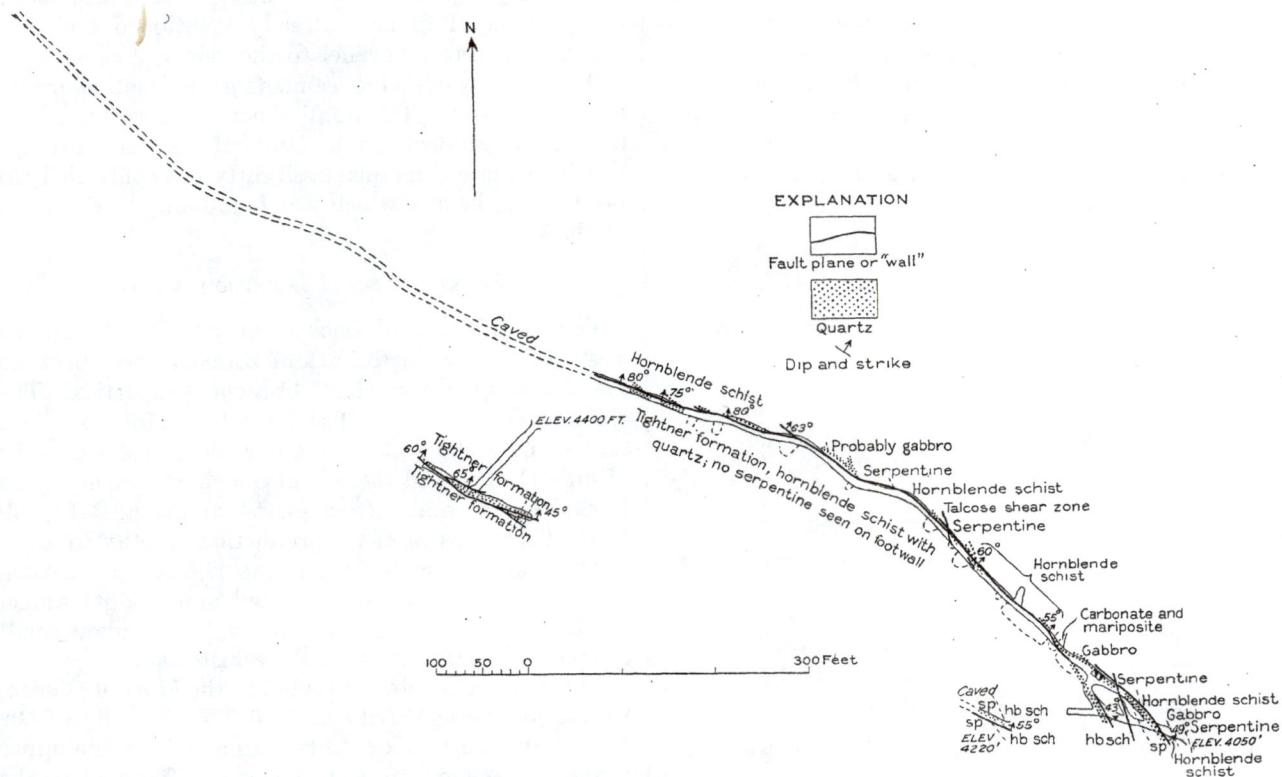

FIGURE 34.—Map of workings, Osceola mine

The country rock is amphibolite schist of the Tightner formation, showing the usual variation in texture and in places heavily silicified. The amphibolite is cut by narrow serpentine and gabbro dikes, in places without sharp contacts.

The displacement of the serpentine on the surface and in the middle tunnel indicates that the vein follows a fault of small normal throw. In the lower tunnel the relations are confused by the splitting of the vein and the presence in the hanging wall of dikes of gabbro and serpentine which were not found on the footwall side. In the productive portion, close to the serpentine, there has also been minor postmineral thrusting nearly along the plane of the vein.

The vein has a very prominent outcrop on the hill west of the creek just below the lava and is easily traceable on the surface as far as the creek. East of the serpentine belt the vein splits, and the two branches have also been developed on the tunnel level. The footwall branch continues to the southeast and is probably the same fissure as that followed by the Colorado vein on the ridge between the forks of the creek. South of the creek what may be the same vein has been developed as the Bullion vein of the Eldorado mine and is traceable uphill to the south at least as far as the 4,100-foot contour, in all a distance of about 4,500 feet. The hanging-wall branch becomes indefinite at its contact with serpentine in the creek bed, but the small Panama vein, 500 feet to the southeast, may be on the same fissure.

The tunnel was visited by Mr. Gannett in the early part of the field season of 1925, before the last cave occurred, but was not studied in detail. His preliminary notes indicate rather less continuity of quartz filling than on the same vein in the Yellow Jacket workings, to the southwest. Minor postmineral faulting nearly parallel to the plane of the vein has in places caused gouge streaks to cross the vein from footwall to hanging wall. The strike of the grooving on these streaks where recorded is N. 60° E.

For the first 400 feet from the portal the vein has an average strike of N. 42° W. and a dip of 55°–60° NE. Beyond this point the sketch map shows an average strike of N. 66° W. and dip of 75°–80° NW. There are stringers in the hanging wall following the strike of the southern segment of the vein. It is possible, therefore, that this change in strike may represent another split in the vein (in the opposite direction from that near the portal) and that an unexplored segment lies to the north. Notes made in 1913,[87] when the full length of the tunnel was open, indicate that near the face the strike is N. 25° W. and the dip 35° E. This may indicate a bending toward the Sixteen to

[87] Ferguson, H. G., Lode deposits of the Alleghany district, Calif.: U. S. Geol. Survey Bull. 580, p. 174, 1915.

One fissure similar to that of the Ophir vein. Quartz is entirely lacking between points about 700 and 1,150 feet from the portal, but thence to the face it is continuous.

The known high-grade stopes are near the junction of the two branches near the portal of the main tunnel and also close to the small serpentine dike cut by the vein. So far as known no high-grade shoots were associated with changes in dip and strike of the vein. Mariposite is abundant only near the serpentine. In part at least the gold occurred in association with crushed arsenopyrite.

PANAMA PROSPECT

The Panama claim, owned by Will Morrison, adjoins the Colorado on the northeast. A 65-foot tunnel has prospected small quartz stringers in much sheared and altered amphibolite close to a serpentine contact. These may represent the south end of the northern branch of the Osceola vein. The serpentine-amphibolite contact strikes N. 25° W. and dips 70° NE. The quartz stringers are within a few feet of the contact and parallel to it but have an average dip of only 30° NE. They do not appear to fault the contact but die out in the serpentine. Quartz is present only in the amphibolite; at the contact the filling changes to a mixture of mariposite and carbonate. The prospect therefore presents an example of a quartz vein fading out into serpentine down the dip. The outcrop of the contact is here marked by gossan derived from iron-bearing carbonate associated with the serpentine.

RED STAR MINE

The Red Star mine (fig. 35) was owned in 1925 by the Tightner Mines Co. It was originally a placer mine worked from a bedrock tunnel whose portal is near the head of the ravine about half a mile north of Alleghany. A vein was encountered and developed to some extent, and one high-grade shoot, reported to have yielded $80,000, was mined. Later work was started from the Tightner tunnel (250-foot level of the Sixteen to One mine), but though a vein was picked up and explored by drifts and raises the results were unsatisfactory and work was abandoned in 1925. A little high-grade ore was obtained in the raise, but no production of any consequence was made. Probably the amount of development on the vein exceeds 2,000 feet, which would give a ratio of about $40 to the foot.

Only a part of the lower tunnel was accessible at the time of visit. So far as could be seen the country rock belongs to the Tightner formation, but it is said that in the high-grade shoot on the upper level the vein adjoined serpentine.

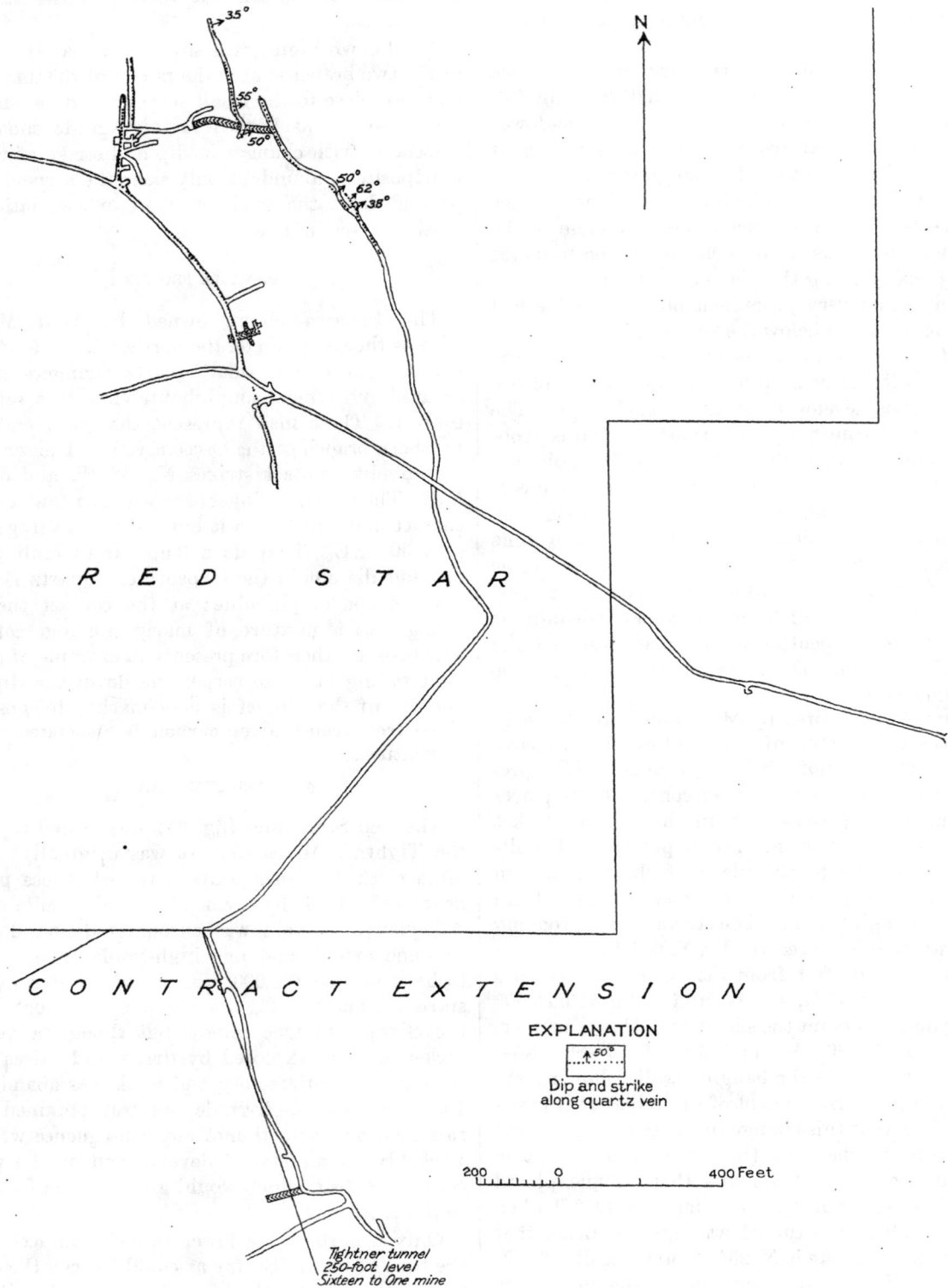

FIGURE 35.—Map of workings, Red Star mine

It has been considered that the Red Star vein was the extension either of the Sixteen to One or the Osceola, but the projection (pl. 8) suggests that it lies to the east of these, about in line with the Amethyst and South Fork veins.

RAO PROSPECT

The Rao claim adjoins the Osceola on the northeast and in 1925 was owned by the Tightner Mines Co. The Tightner formation is cut by a dike of serpentine, which on its northeastern contact is altered to limonitic gossan, resulting from the oxidation of iron-bearing carbonate. This belt of gossan can be followed from the Tertiary gravel near the road almost to the creek. In the upper part there is also a little quartz which has a northwesterly strike and an apparent dip of 85° NE. This vein has been prospected by a tunnel, now caved. Another small vein, also with a northeasterly strike but dipping 60° SW., crops out 300 feet to the southeast.

MAYFLOWER PROSPECT

On the Mayflower prospect (fig. 36), on the North Fork of Kanaka Creek about three-quarters of a mile northeast of Alleghany, a short tunnel and open cut show a small vein with an average strike of N. 35° W. and dip of 55° NE. At the open cut this is cut by a fault which strikes north and dips 80° E., with reverse displacement of a few feet only. Pyrite has been deposited in the fault plane subsequent to movement. The wall rock is amphibolite schist of the Tightner formation.

MARIPOSA MINE

The Mariposa claim extends south from the south bank of Kanaka Creek. The portal of the tunnel is near the creek, 2,200 feet east of the Sixteen to One mine and 1,300 feet southwest of the portal of the Eldorado tunnel. The mine was owned in 1925 by the Ophir Mining Co., but no work had been done for several years. The production is reported to have been more than $50,000. The ratio of production to development work on the vein may be as much as $40 a foot or may be less, for the extent of the work on the vein above and below the adit level is unknown.

The only accessible working is a tunnel 1,350 feet in length. (See pl. 55 and secs. H–H', I–I', pl. 10.) Work has been done both above and below the tunnel level, however.

The country rock consists of amphibolite schist of the Tightner formation, cut on each side of the vein by rather irregular dikes of serpentine. A small dike of gabbro was found on the surface adjoining the

western serpentine mass but was not identified underground. There are no data for measurement of the displacement along the vein fissure, but as the vein is probably the southward continuation of the Ophir it is probably small and in the normal direction. The vein is small but fairly well defined. The average strike is N. 48° W., and the dip 48°–60° NE. The maximum thickness, about 2 feet, is in the stretch from 150 to 300 feet from the portal where the dips are flattest. In the southern part the tunnel leaves the vein and follows a footwall split to a small dike of serpentine, probably a branch from the larger mass to the west. In this part of the tunnel the vein has numerous minor branches on both footwall and hanging wall, which have made it difficult to follow. It is more

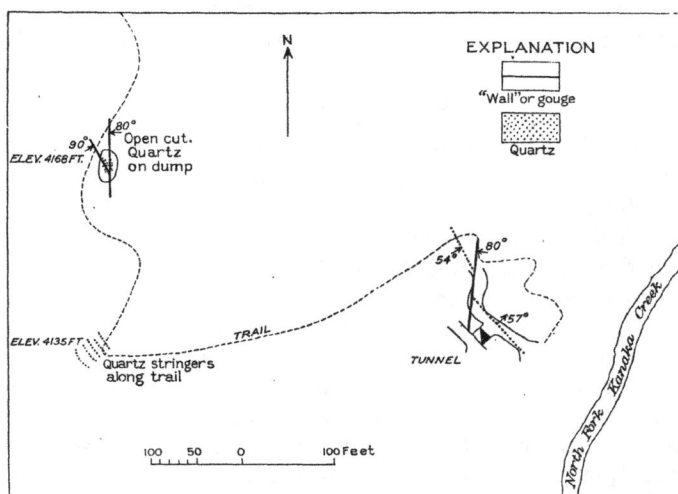

FIGURE 36.—Sketch map of Mayflower prospect

regular at a distance from the serpentine, but one branch was noted in the hanging wall about 200 feet from the portal.

A small westward-dipping vein that crops out on the hillside east of the tunnel (see pl. 55 and sec. H–H', pl. 10) agrees in projection (pl. 8) with the fault on the 2,100-foot level of the Sixteen to One mine. If this follows the same fault, displacement of the Mariposa vein may be expected below the tunnel level. It is assumed that the amount of throw will be less than in the lower workings of the Sixteen to One mine.

Apparently the principal production has been derived from the portion near the portal, where the dip is flattest and the vein thickest.

CENTRAL MINE

The Central group, owned by Colonel Edwards, of Alleghany, consists of four claims on the south side of Kanaka Creek near Chips Flat. A vein, presumably the continuation of the Mariposa and Ophir, is traceable through the Mariposa claim into the Banquet claim of this group. Apparently it lies in a belt of schist between two serpentine masses and seems to follow and perhaps fault a small serpentine dike. Owing to the deep soil cover, however, the relations are uncertain.

On the Central claim, to the west of the Banquet, old dumps show quartz fragments. These workings are said to have produced high grade.

SNOWDRIFT PROSPECT

The Snowdrift claim is a fraction between the Mariposa, Bullion, Eclipse, and Hope Extension.

No well-defined vein is exposed on this claim. A series of stringers run into and along the serpentine contact and into the serpentine. Several bunches worth a few hundred dollars each are said to have been obtained.

ELDORADO MINE

The Eldorado mine is on the south side of Kanaka Creek about 2,000 feet upstream from the mouth of the North Fork and 3,000 feet southeast of Alleghany. The Yellow Jacket property adjoins it on the north and the Plumbago on the south. The property consists of three unpatented claims, the Eldorado, Eldorado Extension, and Terrible, covering an area of about 45 acres. The property was located about 1880 by John Fessler and in 1925 was owned by the Fessler estate and was being worked under bond and lease by the Alleghany Eldorado Mining Co., Harry Engelbright, manager. It was idle in 1928.

Three veins have been developed. (See pl. 55 and secs. H–H′ and I–I′, pl. 10.) On the easternmost, the Eldorado (pl. 56), there are three tunnels, a winze, and sublevels, with a total of about 3,600 feet of drifts on the vein and 1,000 feet of development raises and winzes. The reported production is about $325,-000, equivalent to about $70 per foot of development on the vein. Records of tonnage milled in recent years show a recovery of $16.90 to the ton. The middle vein is developed by a 400-foot drift. It is reported that good prospects were found, but there has been no production from this vein. On the Bullion vein there are two tunnels with about 400 feet of drifts, which have yielded only a small production.

The principal country rock is amphibolite schist of the Tightner formation, for the most part of the more massive type, suggesting much altered flows. At the tunnel portal there is an outcrop of coarsely crystalline marble similar to that found in the North Fork mine and on the bank of the Yuba River, and crystalline limestone is also exposed at the hoist station on the main level. On the surface are dikes of gabbro and serpentine, faulted by the Eldorado vein. The serpen-

tine appears in the underground workings, but the gabbro is recognizable only at the surface, probably owing to the extreme crushing and alteration along the vein. Two small masses of aplite were found near the vein in the southern part of the Eldorado claim but were not encountered in the underground workings.

The Eldorado vein (secs. H–H′ and I–I′, pl. 10) follows an eastward-dipping reverse fault, probably the northern extension of the same fault followed by the Plumbago vein. The displacement is here much less—not quite 50 feet in the northern part of the mine and perhaps about 100 feet in the southern part. The vein is variable in both strike and dip, and there are several minor branches. On the main tunnel and lower levels the vein shows a very sharp bend. From the portal for a distance of about 500 feet the average strike is N. 26° W.; thence for about the same distance it is N. 50° W. At this point there is a sharp change to a strike of N. 30° E., which continues for about 350 feet; thence to the south end of the workings the vein has an average strike of N. 7° W. in the tunnel level and N. 15° W. in the lower level. As the dip flattens greatly above the upper level, this bend does not appear in the course of the outcrop, though its position is marked by a thickening of the quartz. It is thought that, as the change in both strike and dip coincides with the crossing of a serpentine dike by the vein, the deflection may have been caused by the serpentine, but in the Sixteen to One, Ophir, and Osceola mines other serpentine dikes are cut and faulted by vein fissures without such marked deflection either in strike or in dip. This serpentine dike is thicker than those cut by these veins and may have exerted an influence on the fissure intermediate between that of the large mass of serpentine at the Oriental, where the vein turns sharply and is lost in the serpentine, and that of the much smaller dikes that are crossed by the veins without significant change in strike.

The Eldorado vein shows continuous quartz throughout the stretches in which the strike is west of north. In the area of the great bend in the vein quartz is present only in lenses, yet this section has yielded all the high-grade ore so far found. Its maximum thickness, about 6 feet, is reached at the south end of the workings.

Postmineral thrusts occur throughout the length of the vein, but thrusts of intermineral age were noted only in the productive portions. The direction of later movement, indicated by grooving on the later thrust planes, is independent of the changes in strike of the vein and varies from N. 65° E. to N. 85° E., mostly between N. 70° E. and N. 75° E. In the shaft at one point grooving on the hanging-wall slip has a strike of N. 85° E., but directly below on the footwall the grooves strike N. 65° E. The later movements seem to have been most intense in the region near the bend, which is also the region in which the vein faults the serpentine dike, and least intense in the northern part of the mine. In and south of the region of the bend in the vein the effect of pressure is also shown by the development of much "crinkly banding" (pl. 19, B), and reopening of the vein is indicated by the quartz-cemented breccia along the hanging wall and in the vein (pl. 14, A, and fig. 9).

Serpentine occurs on either the footwall or the hanging wall in the productive part of the mine. Several shoots of high-grade ore have been mined. A large amount, said to have been $150,000, was obtained from stopes above the upper tunnel. Here the vein strikes about N. 60° W. and has a footwall of serpentine largely altered to the usual carbonate mariposite aggregate and partly gossanized. The old stopes, now caved, adjoin a well-marked split in the vein. A smaller bunch ($6,000) was obtained near the face of this tunnel at the point where the footwall branch of the vein begins to be again well defined, here with schist footwall and serpentine hanging wall. The second tunnel yielded only one small shoot. Two shoots of high-grade ore have been mined from the lower level. The less productive, which yielded only $6,000, was found on the N. 60° W. segment of the vein about 200 feet from the bend near the shaft, at the point where the footwall serpentine begins. It is also above a distinct split in the vein shown in the level below. The more productive shoot has yielded at least $60,000. Of this about $10,000 came from the large stope extending from the drain level to the intermediate level, with a small amount below the level, and about $50,000 from a much smaller area just above the intermediate level. The zone in which the high-grade ore occurs lies just south of the sharp bend at the shaft and follows the southern contact of serpentine and schist on the footwall of the vein. Just below the best ore on the intermediate level is a well-marked split in the vein, one branch holding the N. 60° W. strike and the other striking about N. 20° W. The high-grade ore is said to have occurred in horizontal bands, and only a small part of the gold is reported to have been associated with coarse arsenopyrite.

The Eldorado mine is one of the few in the district which has yielded ore of mill grade. The body of mill ore occurs near the south end of the drain level, where the vein resumes its N. 20° W. strike, and extends both above and below the drain level. The ore mined from this shoot has a general tenor of $7 to $15 a ton, with patches of higher grade. This ore differs in appearance from the usual type of the district. Sulphides are relatively abundant and carry sufficient gold ($40 to $240 to the ton, averaging about $100) to warrant concentration. The principal sulphide is pyrite, probably of two generations because it occurs both in coarse crystalline aggregates scattered irregularly through the quartz and also, together with a little arsenopyrite, in the irregular "ribbon seams" that cross the quartz of the vein. Small amounts of sphalerite, galena, and tetrahedrite are also present. The gold is only very rarely coarse enough to be visible in the hand specimen. It is usually associated with the pyrite, and the amount of pyrite present gives a fairly reliable indication of the grade of the ore. No isolated spots or bunches of high-grade ore have been found in this ore body.

The two other veins have not yet been found worthy of much development. The Middle vein has been prospected by a tunnel about 400 feet in length. The vein, both walls of which are amphibolite schist, is not well defined and has a strike of about N. 55° W. Near the face the vein appears to split up into stringers which range in strike from north to N. 30° W. The Bullion vein, which is the southern continuation of the Colorado, is well defined, and its outcrop can be traced up the ridge nearly to the south end of the claim. It is developed by a 300-foot tunnel near the creek level and a smaller tunnel in the southern part of the claim. The country rock is amphibolite schist of the Tightner formation. The vein near the portal of the lower tunnel has a strike of N. 40° W., which it holds for about 100 feet. At this point fissuring in a N. 3° W. direction controls the vein to the face. In the upper workings a N. 30° W. strike prevails. The dip is easterly, ranging from 30° to 60°. Postmineral movement is shown in places, with grooving following the same general directions as in the main workings on the Eldorado vein. Near the portal of the lower tunnel grooves of two ages were observed—a younger set striking N. 25° E. superposed on an older set striking N. 50° E. It is possible, however, that this is close enough to the surface for the grooves to be the result of surface movement on the steep hillside.

A minor westward-dipping vein striking N. 20° W. crops out in the gulch in the southwestern part of the Bullion claim and has been prospected by a short tunnel.

YELLOW JACKET (COLORADO) MINE

The property of the Yellow Jacket mine (pl. 55 and sections D–D′, E–E′, F–F′, and G–G′, pl. 10) includes three patented claims, the Yellow Jacket, Colorado, and Colorado Extension, having a total area of 37.7 acres. The group adjoins the Eldorado property on the south at Kanaka Creek and extends northwestward across the point of the ridge between Kanaka Creek and the North Fork, adjoining the Osceola claim near the North Fork. The mine was owned in

1925 by R. W. Gillespie, who was also the owner of the adjoining Osceola mine. It was not being operated at the time of visit.

The production has been small. A little high-grade ore, possibly about $8,000 worth, was obtained in 1924 from the Colorado vein, which in the Yellow Jacket workings has been explored by a 1,000-foot drift and several raises. It is said that about $20,000 was produced many years ago from sluicing on the outcrop of the Yellow Jacket vein. The old workings on the Colorado and Colorado Extension claims are also reported to have yielded a little high-grade ore.

The country rock is chiefly amphibolite schist of the Tightner formation showing the usual variety of amphibolite, amphibolite schist, and slaty schist. A small bed of coarsely crystalline limestone, possibly the same as that cropping out at the portal of the Eldorado tunnel, is cut in the eastern crosscut of the Yellow Jacket tunnel. In places the amphibolite is replaced by quartz along rather indefinite zones parallel to the schistosity. The Tightner formation is cut by several small dikes of serpentine and gabbro roughly parallel to the schistosity. Where serpentine and gabbro are in contact there is an apparent gradation between them.

The Yellow Jacket vein, which forms the northward extension of the Eldorado, follows a thrust fault with a probable displacement of not over 100 feet, though the data for measurement, based on the apparent offset of a small gabbro dike, are not adequate for a close estimate. It is not well defined and is faulted by one or more faults which dip steeply to the northeast.

The Colorado vein, which is essentially continuous with the Bullion vein of the Eldorado mine, to the south, and the Osceola vein, to the north, probably follows a normal fault. The displacement here may be more than 100 feet measured along the dip, but this estimate depends on the apparent offset of a serpentine dike which is cut obliquely by the vein. The small westward-dipping vein appears to lie in the plane of a reverse fault of small throw and faults the fissure followed by the Colorado vein, though the quartz filling of the two is contemporaneous.

The principal workings start from an adit whose portal is at the south end of the Yellow Jacket claim, near Kanaka Creek. Here for about 300 feet a drift follows the Yellow Jacket vein, but it is not as well defined as in its southern continuation, the Eldorado vein, and at its north end it appears to fade out into several different slips with little or no quartz. A crosscut extending 350 feet to the southwest reaches the Colorado vein, which has been followed by a drift for about 1,000 feet, almost the entire length of the Colorado Extension claim. The vein varies somewhat in strike (average N. 20° W.) and dip (average 45° E.) but is much more regular than the Eldorado. In

general the dip is steepest in the upper workings. The maximum width is 10 feet, but over part of the distance quartz is lacking entirely, though a well-marked slip indicates the course of the vein. Post-mineral thrusts within the vein were noted in places. The grooving on the fault planes is very regular, the strike varying only a few degrees to the north and south of west.

The upper tunnels on the Colorado Extension claim show a vein of similar character, though the proportion of the quartz is rather less than in the lower drift. The dip is to the east and is rather steep (54°) in the southern part but flattens greatly (27°) to the north. The uppermost of the three tunnels follows a minor northerly vein which has an average dip of 65° W. and probably belongs to the same group of faults which cut the Sixteen to One vein between the 1,800 and 2,100 foot levels. The middle tunnel shows the major eastward-dipping vein cut off by a shear zone carrying some quartz, which is apparently the continuation of the westward-dipping vein of the upper tunnel. Near this point a small westward-dipping vein joins the major vein without displacement of the vein filling.

The small shoot of high-grade ore mined on the lower level was obtained from the north end of the principal swell in the vein, about 100 feet north of the crosscut. Here the dip (55°) is somewhat steeper than the average. North of this point the quartz is cut off by heavy gouge walls, at a small angle to the strike. A small amount of high-grade ore was also found 100 feet farther north, where quartz comes in again. The high-grade ore was associated with fractured ribbon quartz.

EASTERN CROSS GROUP

The Eastern Cross group (pl. 55) on the south side of Kanaka Creek east of the Eldorado, includes four lode claims—the Eastern Cross, Eastern Cross Extension, Eastern Star, Western Cross—and the Jewel placer claim. The property was owned in 1925 by B. L. Bovee, but no work was being done at the time of visit. According to Mr. Bovee about $3,000 in high-grade ore has been obtained from the veins.

Lack of outcrops on the steep brush-covered hillside prevented detailed study of the surface geology. The principal country rock is the Tightner formation, which is more than usually schistose and nearly everywhere carries abundant quartz stringers parallel to the schistosity. A mass of gabbro, the western prolongation of the great gabbro-serpentine complex that covers much of the southeastern part of the district, crops out in the eastern part of the property. Several small outcrops of serpentine were also found.

There are four veins on the property, known as the Eastern Cross, Eastern Star, Western Cross, and Jewel. The Eastern Cross and Jewel dip to the northeast, the other two to the southwest. The displacement on the Eastern Cross fissure seems to be in the reverse direction but probably does not exceed 50 feet; that of the others is unknown.

The Eastern Cross vein apparently follows the same fissure as the Cedar vein, north of the creek. The only accessible working on this vein is a tunnel near the creek. Other workings higher up the hill are now caved. The tunnel is about 450 feet long and follows the vein for about 350 feet. The vein, which is irregular in detail, has an average strike of N. 65° W. and dips from 30° to 50° and at one place 70° to the northeast. The vein follows a well-defined slip, with gabbro on the hanging wall for most of the distance and schist of the Tightner formation on the footwall. In places quartz is lacking entirely; elsewhere there are small stringers an inch or two wide in the sheared rock. Only for the 50 feet in the western part of the drift is the vein at all well defined; here it ranges in width from 6 inches to 2 feet. Where gabbro is the country rock extensive carbonate alteration of the wall rock has occurred, with the formation of a little mariposite.

Two small reverse faults with steep westerly dips cut the vein at distances of 115 and 75 feet from the face of the tunnel but displace it only a few feet. The unusually steep dip of the vein to the east of the fault, 70° as compared with 50° at the face, may be due to the drag on the fault. Minor thrusts along the plane of the vein were noted in places.

Higher up the hill the vein cuts and displaces slightly a small serpentine dike. It is reported that the caved workings near this point yielded high-grade ore amounting to about $2,500.

The Western Cross vein is exposed in a small cut in the ravine south of the Eastern Cross, just above the Eldorado pipe line. It is traceable on the surface for about 250 feet. Its average strike is N. 25° W. At the cut it consists of a 3-foot zone of much-crushed quartz stringers dipping 45° SW. It is possible that this vein follows the same fissure as the Eastern Star vein, which crops out higher up the hill to the southeast.

The Eastern Star vein has been prospected in several tunnels between altitudes of 4,300 and 4,400 feet on the eastern bank of the same ravine. These are now caved except for one large straight tunnel apparently run for gravel but discontinued. This appears to cut the vein, which is here not well defined. Another tunnel 35 feet in length higher up the hill connects with an old raise from one of the lower workings. Here $520 in high-grade ore was obtained. Between this and the tunnel next below is a small serpentine dike, and at the portal of this tunnel the footwall is completely altered for a few feet to the carbonate-mariposite mixture commonly characteristic of serpentine, though only silicified schist is found in the tunnel itself. The vein here consists of a crushed zone in the

schist 3 feet wide with a varying amount of quartz. The strike is N. 12° W. and the dip 60° W.

The Jewel vein crops out in the western part of the property, close to the Eldorado side line, and is traceable by float to the edge of the Tertiary gravel on Balsam Flat. The only development work consists of a 40-foot tunnel on the ridge about 700 feet northwest of the cabin on Balsam Flat. Here a vein averaging 2 feet in width strikes N. 55° W. and dips 75° NE.

It is said that some very rich angular gravel, a single fragment of which carried $1,100 in gold, was found near Balsam Flat. This is considered by local prospectors to have been derived from the Jewel vein.

RISING SUN MINE

The Rising Sun property (pl. 55) includes four patented claims—the Cedar, Baltimore, Rising Sun, and Monroe—and extends from Kanaka Creek north of the Yellow Jacket northward across the ridge between Kanaka Creek and its North Fork. It was noted as a producer as early as 1881.[88] In 1925 the property was owned by the Mercantile Trust & Deposit Co. and the Safety Trust & Title Guarantee Co., of Baltimore, and was under lease to Frank Nichols, of Alleghany. The reported production is $58,000, all from the Rising Sun vein. The ratio of production to development work can not be given accurately, as the extent of the old workings is unknown. It must be less than $100 to the foot of work on the vein.

The country rock is the Tightner formation, cut here and there by small dikes of serpentine and gabbro, which, however, do not crop out in the vicinity of the veins. The Tightner formation here shows greater lithologic variety than in the region to the west and contains, besides amphibolite schist of the usual type, bands of dark quartz-chlorite schist and hornblende-mica schist. The more massive phases found farther to the south are lacking here. The schist contains abundant small quartz stringers parallel to the schistosity. No data are available on which to base an estimate of the amount or direction of displacement along the fissures followed by the different veins.

The principal vein is known as the Rising Sun and has been developed by two tunnels on the ridge north of Kanaka Creek. The upper tunnel is now inaccessible, but the principal tunnel, at an altitude of 4,470 feet, though caved at both ends, is open for about 500 feet along the drift. The strike is about N. 8° E. and the dip 60°–85° W., averaging about 75°. The vein is irregular in detail, and small splits are numerous, particularly on the hanging-wall side. The principal split is near the point where the crosscut meets the

[88] Burchard, H. C., Report of the Director of the Mint for 1881, p. 98, 1882.

vein. Here two distinct strands are present, separated by a horse 25 feet wide. The vein varies in width and reaches a maximum of 6 feet, though for most of the drift it does not exceed 2 or 2½ feet. Quartz is nowhere entirely lacking, but at a few points it is present only as small stringers in the crushed schist. There is evidence in places of successive reopenings of the vein by minor reverse faults approximately parallel to the vein.

The Rising Sun vein has been one of the most productive of the westward-dipping veins. The ore mined was said to have been largely of mill grade, with small but fairly numerous patches of high grade. There were two ore shoots, one in the region where the vein splits and reunites and the other in the southern part of the tunnel, now caved. At the former there are sharp but minor changes in strike and dip and an unusual number of quartz stringers and minor gouge walls in the wall rocks.

The tunnel on the Cedar claim follows for about 150 feet a small vein which is the western extension of the Eastern Star vein. This is faulted by a westward-dipping vein parallel to the Rising Sun but 900 feet farther east. The amount of displacement is unknown, though if the slip with northwesterly dip at the face of the tunnel represents the same fissure, it can not be more than a few feet. What may be the same northerly vein as that which faults the Cedar is exposed in small prospect pits east of the portal of the Rising Sun tunnel.

IROQUOIS PROSPECT

The Iroquois prospect (pl. 55) is on the north side of Kanaka Creek about 200 feet northeast of the Cedar tunnel. A small vein with a strike of N. 60° W. and a dip of 65° NE. has been prospected by a tunnel about 100 feet in length. At 60 feet from the portal there is a hanging-wall branch with northerly strike and steep easterly dip.

DREADNAUGHT MINE

The Dreadnaught property, which includes the Dreadnaught and Seneca claims, is on the north bank of Kanaka Creek about three-fourths of a mile east of Alleghany. It has been idle for many years, but in 1928 was under bond to Pete Fleurers, of Alleghany. The total output according to MacBoyle[89] is between $50,000 and $100,000. The mine was in operation at least as early as 1870,[90] and the last production recorded in Mineral Resources was made in 1893.

There are several tunnels on the property, only one of which is now in part accessible. (See fig. 37.)

The country rock is amphibolite schist of the Tightner formation, which is highly sheared at small angles to the original schistosity. These shear planes commonly strike a few degrees west of north and dip steeply to the west, though a few with more gentle easterly dips were seen. Here and there discontinuous quartz seams are present along these shears. The vein, though irregular in detail, possibly owing to the influence of these northwesterly shears, strikes nearly north and dips about 60° W. Near the end of the

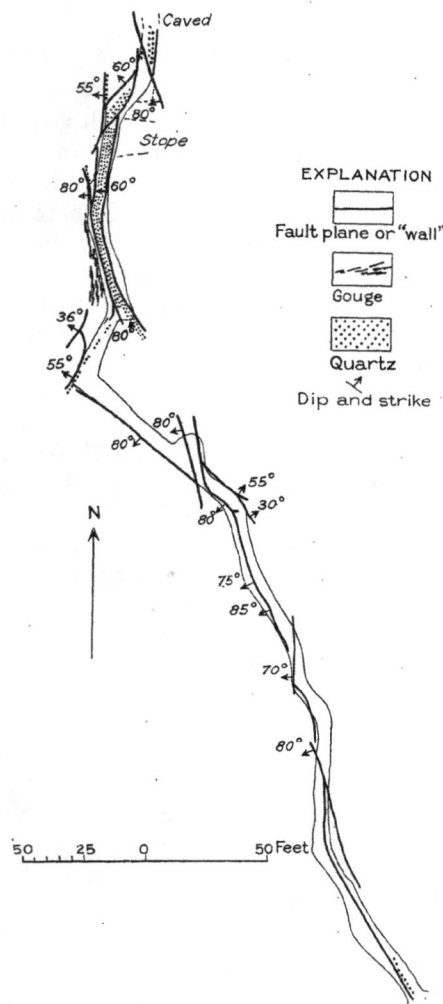

FIGURE 37.—Plan of Dreadnaught tunnel, from compass and tape survey

accessible portion of the drift there is a marked split in the vein, and just to the south of this a small stope from which, according to Mr. Fleurers, $20,000 in high-grade ore was taken.

A similar parallel vein crops out on the Seneca claim and has been prospected by a short tunnel.

BELMONT PROSPECT

The Belmont and Annex claims, owned by Pete Fleurers, lie between the Rising Sun and the Dreadnaught. Three tunnels on the Belmont claim prospect

[89] MacBoyle, Errol, op. cit., p. 83.
[90] Min. and Sci. Press, vol. 20, p. 212, Apr. 2, 1870.

a well-defined vein, which strikes on the average N. 73° W. and dips 50°–70° SW. The longest of these tunnels is 300 feet long. Coarse pyrite is found in the vein here and there but is not common. In places a little gold has been obtained from quartz showing irregular ribbon structure. The total production has not exceeded a few hundred dollars.

Other veins have been prospected by cuts and short tunnels.

A vein with a strike of a few degrees north of west and a northeasterly dip may follow the same fissure as the Docile vein and is apparently cut off by the Belmont. There are also two veins with northerly strike and steep westerly dip. The western one crops out 100 feet south of the Belmont vein and may end against that vein. Another outcrop 700 feet to the south on the trail at the 4,400-foot contour is probably the same vein. The other northerly vein also crops out south of the Belmont but may limit the southern extension of the Belmont and continue northward to the Docile. It is traceable on the surface for a distance of 400 feet, and what may be the same vein has been prospected by a tunnel on the north bank of Kanaka Creek, 1,300 feet south of its most northerly outcrop.

GOLDEN KING MINE

The Golden King mine is in the valley of Kanaka Creek a mile east of Alleghany. The property includes two claims—the Golden King, crossing the creek, and the Golden Queen, on the southern bank. The mine has long been idle and the workings are now inaccessible. It is reported to have produced over $100,000, mostly from ore of the mill grade. According to MacBoyle,[91] however, the production from 1890 to 1895 was $250,000.

At the outcrop near the creek the strike is due north and the dip 80° W. The position of the tunnels on the north side of the creek also indicates a steep westerly dip. Both walls are here in amphibolite schist of the Tightner formation, but the vein is not far from the serpentine contact, and the presence of gossan on the dump suggests that serpentine was encountered underground.

A report by George F. Taylor contains information derived from an old miner familiar with the workings in 1895, when the mine was last worked, and the following statement is condensed from this information:

The ledge was first developed by a shaft sunk vertically in the footwall (possibly mistake for hanging wall; see below). A 100-foot drift (level No. 1) north from the shaft found no ore. Level No. 2, the drain tunnel, was driven north from a point just above the creek and considerable ore stoped from above this level. (According to MacBoyle[92] this drift is 800 feet in length, and the ore stoped yielded $25 a ton.) The vertical shaft crosses the dip of the ledge[93] at the drain tunnel and was continued down 98 feet to the No. 3 level. Here a 36-foot crosscut encountered a 10-foot ledge of barren quartz. A long drift to the south found no ore. To the north a split was encountered and the drift was continued along the hanging-wall branch, though it was only a seam. A large area of ground between the No. 3 and No. 2 levels was stoped, and the stope on the footwall side, beyond this split, had a width of 16 feet of $10 ore and still showed a good width when the mine was closed down.

The No. 4 level was 98 feet lower, and the vein was reached by a crosscut 126 feet long. (MacBoyle states that this level was driven 600 feet northwest from the shaft.) The vein here was found to be only 18 inches wide, though the ore was of good value. A short drift was driven south, but the vein pinched to a few inches in width. In driving north the ledge widened to 4 feet, and the raise to the No. 3 level opened up 9 feet of $20 ore. A large amount of this has been stoped, but when the mine closed there was ore in the breast of the stope, just below No. 3 level, in the north face of No. 4, and along the bottom of the drift. From this level the vein was followed by an incline.

Twenty feet below the floor of No. 4 the ore gave out entirely, but at 50 feet a 12-inch ledge of solid sulphurets assaying $175 a ton was encountered. Then quartz came in successive layers till at 110 feet, where a station was cut, the quartz was 4½ feet wide, but sandy and with a flood of water, and very unstable and broken ground. No. 5 level was driven north and south in this but showed quartz of no value, leached, pulverized, and with water-deposited iron pyrites. A careful survey showed that this broken ground was caused by the proximity of the "Tight Squeeze" vein, less than 25 feet to the west, and the intention was to crosscut to it rather than take the great expense of increasing the plant for further sinking.

DOCILE MINE

The Docile property consists of two patented claims, the Docile and Loosner, extending up the north bank of Kanaka Creek about a mile east of Alleghany. The mine has been idle since 1894. It has recently been taken under option by Pete Fleurers, of Alleghany, and plans are being made to reopen it.

According to MacBoyle[94] the total production was about $100,000; some reports give $200,000. As far as can be determined from the old maps the total of drifts and raises on the principal vein may have exceeded 700 feet, which on the basis of a production of $100,000 would give a production ratio for the development on this vein of about $140 a foot. The

[91] MacBoyle, Errol, Mines and mineral resources of Sierra County, p. 90, California State Min. Bur., 1920.

[92] MacBoyle, Errol, op. cit., p. 90.
[93] Apparently inconsistent with first statement or implies a change in dip of the vein.—H. G. F.
[94] MacBoyle, Errol, op. cit., p. 83.

country rock consists of amphibolite schist of the Tightner formation. A large mass of serpentine lies immediately east of the workings, and there is a small

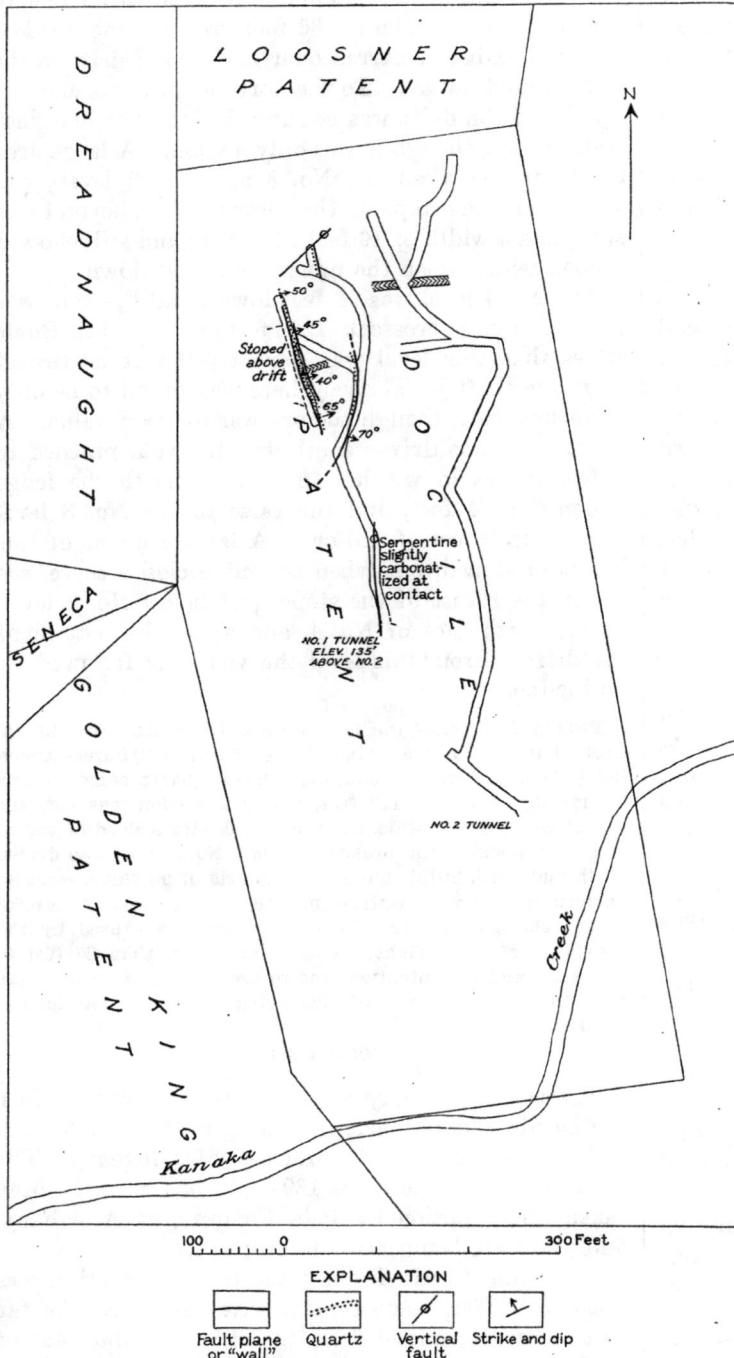

FIGURE 38.—Map of workings, Docile mine. Lower tunnel from map by G. F. and R. F. Taylor

outcrop of serpentine in the ravine west of the tunnel, apparently trending about across the course of the outcrop of the principal vein. On the surface the trace of the vein outcrops can be followed on a nearly

west-northwest course for about 200 feet from the ser pentine contact and doubtfully for about 500 feet to the Belmont property. Here a vein with similar strike continues for about 300 feet to the Belmont vein. At the outcrop near the serpentine the dip is 60° N. A vein with northerly strike and easterly dip south of the principal vein crops out 100 feet east of the serpentine, on the ridge above the tunnel, but does not cross the northward-dipping vein. There are also outcrops of quartz close to the serpentine contact between the two tunnels. The amount and direction of displacement along the veins are unknown.

There are two tunnels, connected by a raise, the lower near the creek at an altitude of 4,020 feet and the upper on the ridge to the north at about 4,240 feet. Some work has been done from a shaft below the lower tunnel. Only the upper tunnel (fig. 38) was accessible at the time of visit. Here the principal vein has been followed for about 150 feet. The strike, N. 29° W., is much more northerly than on the surface. The dip is about 40° NE. where the quartz is thickest but is steeper at the ends of the drift, where the quartz pinches to stringers. A nearly parallel but steeper vein containing at its maximum only a few inches of quartz has also been explored. This is cut off at the north end of the drift by a vertical fault with northeasterly strike. The serpentine contact near the portal of the tunnel dips vertically, but the average dip below the tunnel level must be steep to the east.

It is not known whether all the production came from the stope in the upper tunnel. The discordance in strike of the vein between the tunnel and outcrop suggests a bend toward the serpentine contact similar to that of the Oriental vein, and it is possible that this bend controlled the position of the high-grade ore. Although the vein pinches to the north in the drift, further exploration in this direction might be desirable.

CONTINENTAL MINE

The Continental claim, owned by J. A. Casserly, adjoins the Loosner claim of the Docile group on the north. There are several tunnels on it, now caved. It is said that some production was made about 1899. The vein appears to strike N. 40° E., but the dip is unknown, and consequently the vein is not shown on the projection (pl. 8).

MAMMOTH SPRINGS MINE

The Mammoth Springs mine consists of three claims, the Contact, Contact No. 1, and Contact No. 2, which adjoin the Kinselbach property on the north and extend northward from the vicinity of Little Kanaka Creek to the ridge south of Oregon Creek, covering a part of the Mammoth Springs placer claim. The production has probably been very small.

At this mine, as at the Kinselbach, the lode follows the fault that limits the Blue Canyon formation. Here, however, the rocks on the west are gabbro and serpentine. The gabbro is more basic than usual, and there is no sharp boundary between it and the serpentine. The Tightner formation crops out a short distance to the west of the mine. Development work consists of two tunnels, the lower (fig. 39) about 750 feet and the upper 150 feet in length.

The lode lies along the fault contact, which in the lower tunnel strikes a few degrees west of north and has a dip ranging from 80° W. to vertical, but the positions of the upper and lower tunnels show that the average dip is to the east, though nearly vertical.

The Blue Canyon formation here consists of dark fissile slate varied by a few beds, as much as 6 inches thick, of light-gray sandy slate. The average strike is N. 20° W. and the dip 80° W. The slate is crushed and sheared close to the contact, but a few feet away the only effect of alteration is the presence of flakes of pyrite on the cleavage surfaces. The gabbro, on the other hand, is altered, though irregularly, to crushed talcose material containing lenses of carbonate. At one place small irregular veinlets of cross-fiber amphibolite asbestos, 6 inches in maximum width, are crossed by irregular quartz stringers, and the amphibolite is replaced by quartz. In spite of the intense alteration of the gabbro the usual mariposite-carbonate alteration is lacking. Quartz, however, is almost completely lacking, and the gold obtained is said to have been found in thin flakes on the cleavage surfaces of the dark slate.

KINSELBACH (ACME) MINE

The Kinselbach mine, also known as the Acme, is in the extreme eastern part of the district, 1½ miles east of Alleghany. The property comprises four unpatented claims extending from Little Kanaka Creek across the ridge south of Kanaka Creek and is owned by Langdon Smith, of Forest, and associates. Production from small pockets on the surface is said to have exceeded $8,000. The principal development work consists of two tunnels running north and south from Kanaka Creek. The northern tunnel, now caved, is 165 feet in length and the southern one 225 feet.

The lode follows the fault that forms the western boundary of the Blue Canyon formation. At the mine the rock west of the fault belongs to the Tightner formation. As exposed in the tunnel the fault is marked by a heavy gouge at the contact and parallel fault planes a few feet to the east, within the amphibolite schist of the Tightner formation. The fault here strikes N. 18°–26° W. and dips 74° E. to vertical. There has been no mineralization at the fault contact, but small quartz seams occur in the crushed schist parallel to the fault. In places near the quartz the schist is altered to carbonate with a little mariposite.

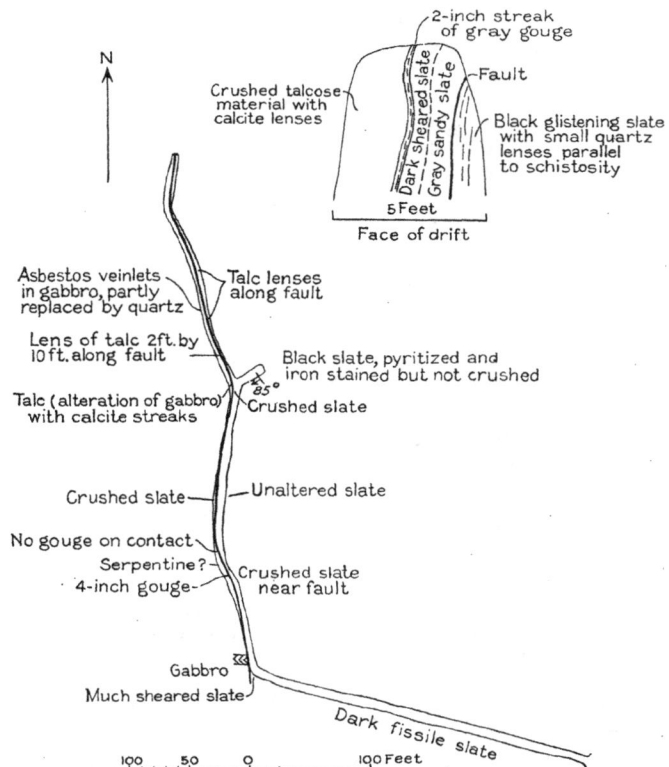

FIGURE 39.—Sketch map of lower tunnel, Mammoth Springs mine

The lode has been traced up the ridge to the lava contact in a series of small cuts, now caved. South of the lava were found outcrops that indicate veining within the Blue Canyon slate.

MINES IN THE DRAINAGE BASIN OF THE MIDDLE FORK OF THE YUBA RIVER

MOUNTAIN VIEW PROSPECT

The Mountain view prospect is in the southern part of the district on the south side of Lafayette Ridge, just above the Moores Flat trail, about 3,500 feet west of the Irelan mine.

The veins are in gabbro, though close to the serpentine contact. The workings consist for the most part of shallow surface cuts, but one tunnel about 140 feet in length (fig. 40) has developed the junction of two veins. One of these, which strikes N. 22° W. and dips about 50° SW., is cut off by a vein which at the junction has a northerly strike and dips 82° W. and near the face strikes N. 5°–10° E. and dips 55° W.

The flatter vein has a maximum thickness of 1½ feet. The footwall is free but carries no gouge, and the hanging wall is frozen. The steeper vein ranges from a mere stringer of quartz against the hanging wall to a vein over 3 feet thick. There is well-marked shearing, particularly along the hanging wall. The flatter vein is not cut off abruptly by the steeper one

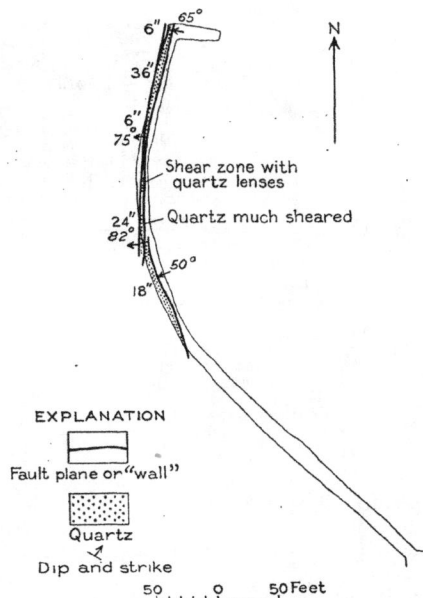

FIGURE 40.—Sketch map of Mountain View tunnel (pacing and compass). Upright figures show thickness of quartz

but bends sharply toward it in both strike and dip, the quartz of the flatter vein apparently forming the footwall strand and grading into the steeper vein.

MARYLAND PROSPECT

The Maryland prospect, on the steep hillside west of Minnesota Flat, is about a quarter of a mile west-southwest of the Irelan mine.

The lower workings, consisting of surface cuts and pits and a 20-foot tunnel, prospect a poorly defined vertical northerly vein which apparently faults the contact of the Kanaka and Relief formations. The horizontal displacement on the surface is less than 100 feet to the south on the east side, but as the contact between the two formations is gradational the amount of displacement is uncertain. This vein has a maximum thickness of about a foot and is much sheared.

The upper tunnel, near the trail at an altitude of 4,000 feet, is about 100 feet long. The vein exposed here strikes N. 20°–37° W. and dips 37° W. It con-

sists of a series of lenses and stringers of quartz about 1½ feet in maximum width, in much sheared schist of the Kanaka formation. At an average distance of about 2 feet below the quartz is a strongly marked fault plane separating the shear zone that carries the quartz from the less altered schist to the east. On the hanging-wall side the boundary of the shear zone is less sharp. The quartz carries a little pyrite, and the schist, which near the quartz is altered to talc or sericite, is pyritized in places.

ORIFLAMME MINE

The Oriflamme mine is on the west side of Minnesota Ridge about 2,000 feet west of the Irelan mine.

Development work consists of a crosscut tunnel 440 feet in length that cuts three veins (fig. 41), on one of which there is a drift now accessible for about 150 feet. An upper tunnel close under the lava capping is caved. No production is known to have been made, though promising prospects are said to have been obtained.

The country rock belongs to the Kanaka formation and is chiefly green schist, regarded as altered tuff, with bands of characteristic black slate. At the portal is a more quartzitic slate that is considered the base of the Relief formation. Gabbro was identified at one point in the footwall of the eastern vein. No serpentine has been encountered in the workings.

At 90 feet from the portal a small vertical vein striking N. 10° W. has been cut and explored by a drift for a few feet. Prospects showing arsenopyrite and coarse gold in the pan are said to have been obtained. This is probably the same vein that has been prospected by shallow surface workings near the point where the trail crosses the ridge about 600 feet to the south. The next vein, 380 feet from the portal, strikes N. 65° W. and dips 50° NE. There has been a small amount of stoping just north of the crosscut. The third and easternmost vein is explored by a drift, of which about 150 feet was accessible at the time of visit. The vein near the crosscut strikes N. 64° W., but at the north end of the open portion of the drift the strike changes to N. 15° W. For most of the distance there is a thickness of 2 to 3 feet of quartz; near the north end this fades to small stringers. The vein shows banding, quartz strands 6 inches to 2 feet in width being separated by ribbons of slaty material. Adjacent to and along these ribbons are small crystals of pyrite and arsenopyrite.

The projection (pl. 8) suggests that the two northeastward-dipping veins are not, as has been supposed, the northward continuations of the Irelan veins, though they probably belong to the same system. However, the small amount of data for dips of these veins and the considerable distance (500 feet) above the datum plane makes their correlation doubtful.

There is little evidence of postmineral shearing except along the walls of the vertical vein. The eastern vein has on the whole a well-defined hanging wall but less sharp footwall. In places small vertical quartz stringers enter the vein from the footwall side. No evidence of marked replacement of the wall rocks either by carbonates or by quartz was seen along any of the veins.

IRELAN MINE

The Irelan mine is on the east side of Minnesota Flat, south of Lafayette Ridge, nearly 2 miles southeast of Alleghany. The property consists of a group of claims extending southeastward across Minnesota Flat and down the ridge west of Jackass Ravine.

Work is reported to have begun as early as 1860,[95] and the mine has been worked at intervals under different owners until 1924, when it was closed down by its present owner, the Durston Mining Co., of San Francisco. The production is said to have been in excess of $300,000, largely from shallow surface workings and stopes above the adit level. The workings are in part inaccessible, and the only available maps are incomplete, so no close estimate of the ratio of production to development can be made. It can not be over $100 a foot, and may be much less.

The mine (pl. 57) is developed by an adit level near the mill at an altitude of 4,120 feet. From this a 200-foot winze has been sunk, and some drifting has been done from the winze, but the workings below the adit level were not accessible. Another tunnel 1,500 feet to the southeast at an altitude of 3,700 feet followed the vein for nearly 1,000 feet. This is now caved, and the authors are indebted to a map made by Mr. W. R. Simkins and reproduced in Plate 57 for information concerning this part of the mine.

The country rocks include various phases of the Kanaka formation complexly intruded by dikes of

schistose gabbro. No serpentine was found in the vicinity of the mine, the nearest known mass being a quarter of a mile to the northeast, near the head of Jackass Ravine. The intense shearing and alteration close to the vein have made it impossible to identify the rocks throughout the greater part of the mine workings, but apparently both gabbro and the slate

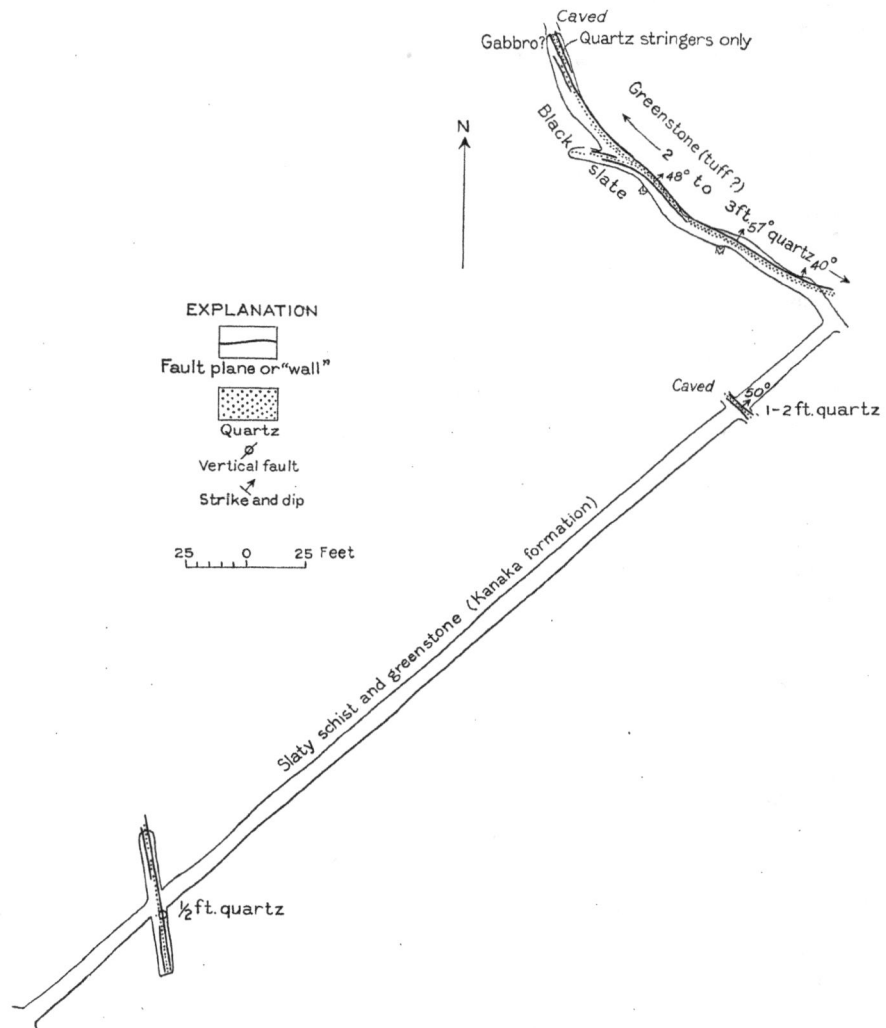

FIGURE 41.—Sketch map of Oriflamme tunnel

and tuff of the Kanaka formation are present on both walls of both veins. On the surface, south of the area covered by gravel, there seems to be dominant gabbro between the two veins, with the Kanaka formation principally on the hanging-wall side of the eastern vein and the footwall side of the western vein. These relations suggest reverse movement on both veins, but the exposures are so poor that it is impossible to be certain.

Two veins have been developed in the upper workings. The western vein is poorly defined in the tunnel, though traceable for a considerable distance on the surface. The eastern vein, which is fairly well

[95] It is not known whether the Copper Hill mine reported by Trask (Report on the geology of northern and southern California, p. 63, Sacramento, 1856) as having a stamp mill in operation in 1855 is the same as the Irelan. The first reference seen to the mine under its present name is in the Mining and Scientific Press for Apr. 12, 1879, in which it is noted that the tunnel of the Irelan mine (Sam Irelan, owner) is in 550 feet.

defined throughout the accessible portion of the tunnel level, has yielded all the gold so far obtained. Outcrops which indicate a third vein are found on the ridge between the two tunnels.

The average strike of the major vein is about N. 60° W. in the southern part of the main tunnel but changes sharply to N. 30° W. at a distance of about 500 feet from the crosscut. The vein maintains this course for another 500 feet, to a point near the face, where it turns toward its original direction. The first change in strike approximately corresponds to the projected junction with the western vein, and a footwall branch of the eastern vein at the beginning of the bend may represent this junction. There is also another footwall stringer 150 feet to the southeast. The map of the lower tunnel shows a vein holding a fairly constant strike of about N. 40° W. for most of its distance but turning sharply to a N. 75° W. course near the face. At about 500 feet from the portal two branch veins diverge toward the north, one from the footwall and one from the hanging wall.

The dip is 30°–70° NE. It is flatter than the average in the upper tunnel in the section of the vein that strikes N. 30° W. and steeper than the average in the lower tunnel. Dips of 50° to 60° were observed on the western vein in the lower tunnel, but as the distance between the two veins is greater on the surface than in the tunnel the western vein above the tunnel level must have an average flatter dip than the eastern.

Quartz is lacking in places, but over the total of accessible drifts on the main level quartz is fairly continuous. The maximum width of 5 feet is reached in the small lens just north of the crosscut in the upper tunnel and in the bottom of the winze.

Conspicuous carbonate alteration of the wall rock seems to have been confined to the part of the vein that has been productive. Mariposite was observed, but, as would be expected from the lack of serpentine, is rare.

Postmineral movement along the veins has produced only minor effects. Here and there are wall-rock slips with a little gouge, but these are not well defined nor continuous, nor was any evidence seen of postmineral thrusting along the vein, though the presence of quartz-cemented quartz breccia indicates movement prior to the completion of vein filling. A small transverse fault of postmineral age cuts the western vein but apparently does not affect the eastern, and a similar fault crossing the eastern vein is shown in Simkins's map of the lower tunnel. (See pl. 57.)

There has been some stoping on the lens of quartz just north of the crosscut, but the portion of the vein that has yielded nearly all the production is that north of the probable junction with the eastern vein, in which the strike is about N. 30° W. instead of the usual N. 60° W. and the dip is flatter than the average.

So far as known no gold was obtained in the lower tunnel. In the zone that yielded high-grade ore the vein has been stoped above the tunnel level for a distance of 400 feet and below the level for at least 250 feet.

The high-grade pockets are said to have been found where the quartz was thickest, and there is a possible correlation of high-grade bunches with the stringers branching from the vein on the hanging-wall side. The largest high-grade bunch was 15 feet in drift length by 25 feet on a steep southerly pitch. The richest shoot, from which picked specimens are said to have assayed as high as $300,000 a ton, was a narrow streak crossing the vein diagonally to the north from footwall to hanging wall. The ore on the whole, however, was not of the sensational richness sometimes found in the district.

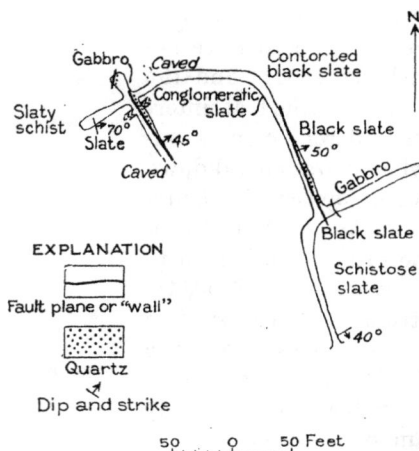

FIGURE 42.—Sketch map of upper tunnel, Arcade mine

The high-grade ore seems to have been largely associated with arsenopyrite, although serpentine is lacking. Specimens of high-grade ore show crushing of the arsenopyrite prior to the introduction of the gold. (See pl. 41, B.)

ARCADE MINE

The Arcade mine, near the head of Jackass Ravine, is reported to have produced a small amount of high-grade ore, probably not more than a few hundred dollars' worth, at most. The mine is developed by two tunnels, the upper (fig. 42) just below the gravel and about 1,200 feet north of the Irelan mine, and the lower (fig. 43) in Jackass Ravine near the 4,000-foot contour.

In the upper tunnel the country rock is dark slate and schist of the Kanaka formation cut by gabbro. The lower tunnel shows no Kanaka formation, except possibly at the extreme west end of the workings, but is chiefly within the amphibolite schist of the Tightner formation, which shows in places slaty phases and is also cut by dikes of gabbro and diorite.

The lower tunnel develops two rather weak parallel veins which have a northwesterly strike and northeasterly dip. A third vein with eastward dip and strike of a few degrees west of north seems to unite with the eastern of the two northwesterly veins at the adit. The long southerly crosscut above the raise crosses a steeply dipping gouge, which may mark the boundary between the Kanaka and Tightner formations, inasmuch as the rock to the west is more slaty, but this could not be certainly determined.

Only one distinct vein is present in the upper tunnel, though a well-marked eastward-dipping slip along which there is a little quartz in places may represent the eastern vein of the lower tunnel. The distinct vein, which has yielded the only production from the mine, has been developed by a drift of which about 110 feet is now open and a 50-foot winze from which high-grade ore is said to have been obtained. The quartz has a maximum width of about a foot at the winze and pinches out entirely a short distance to the south.

Evidence of postmineral movement was seen, particularly in the lower tunnel, where the eastern vein seems to be displaced by a fault at a very small angle to the vein.

COOPER PROSPECT

The Cooper prospect, in Jackass Ravine north of the Arcade mine, is developed by two tunnels, of which the upper and more extensive is caved. In the lower tunnel 30 feet of drifting has been done on a vein in gabbro, which strikes N. 20° W. and dips 50°–60° W.

The dump of the upper tunnel shows white vuggy quartz, quartz with spots of iron oxide, containing also little veinlets of chalcedony, and the peculiar " ghost quartz " derived from the replacement of brecciated country rock. Fragments of gossanized serpentine indicate that exploration extended as far as the serpentine mass to the north.

PLUMBAGO MINE

The property of the Plumbago mine includes eight patented claims—the Hope, Crafts, Clute, Gold Beater, Standard, Standard Extension, Enterprise, and Marion—extending southeastward across Lafayette Ridge from the Eldorado property to the claims of the Independence group at Hope Ravine. There is also a patented mill site at the Middle Fork of the

Yuba River near the mouth of Wolf Creek, on which is situated the power house. Water for power is obtained from the Middle Fork and carried through a 6,000-foot pipe line.

The mine has been owned for many years by the Croesus Gold Mining & Milling Co. It was recently worked under bond by the Ante-Up Mining Co., in which the North Star Mine of Grass Valley had a controlling interest, but in 1925 work, except by a few lessees, was abandoned and in 1927 the option was given up. The mine was idle in 1928.

The vein crops out along the southward extension of Lafayette Ridge, which separates the valleys of

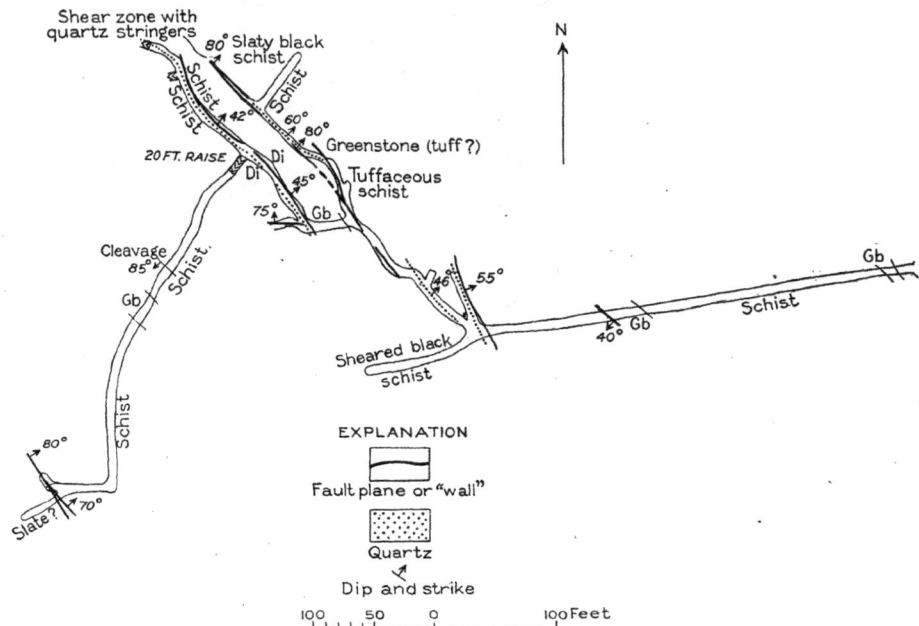

FIGURE 43.—Sketch map of lower tunnel, Arcade mine

Jackass and Hope Ravines. The camp, at the site of the now inaccessible No. 3 tunnel, is on the west bank of Hope Ravine at an altitude of 4,200 feet. The mill and present working tunnel, the No. 4, is farther down the same slope at an altitude of 3,700 feet, a short distance above the creek. Although the camp is less than 2 miles southeast of Alleghany in an air line, the rugged topography makes the distance by the road, which skirts the head of Kanaka Creek, over 17 miles.

The first workings, dating from the early days of the camp, were along the outcrop, and it is said that a large production was obtained by shallow mining and surface sluicing. Later the claims were consolidated and mining was carried on from tunnels on the east slope of the ridge. The Croesus company began operations about 1899 and established the present No. 4 tunnel, from which all recent work has been carried on. Operations have now extended to the No. 10 level, at a depth of about 800 feet down the dip from the No. 4

tunnel level. The total vertical depth of exploration between the lowest level and the outcrop on the ridge is about 1,200 feet. (See pl. 58, fig. 44, and sec. J–J', pl. 10.)

The production since 1896 has been slightly greater than $2,000,000; the production of which there is certain record exceeds $2,500,000; and the total, if the stories of the immense richness of the outcrop work-

the other mines, and for the last 20 years the ratio of production to tonnage mined has been about $8.40 a ton.

The principal country rock is gabbro, which is on the whole less schistose than usual. On the outcrop along the ridge the footwall rock of the vein is much weathered hornblende schist of the Tightner formation. Similar schist is found on the footwall side at least as far as the No. 2 tunnel, 200 feet below the

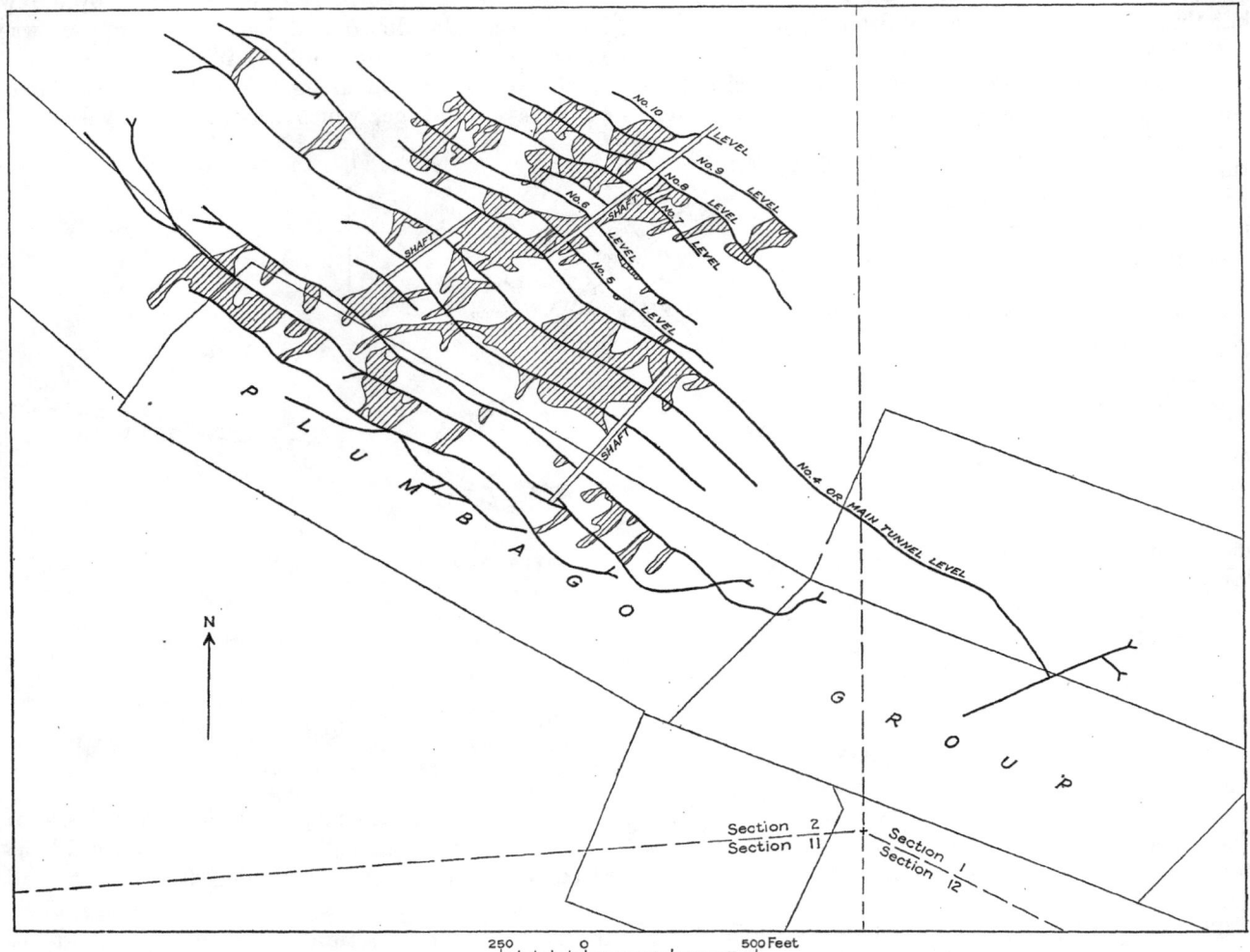

FIGURE 44.—Map of Plumbago mine

ings and surface sluicing are well founded, may be greatly in excess of this. It is estimated that prior to 1875 the production from the mine amounted to $100,-000, all obtained from hand mortars.[96] On the basis of an estimated production of $2,500,000 the ratio of production to development work, approximately 20,000 feet of drifts and 9,000 feet of raises and winzes, is about $86 a foot. Owing to the fact that much of the quartz between the high-grade bunches is of a high enough tenor to constitute mill rock, there has been a relatively greater amount of stoping than in most of

outcrop (pl. 58 and sec. J–J', pl. 10), indicating reverse faulting along the vein. Two masses of serpentine crop out north and south of the vein. The eastern mass is also found underground on the hanging wall of the vein at the tenth level near the shaft. The vein appears to pinch out before meeting the serpentine on its strike on the surface near the No. 4 tunnel. The western mass is found in the western part of the workings, where it occurs on the footwall of the vein at least as far as the No. 4 tunnel and has probably caused the fraying out of the vein in this direction.

---

[96] Min. and Sci. Press, vol. 31, p. 378, 1875.

On the levels above the A drift the vein is cut by the basalt plug described in detail on page 18. Apparently the basalt enters from the footwall and follows the vein from the B drift up to at least the No. 1 tunnel, where according to the map it cuts out both branches of the vein. The vein is not in any way distorted by the intrusion, nor has there been any apparent shearing or faulting at its margin. A section of the vein from 60 to 150 feet in length on the different levels seems to have been replaced by the basalt.

The vein is well defined throughout most of the area in which it has been developed. In the productive part of the mine it has been followed for a distance of over 2,000 feet along the strike, and only here and there for short distances is quartz lacking. In the southeastern part of the No. 4 tunnel the vein fades out to a small shear zone in the gabbro and does not reach the eastern serpentine mass. There is one large lenticular swelling where the vein reaches a thickness of over 30 feet, but as a rule the width of the quartz ranges between 2 and 4 feet. The "walls" are usually well defined and either footwall or hanging wall is commonly separated from the quartz by a greater or less thickness of crushed and altered gabbro.

The average strike of the vein is N. 49° W. and the dip about 38° NE. There are, however, considerable variations in both dip and strike. A well-marked outward bend is well shown on No. 8 and No. 9 levels, where it is associated with a flatter dip and greatly increased thickness of the quartz. The same bend is observable, though less pronounced, on the upper levels as far as the A drift. The rake is to the southeast, down the dip. Bends in the opposite direction—that is, inward from the hanging wall—are indicated on each side of this outward bend but are less prominent.

Along the southern part of the outcrop there are two veins, close together and parallel. The map of the No. 1 tunnel shows a drift to the west of the main workings, apparently on the parallel vein. As no crosscutting has been done below this level, it is not known whether this parallel vein is persistent in depth. There are several distinct splits in the vein. On the No. 1 tunnel there is a footwall branch whose junction with the main vein pitches to the east and is also found on the No. 2 tunnel just east of the No. 1 main raise. A well-marked branch with easterly pitch occurs at the west ends of the No. 2 and No. 3 tunnels. In the western part of the No. 4 tunnel there is a well-defined split in the vein, not identified in the lower levels.

The vein quartz contains fragments of gabbro embayed by quartz and irregular patches whose slightly darker color suggests complete replacement of gabbro fragments, though there is nothing in the texture of the quartz to confirm this suggestion. Elsewhere there is the usual banded structure, with the strands of quartz separated either by septa of altered gabbro or by mere dark lines of later minerals, which may in part be the result of thrusting within the vein.

Later quartz veins transverse to the major structure are prominent in places, particularly in the productive portion of the mine. Along the hanging wall of the vein there is in places breccia containing abundant angular fragments of wall rock altered to carbonate, mariposite, and vein quartz, cemented by fine-grained quartz which is dark from the presence of numerous minute inclusions of foreign material.

Altered wall rock of the carbonate type was observed in many places. As little work has been done outside the vein, it is impossible to say how extensive such alteration may have been. Mariposite is nearly everywhere present in small amount in the altered wall rock. As noted above, the vein, though within the gabbro, is no where far from the serpentine, a fact which may account for the widespread presence of the mariposite. In places inclusions in the vein are unaltered or only partly silicified where inclosed only in the older quartz but completely changed to the carbonate-mariposite aggregate where crossed by veinlets of later quartz, indicating that in part at least the carbonatization was associated with the later quartz.

The metallic constituents of the vein include arsenopyrite, pyrite, galena, and very rare tetrahedrite, as well as gold. The high-grade ore seems to be more closely associated with the coarse arsenopyrite of the vein than in most of the other mines, except perhaps the Oriental. In places the gabbro close to the vein contains small arsenopyrite crystals. The arsenopyrite in the vein is commonly in large bladed clusters. These radiate from nuclei on the hanging wall but commonly are separated from the wall by a film of quartz and are veined by quartz. Pyrite is fairly widespread in the vein and in small crystals in the adjoining country rock. The "mill rock" usually contains abundant pyrite, though pyrite is also found in barren portions of the vein. Galena and tetrahedrite are found only in small amount and only in or near the high-grade bunches. The mine is one of the few in the district where the concentrates contain enough gold to warrant their being saved. The proportion is small, but the average tenor is about $150 a ton.

A mineral not found elsewhere in the district is the so-called "leather gouge," consisting of tough felt-like sheets of palygorskite. This mineral occurs in the neighborhood of the outward bend of the vein as the filling of small seams, suggesting gash veins, extending downward from the hanging wall for distances of a few inches to a foot, and less commonly along joints near and parallel to the hanging wall. These veins are nowhere over a tenth of an inch in thickness and are widest near the top, but sheets of this tough leathery material as much as a foot in diameter can be obtained.

Postmineral thrusts are prominent in places, particularly in the region between the No. 1 and No. 2 raises and in the lower levels, which have also been the most productive parts of the mine.

The high-grade ore of the Plumbago mine occurs over a large area, but the individual bunches have generally been small. For the most part the mill rock seems to have been obtained where small bunches of high-grade ore were spaced closely enough to make it profitable to mine considerable areas of the vein, but elsewhere, particularly in the upper levels, there was mill rock without high-grade ore, of the type mined in the Eldorado mine. Also the quartz outside the high-grade shoots contains more gold than usual in the district, so that stoping large areas of the vein has been, if not profitable, at least less expensive than in other mines.

Possible controls for the high-grade ore are less evident in the Plumbago than in other mines. Apparently some high-grade ore has been found wherever the vein carries considerable quartz. Plate 58 indicates two rather ill-defined productive zones pitching gently to the southeast. In the barren zone between them the quartz pinches greatly and is in places entirely lacking.

Although the vein at its north end has a serpentine footwall, the serpentine contact as far as prospected has not proved highly productive except for one high-grade stope above the No. 4 tunnel. The vein itself, however, crosses the space between the two serpentine masses and is nowhere at a great distance from the serpentine. The high-grade zones in the upper workings, above the No. 3 tunnel, appear to lie close to the junction of minor veins that branch off on the footwall side.

Should further exploratory work be undertaken attention might well be given to the portions of the vein close to the serpentine masses—extensions of the levels below the No. 4 tunnel northwestward toward the western serpentine mass and development at depth of the contact of the vein with the eastern serpentine. Exploration to determine the possible existence in the lower levels of the footwall split or splits indicated in the southeastern part of the upper tunnels might also be desirable.

GERMAN BAR MINE

The German Bar mine is on the south bank of the Middle Fork of the Yuba River, about 500 feet north of the mouth of Jackass Ravine, in the peninsula formed by the sharp northward bend of the river. The mine is probably the first in the region to have been systematically worked and is said to have been in operation as early as 1851. There is record of a stamp mill at the mine in 1853.[97] It has been idle for many years but has been reopened since the writer's last visit to the district.

The total production is reported to have been about $200,000. There are in all about 2,000 feet of drifts and 500 feet of raises, indicating that the ratio of production to development is about $80 a foot.

The principal vein strikes N. 70°–80° W. and dips 40°–80° N. (See fig. 45.) It follows a fault, as there is gabbro on the hanging-wall side and conglomeratic black slate, schist, and tuff of the Kanaka formation along the footwall in the western part, and it is believed that the movement was in the reverse direction. The vein can be traced uphill to the southeast for some distance, though the intersection with the serpentine, which lies farther east, has not been explored. Its continuation on the north side of the river was not found. The projection (pl. 8) suggests that it may belong to the same zone of fissuring as the west vein of the Irelan mine.

The workings consist of four tunnels, the lowest close to the river bank, two others 67 and 133 feet above, and a fourth on the opposite side of the peninsula still higher. The lower tunnel was studied, but the upper tunnels, which together with shallow surface work and surface sluicing yielded by far the greater part of the production, were inaccessible at the time of visit. It is reported that much of the production was derived from "mill rock" having a tenor of about $10 a ton. The "mill rock" stopes also yielded small bunches of high-grade ore in which the gold was associated with coarse arsenopyrite. Two veins have been developed—the German Bar vein, followed from the portal of the lower tunnel, which is said to have yielded the ore taken from the two upper tunnels, and the Wheeler vein, which is a short distance to the north and is apparently joined by the German Bar vein toward the east.

A zone of quartz stringers, accompanied by some gouge, connects the two veins. The junction of this zone with the footwall of the Wheeler vein has been explored between the two lower tunnels by an inclined raise. This has yielded nearly all the high-grade ore found outside the ore bodies of the upper levels.

Postmineral movement seems to have been of comparatively minor extent, though postmineral gouge is present fairly continuously along the German Bar vein, generally along the hanging wall, and is less noticeable on the Wheeler vein. At the portal of the lower tunnel a steep postmineral fault cuts the vein at a very acute angle. The same fault is also observable above the portal of the next tunnel (No. 3).

As the surface exposures were not studied in detail, no opinion can be offered as to whether development

---

[97] Trask, J. B., Report on the geology of northern and southern California, p. 63, Sacramento, 1856.

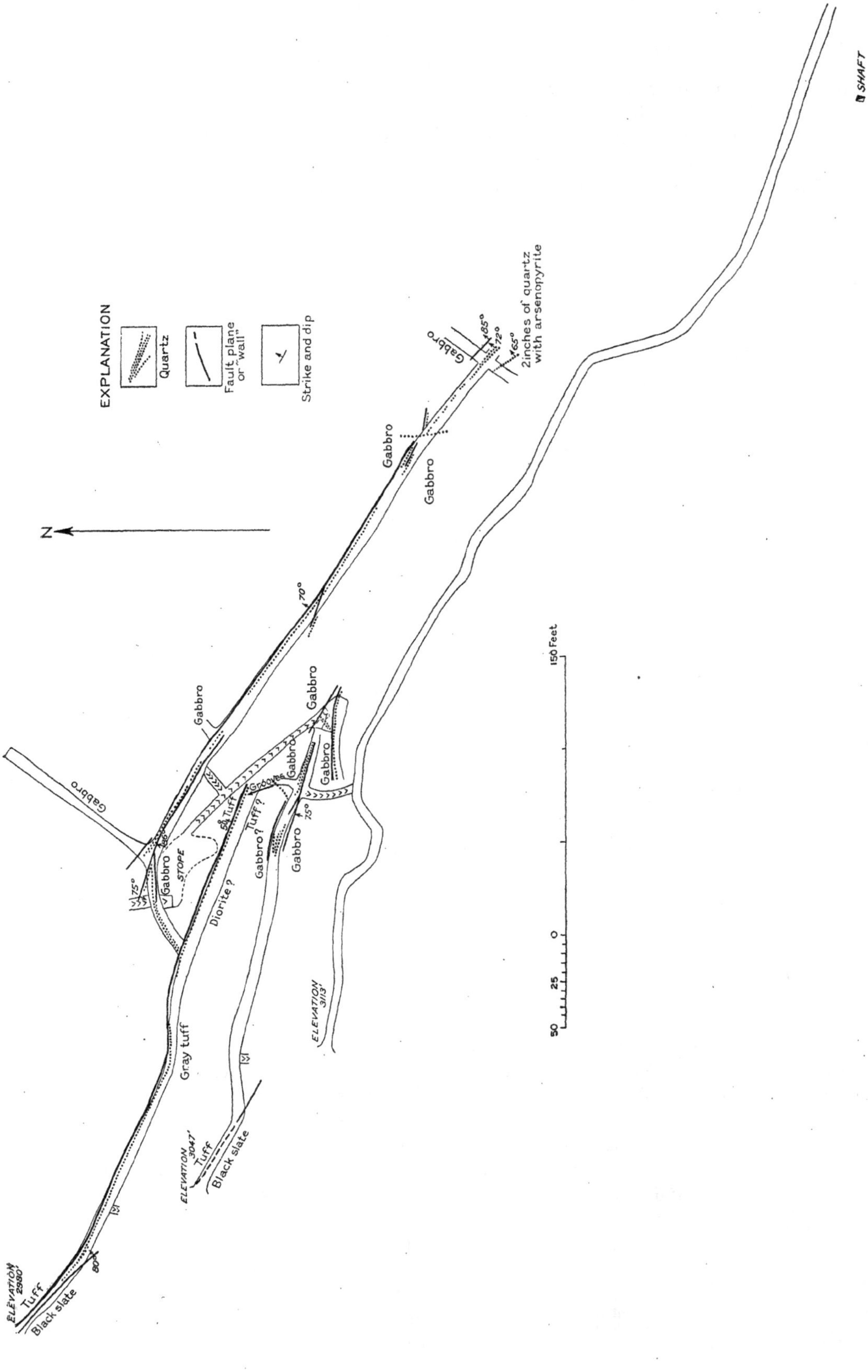

FIGURE 45.—Map of German Bar mine

EXPLANATION

Quartz

Fault plane or "wall"

Strike and dip

eastward to meet the serpentine would be likely to prove productive.

GOLD CANYON MINE

The Gold Canyon mine (fig. 46) is on the north bank of the Middle Fork of the Yuba River just north of the sharp bend. The mine has been worked at intervals since about 1860 but has not been operated for several years. The production, according to Mac-

The principal vein, which has an average strike of N. 60° W. and dips about 55° NE., is entirely within the gabbro and does not enter the adjoining serpentine. It presumably occupies the continuation of the Irelan fissure. Reverse faulting along the vein is indicated by the fact that at the east end of the workings serpentine forms the hanging wall. In the direction of the strike the vein pinches out against the serpentine, breaking up into irregular stringers. The

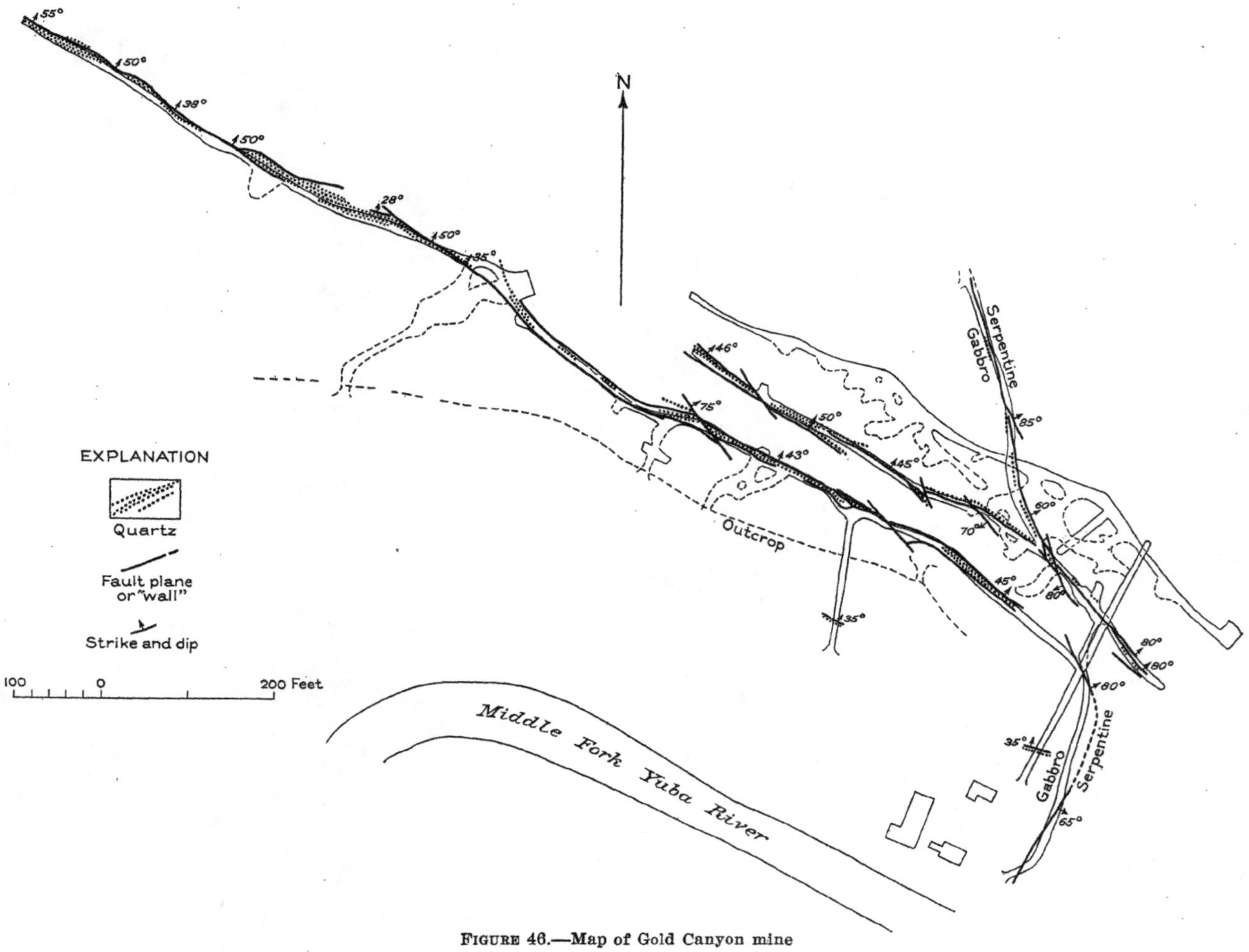

FIGURE 46.—Map of Gold Canyon mine

Boyle,[98] has been $735,000, almost all of which has come from one small ore shoot. The workings include between 5,000 and 6,000 feet of drifts and perhaps 2,000 feet of raises and winzes on the vein, giving a production ratio of $92 to $105 a foot. If 1,000 feet of barren development on the Contact vein is excluded the ratio for the main vein is $105 to $122 a foot.[99]

[98] MacBoyle, Errol, op. cit., p. 83.
[99] These figures are taken from Mr. Gannett's notes and are presumably based on information obtained by him as to the extent of workings not shown on the only map available to the present writer (fig. 46), which is known to be incomplete. The drifts on this map aggregate about 3,300 feet and the raises about 400 feet, which would give a ratio of nearly $200 to the foot for the mine as a whole, and about $270 for the development of the principal vein.

serpentine adjoining these is altered near the surface to the brown cellular mass resulting from the oxidation of iron-bearing carbonate. Besides the main vein, which has yielded the entire production, there is a small vein, well defined only near the main vein, which follows the serpentine contact. Another small vein less than a foot in width 130 feet south of the main vein and approximately parallel is said to yield fine gold in the pan.

Evidence of successive thrusts during the development of the vein is obtainable in places. Such faulting appears to have been more intense in the neighborhood of the serpentine. The vein that follows the serpentine contact has been prospected principally on the drain level and consists for the most part of a mere stringer of quartz separated from the serpentine on the hanging wall by a well-marked postmineral slip.

Apparently the vein at a distance from the serpentine did not repay development. Development near the serpentine contact has been carried on below the lowest level shown on the map, apparently with no great success, but the amount of work done is unknown. Presumably this is still the most hopeful region for future work. The only other region of possible promise seems to be in the neighborhood of the junction of the hanging-wall split referred to above. This has apparently yielded some gold between the upper level and the outcrop but has not been prospected in depth.

The main vein is well defined throughout the length of the main drift, which exceeds 1,500 feet, but it has been productive only near its east end, close to the serpentine contact. Here the vein is cut by a number of nearly parallel small faults. These are most closely spaced within a zone of 150 feet west of the point where the vein is lost against the serpentine. The principal movement on these faults was clearly postmineral and the displacement small—from a few inches to a few feet. There is evidence, however, that these fissures were formed prior to the phase of mineralization that brought the major part of the gold, for in this zone the older vein quartz is crossed by small veinlets which contain quartz, carbonate, mariposite, galena, and sphalerite. Similar later veining was observed in the central part of the upper level, where there is a split in the vein on the hanging wall. Postmineral movement was here nearly parallel to the vein, the gouge streak on the hanging wall to the west crossing the vein diagonally and becoming the footwall. There has been stoping above the level at this point, but the amount obtained is unknown. The high-grade ore was associated with coarse arsenopyrite. (See fig. 18.)

INDEPENDENCE MINE

The Independence mine is on the east bank of Buckeye Ravine about 3,000 feet east-northeast of the Gold Canyon mine and 3,800 feet southeast of the Plumbago mine. The mine was worked many years ago and has long been idle. The work was done from a shaft, now inaccessible, on the bank of Buckeye Ravine.

The vein follows the fault that bounds the district on the east. On the west is a large mass of serpentine, but at the shaft the rock west of the fault is a chloritized schist, probably derived from gabbro. This forms a belt 30 to 40 feet in width between the fault and the serpentine. The schist is much contorted, but the average strike of the schistosity is N. 20° W. and the dip 80° W. to vertical. East of the fault the rock

is smooth dark slate of the Blue Canyon formation, which near the shaft contains a lens of impure limestone 5 feet wide and traceable along the strike for about 60 feet. The strike of the slaty cleavage, which is close to that of the slate and limestone contact, is N. 8° E. and the dip 85° E., flattening slightly close to the fault.

The vein follows the fault plane and in the short distance through which the outcrop can be recognized has a strike of N. 1°–20° E. At the shaft the dip is 80° E. The maximum width, where exposed at the shaft, is 4 feet.

The quartz on the dump contains abundant inclusions of country rock altered to carbonate and mariposite. The schistose gabbro on the surface has also been replaced close to the vein, but the slate on the hanging wall is unaltered.

FUTURE OF THE DISTRICT

It must be evident to anyone who has studied the body of this report, particularly the section on ore shoots, that the authors consider that the veins of the Alleghany district hold promise of considerable future production, although owing to the peculiar type of ore prevalent in the district production will continue to be irregular.

There is nothing in the mineralogy of the veins which suggests that the gold of the high-grade shoots is of supergene origin or in any way related either to the present surface or to that existing at the time of deposition of the Eocene auriferous gravel. In spite of the long history of mining in the district, development of the veins has so far reached only comparatively shallow depth, but if the smaller amount of development work at the lower altitudes is taken into account, there has been no decrease in production with depth. It is of course beyond the ability of the writer to make any predictions as to the ultimate depth to which the veins will prove workable, and as the ore of Alleghany differs so greatly from that of other California districts that have been developed successfully to much greater depth, no prediction based on comparison with these districts can be made. It is reasonable, however, to infer a productive depth at least as far below the present average depth of development as the present depth is below the lava overlying the veins.

It may be that the deposition of the coarse gold of the high-grade shoots was dependent upon a delicate adjustment of physical conditions such as temperature and pressure and that at much greater depth a wider distribution of gold into lower-grade shoots will be found. The increase in the amount of gold outside the high-grade shoots in the lower levels of the Sixteen to One mine and the "mill ore" of the German Bar mine suggest that this is possible. But high-grade rather than mill ore is by far the principal source of gold in the lowest workings of the district, such as the lower parts of the Sixteen to One and Plumbago mines, and the workings of the Gold Canyon and German Bar mines. It therefore seems reasonable to suppose that no significant change in the general type of the ore need be expected within moderately greater depth on the veins.

As shown in the table on pages 54–56, the veins that follow the eastward-dipping reverse faults have been the most productive, and it is thought that they are likely to yield the best returns in the future. Nevertheless, a considerable production has been obtained from the veins belonging to the other groups, and if other indications are favorable they should not be neglected.

The most cursory inspection of Plates 8 and 9 shows that by far the greater part of the development work has been done on veins that crop out in areas from which the lava has been eroded. There is no reason to suppose that veins which lie beneath the lava may not prove equally productive. Naturally the exploration and development of such veins, even those which have been definitely located by the placer tunnels, will be a far more expensive process than the further exploration of the veins already in part developed and should hardly be undertaken without capital ample to carry on a long campaign of unproductive exploration. To give quantitative expression to this opinion of the future possibilities the estimates made on page 68 are here repeated, with an additional figure based on the expectation that for at least another 800 feet of vertical depth the veins will prove equally productive.

Possible recoverable gold content of veins concealed beneath the lava, above an altitude of 3,700 feet_____ $10,000,000

Possible recoverable gold content of undiscovered shoots in and near veins now being developed above 3,700 feet_____ $6,000,000

Possible recoverable gold content of all veins between 2,900 and 3,700 feet_____ 40,000,000

Such estimates should not be taken too literally,[1] but they nevertheless reflect the writer's considered opinion as to the productive possibilities of the district.

The chief period of placer production is believed to be over. No study of the auriferous gravel was made, however, and no opinion is offered as to the possibility of future placer production, either by drift mining or by hydraulic mining, should its resumption be permitted.

In closing there may perhaps be permitted a word of warning, applicable alike to active operators and to intending investors. The many failures in the district are believed to have been in large part due to the peculiar occurrence of the ore. It seems to be almost inevitable that the discovery of high-grade ore causes such a feeling of elation that the operator is too strongly tempted to cash in on his good fortune and does not reserve adequate funds for the inevitable period of unproductive exploration which follows the exhaustion of a high-grade shoot. Figures have been cited to show that the reward at Alleghany per ton of quartz mined or per foot of development driven is probably greater than for other California districts, but it must be kept in mind that in exploring veins of this type adequate initial capital is essential and that when the operator's perseverance is rewarded by the discovery of high-grade ore self-restraint must be exercised if a repetition of the first success is desired.

---

[1] "Merely corroborative detail added to give verisimilitude to an otherwise bald and unconvincing narrative."—W. S. Gilbert, Mikado. Act 2.

INDEX.

A

	Page
Acknowledgments for aid	3–5
Acme mine, description of	125
Albite, sections showing	pls. 20, A, 26, 27, A, 30, B, 31, B, C, 32, C, 33, A, B
Alleghany, early mining at	25
Alta mine, description of	100–101
Alta vein, junction of, with Oriental vein, section showing	pl. 19, A
Altitude of peaks in the adjacent region	62
Amethyst mine, description of	94–95
Amphibolite, occurrence of	125
Andesitic breccia, occurrence and character of	18
Ankerite, formation of	74
section showing	pl. 34, B
Annex claim, location of	122
Apatite, occurrence of	45
Aplite, occurrence and character of	16
Arcade mine, description and sketch maps of	128–129
Arsenopyrite, association with gold	50, 59, 82–83, pls. 21, 41
general occurrence and character of	40, 49, 75
sections showing	pls. 33, D, E, 39, A, C, 40, D, 41, 46, C–F
Asbestos, occurrence of	125

B

Bald Mountain, rocks capping	pl. 5, D
Bald Mountain mine, early history of	25
Balsam Flat, early mining at	25
Baltimore claim, location of	121
Barite, occurrence of	44
section showing	pl. 39, C
Basalt, occurrence and character of	18–19, pl. 5, D
"Bedrock series," age of	16–17
description of formations of	6–16
structure of	19–21
Beidellite, occurrence of	45
Belmont prospect, description of	122–123
Bennett, C. A., acknowledgments to	5
Bibliography	5
Blue Canyon formation, age of	19–20
occurrence and character of	6, pl. 2, B
"Blue jay," nature and significance of	46–47, 57, 60
Bob Evans claim. See Evans prospect.	
Bradbury, T. J., acknowledgments to	5
Brush Creek mine, description and sketch maps of	87–89
production from	26, 87
veins in, details of	pls. 13, D, 24, A, 29, C, D, 33, C, D, 35, C, D, 42, A, B, 45, B, C
Bullion vein, development on	118, 119

C

Calaveras formation, age of	16, 20
Cape Horn slate, conditions of deposition of	20
occurrence and character of	12–13, pl. 13, D
Carbon, occurrence of	47–48
Carbonate, general occurrence and character of	45–46
sections showing	pls. 28, D, E, 33, A–C, 34, 42, A, B
source of	74–75
Carbonate stage of mineralization	45–51, 74–75, 85–86
Cedar claim, description of	121–122
Central mine, description of	117–118
Chalcedony, occurrence of	44–45, 48–49
sections showing	pls. 32, D, 36, A, B, 37, C, D
Chalcopyrite, occurence of	49
Chert member of the Kanaka formation	11
Chips Flat, early mining at	25
Chlorite, general occurrence of	47
sections showing	pls. 20, A, 33, C, 34, D
Chlorite stage of mineralization	39–40, 74, 77
Chromite, occurrence and character of	75, pl. 20, B
Clinton vein, character and relations of	37, 79, 103–105, pls. 13, B, 24, B–D, 29, B, 37, A, B
Clipper Gap formation, age of	20
Clute claim, location of	129
Colorado claim, location of	119

D

	Page
Colorado Extension claim, location of	119
Conglomerate member of the Kanaka formation, occurrence and character of	8–10, pl. 4, B
Contact claims, location of	125
Continental mine, location of	124
Cooper prospect, description of	129
Copper Hill mine, early stamp mill at	127
Crafts claim, location of	129

D

Dead River claim, workings on	27, 102
Denudation, estimates of rate of	62–63
Derelict claim, location of	89
Diadem mine, description of	95
Diamond Peak mine. See Mugwump mine.	
Diorite, occurrence and character of	14–16
Docile mine, description and map of	123–124
Drainage of the district	3
Tertiary	18, 23, pl. 43
Dreadnaught mine, description and map of	122
Drift mines, list of some not described	87
Duggleby, A. F., acknowledgments to	5

E

Eastern Cross Extension claim, location of	120
Eastern Cross mine, description of	120–121
Eastern Star claim, description of	120, 121
Eclipse mine, description of	110–111
map of	pl. 54
Eclipse vein, relations of	109
Eldorado Extension claim, location of	118
Eldorado mine, description of	118–119
maps of	pls. 55, 56
veins in, details of	34, pls. 12, A, 14, 17, A, 19, B, 27, B, C, 30, A, 31, A, 32, D, 40, D, 42, C, 44, C, D, 45, A, 46, C
Enterprise claim, location of	129
Eocene gravel, estimates of gold content of	64–68
view of	pl. 5, C
Erosion, estimates of amount of	61–70
Eureka Extension claim, location of	92
Eureka mine, description of	92
Evans prospect, description of	92
sketch map of	93
Extension of the Minnie D mine, description and maps of	113, pl. 54
vein in, section of	35

F

Faulting in the district, general character of	21–22
relation of ore deposits to	33–36, 57–58, 70–72
sections showing	pls. 6, 7, B, C, 11, B, D, 12 A, C, 13, C, D, 15, 16, 17, 18, B, C, 34, B
Federal mine, description of	94
Field work	3–5
Finan prospect, description and sketch map of	89
Fissure systems, origin of	70–72
Folding of rocks in the district	20–21
Forest, early mining at	25
section of Kanaka formation near	11
view of part of	pl. 5, D
Frances D prospect, description of	101
French Ravine, section showing relations of granite and gabbro in	16
Future of the district	135–136

G

Gabbro, occurrence and character of	14–15
section showing	pl. 34, D
Galena, occurrence of	49–50, 109, 119, 131, 135
section showing	pls. 30, A, 41, A
Gaylord, H. M., acknowledgements to	27
General Sherman prospect, description and sketch map of	101–102

Page

German Bar mine, description and map of_____ 132–134
    veins in, details of_____ 78, pls. 18, A, 46, D, F
Gold, concentration of_____ 52–53, 85
    mode of occurrence of_____ 50–51
        sections showing_____ pls. 21, 39, B, 41, 42, A, B
    production of, in California, estimates of_____ 64
        in the district_____ 25–28
    source of_____ 75–76
    total amount in the district, estimates of_____ 64–70
Gold Beater claim, location of_____ 129
Gold Bug prospect, description of_____ 94
Gold Canyon mine, description and map of_____ 134–135
    veins in, details of_____ 59, 78, pls. 33, D, E, 39, A
Gold Star mine, description of_____ 92–93
Golden King mine, description of_____ 123
Golden Queen claim, location of_____ 123
Goldsworthy, W. H. J., acknowledgments to_____ 5
Granite, occurrence and character of_____ 16
Graphite, occurrence of_____ 47–48, 76–77, pl. 42, C
Gravel, early, occurrence and character of_____ 17
    intervolcanic, gold in_____ 26–27
        occurrence and character of_____ 17–18
    Pleistocene and Recent, occurrence and character of_____ 19
"Great Blue Lead," gold content of_____ 64–65
Greenstone, in the Kanaka formation_____ 10–11

H

Hardy Kate prospect, description of_____ 91
"Headcheese" breccia, occurrence of_____ 48, 59–60, 109
    sections showing_____ pls. 14, B, 36, C
Heikes, V. C., acknowledgments to_____ 27
Highland and Masonic mine, production from_____ 25
Hill, J. M., acknowledgments to_____ 27
History, geologic_____ 19–24
    of mining in the district_____ 25–28
Hope claim, location of_____ 129
Hornblende schist of the Tightner formation, occurrence and character of___ 7,
                                                                pls. 2, A, C, 4, A
Hydraulic mining in the district_____ 25, 26

I

Independence mine, description of_____ 135
Irelan mine, description of_____ 127–128
    map of_____ pl. 57
    section along prospect tunnel south of_____ 14
    vein in, details of_____ pl. 29, A, 41, B
Iroquois prospect, description of_____ 122

J

Jackass Ravine, sketch map showing outcrops of rock near_____ 8
Jamesonite, occurrence of_____ 49, 109
    sections showing_____ pl. 40, B, C
Jewel claim, description of_____ 120, 121
Johnston, W. D., jr., acknowledgments to_____ 5

K

Kanaka Creek, sections near_____ 12
    views in valley of_____ pls. 2, D, 5, B, 7, A
Kanaka Creek drainage basin, mines of_____ 97–125
Kanaka formation, areal extent of_____ 7–8
    chert member of_____ 11, pl. 4, C, D
    conglomerate member of_____ 8–10, pl. 4, B
    geologic relations of_____ 20
    sections of_____ 11–12
    state and greenstone member of, lower_____ 10–11
        upper_____ 11
Kate Hardy mine, description of_____ 89–91
    geologic maps of_____ 90, pl. 47
    rock in, character of_____ pls. 12, B, 13, C, 18, C, 30, C, D
Kenton mine, description and sketch map of_____ 97–98
    vein in, details of_____ 35, pl. 28, A, B
King Solomon claim, early history of_____ 97
Kinselbach mine, description of_____ 125
Knowlton, F. H., fossil plants identified by_____ 17–18

L

"Leather gouge," occurrence of_____ 48, 131–132
Lindgren, Waldemar, quoted_____ 61, 75
Live Yankee mine, production from_____ 25
Location of the district_____ 3
    map showing_____ 4

Page

Lode mines, production from_____ 26–28
Loosner claim, location of_____ 123
Loughlin, G. F., acknowledgments to_____ 5
Lucky Larry mine, description of_____ 113–114
    map of_____ pl. 54

M

Mammoth Springs mine, description and sketch map of_____ 125
Map, geologic, of Alleghany district_____ pl. 1 (in pocket)
    geologic, of region surrounding Alleghany district_____ pl. 3
    showing location of fault east of Alleghany with reference to Mother Lode
        system_____ pl. 6
    showing mine workings and claims_____ pl. 9 (in pocket)
    See also names of mines.
Marcasite, occurrence of_____ 51
Marion claim, location of_____ 129
Mariposa formation, geologic relations of_____ 20–21
Mariposa mine, description of_____ 117
Mariposite, occurrence and significance of_____ 46–47, 57, 60, 75
    sections showing_____ pls. 33, D, E, 34, B, 35, A
Maryland prospect, description of_____ 126
Mayflower prospect, description and map of_____ 117
Mica, occurrence and character of_____ 46–47
Microbrecciation, sections showing_____ pls. 24, D, 25, A, B, 34, C, 38, A, B, 44, A
Middle Fork of Yuba River. See Yuba River, Middle Fork of.
"Mill rock," occurrence and character of_____ 60, 119, 131, 132
Mineralization, carbonate stage of_____ 45–51, 74–75, 85–86
    chlorite stage of_____ 39–40, 74, 77
    depth of vein formation_____ 61–70
    final stage of_____ 51
    fissure systems in relation to_____ 70–72, 80–84
    oxidation products_____ 51–52
    pressure and temperature in relation to_____ 72–73
    quartz stage of_____ 40–45, 74–75, 77–85
    summary of_____ 38–39, 86–87
    See also individual minerals and mines.
Mineralogic guides in ore shoots_____ 58–60
Minerals, source of_____ 73–77
    See also names of minerals.
Mines, descriptions of_____ 87–136
    map showing_____ pl. 9 (in pocket)
    See also names of mines.
Minnesota, early mining at_____ 25
Minnie D Extension. See Extension of the Minnie D mine.
Minnie D mine, description of_____ 113–114
    map of_____ pl. 54
    vein in, details of_____ pl. 26, A–C
Monroe claim, location of_____ 121
Morning Glory mine, description of_____ 111–112
    map of_____ pl. 54
Morning Glory vein, relations of_____ 109, 111
Mother Lode system, relation of faults to_____ pl. 6
Mountain View prospect, description and sketch map of_____ 125–126
Mugwump mine, description of_____ 92–94
    sketch map of_____ pl. 48

N

Nolan, T. B., acknowledgments to_____ 5
North Fork mine, description of_____ 27, 95–97
    plan and section of_____ 96
    production from_____ 27, 95
    vein in_____ pl. 11, A
North vein, relations of_____ 103–104

O

Oak prospect, description of_____ 92
    sketch map of_____ 93
Oligoclase, section showing_____ pl. 35, B
Opal, occurrence of_____ 48–49
    sections showing_____ pls. 36, A, 37, C, D, 38, D, 39, B, C
Ophir mine, description and sketch maps of_____ 110, 111
Ophir vein, relations of_____ 33, 108–109
Ore deposits, origin of_____ 60–87
Ore mined_____ 26–28
Ore shoots, character of_____ 52–53
    favorable structural features of_____ 56–58
    mineralogic indications in_____ 58–60
    persistence in depth of_____ 60
    productivity of_____ 53–56
Oregon Creek, physiographic history revealed in valley of_____ 18
Oregon Creek drainage basin, mines of_____ 87–97

Page

Oriental mine, description of_____ 98–101
    early history of_____ 26
    maps and sections of_____ 32, pls. 49–51
    rock in, details of_____ pls. 17, B.
        19, A, 20, A, 22, A, B, 30, B, 31, B, C, 32, C, 33, A, B, 34, C, 35, B
Oriflamme mine, description and sketch map of_____ 126–127
Osceola mine, description and map of_____ 114–115

P

Palygorskite, occurrence of_____ 48, 131–132
Panama prospect, description of_____ 115
Parallel vein, relations of_____ 103–104
Phillips, J. A., quoted_____ 73
Placer mining in the district_____ 25, 26
    gold produced from_____ 26, 27–28
Plagioclase, occurrence of_____ 44
    source of_____ 77
Plumbago mine, description and maps of____ 129–132, pl. 58
    production from_____ 26, 54, 130
    veins in, details of_____ pls. 12, C,
        15, A, B, 16, A, 25, 27, D, 34, A, B, D, 36, C, 38, C, 40, C, 44, A, B, 46, A, B
    relations of_____ 29, 36
Pressure, relation of, to ore deposition_____ 72
Pyrite, general occurrence of_____ 40, 49, 51
    sections showing_____ pls. 22, A, B, 40, A, C
Pyrrhotite, occurrence of_____ 50

Q

Quartz, crinkly banding in_____ 36, 78–80, pls. 19, B, 40, D, 42, C, 44, 46, A, B
    general relations and character of_____ 31–32,
        40–44, 48–49, 74–75, 77–84, pls. 11–42, 44–46
    microbrecciation in_____ 41, pls. 24, D, 25, A, B, 34, C, 38, A, B, 44, A
    microphotographs showing_____ pls. 22–42, 44–46
    origin of_____ 74–75, 83–84
    ribbon structure in_____ 36, 78, pl. 18
    strain twinning in, sections showing_ pls. 24, D, 29, C, D, 31, A, 37, A, B, 46, A, B
    vacuoles in_____ 41–44, pls. 22–30
Quartz diorite, occurrence and character of_____ 15–16

R

Rainbow Extension mine, description of_____ 105–106
    map of_____ pl. 52 (in pocket)
Rainbow mine, description of_____ 102–105
    map and sections of_____ 37, pl. 52 (in pocket)
    production from_____ 26–27, 54, 102
    veins in, details of_____ 79,
        pls. 11, C, 13, A, B, 24, B–D, 28, D, E, 29, B, 37, A, B, 40, B
Rao prospect, description of_____ 117
Red Star mine, description and map of_____ 115–117
Relief quartzite, conditions of deposition of rocks of__ 20
    occurrence and character of_____ 12, pl. 5, A
"Ribbon quartz"_____ 36, 78, pl. 18
Rising Sun mine, description of_____ 121–122
Roads, difficulties of construction and maintenance____ 3
Rock formations, descriptions of_____ 5–19
Ross, C. S., acknowledgments to_____ 5
Roye-Sum prospect, description of_____ 98
    quartz from sections of_____ pls. 36, A, B, 37, C, 38, D
Rutile, occurrence of_____ 48, pl. 35, B

S

Sailor Canyon formation, age of_____ 20
Schaller, W. T., acknowledgments to_____ 5
Scope of the report_____ 3
Seneca claim, location of_____ 122
Sericite, occurrence of_____ 45, 46–47
    sections showing_____ pls. 32, A, B, 40, C, 45, A
Serpentine, occurrence and character of_____ 13–14, pl. 5, B
    relation of, to ore deposits_____ 31–32, 56–57, 71, 74–76
Short, M. N., acknowledgments to_____ 5
Sierra Nevada, Tertiary drainage of, map showing_____ pl. 43
Simkins, W. R., acknowledgments to_____ 127

Sixteen to One mine, description of_____ 106–109
    mill of, view showing_____ pl. 7, A
    production from_____ 54, 166
    veins in, details of_____ 31, 33, 35, 38, pls. 7, B, C, 11, B, D,
        16, B, 18, B, 21, 22, D, E, 23, 26, D, 28, C, 35, A, 39, B, 40, A, 46, F
    workings of, plans and sections showing_____ 108, 110, pl. 53
Slate, in the Blue Canyon formation_____ 6, pl. 2, B
    in the Kanaka formation_____ 10–11
Snowdrift prospect, description of_____ 118
Sosman, R. B., quoted_____ 82
South Fork mine, description of_____ 94
    map of_____ pl. 48
Sphalerite, occurrence of_____ 109, 119, 135
Spiritualist tunnel, description of_____ 105
    map of_____ pl. 52 (in pocket)
Spoohn mine, description of_____ 101
Standard claim, location of_____ 129
Standard Extension claim, location of_____ 129
Structure in the district, general features of_____ 19–23
    relation of veins to_____ 33–38, 57–58
    sections showing_____ pl. 10 (in pocket)
Sulphides, occurrence of_____ 49–50
"Superjacent series," notes on the formations of_____ 17–19

T

Talc, occurrence and character of_____ 47
    sections showing_____ pl. 35, C, D
Taylor, G. F., cited_____ 123
Taylor, G. F. and R. F., acknowledgments to_____ 124
Temperature, relation of, to quartz crystallization____ 72–73
Terrible claim, location of_____ 118
Tertiary history of the district_____ 22–23
Tetrahedrite, occurrence of_____ 49, 109, 119, 131
    section showing_____ pl. 40, A
Tightner formation, occurrence and character of____ 6–7, pls. 2, A, C, 4, A
    origin of_____ 20
Tightner mine, description of_____ 106–109
    veins in, details of_____ pls. 18, B, 21, A, 38, A, B
Tillite (?), possible occurrence of_____ 10, pl. 4, B
Tomboy prospect, description and sketch map of_____ 91–92
Topography of the district, development of_____ 23–24
    general features of_____ 3, pl. 2, D
Transportation in the district_____ 3
Twenty-one vein, description of_____ 105–106

U

Uncle Sam mine, early workings at_____ 25

V

Vacuoles in quartz and other minerals_____ 41–44, pls. 22–30
Veins, age of_____ 61
    distribution of_____ 29–31
    gold contained in, estimates of total amount of___ 68–69
    gold produced from_____ 53–56
    origin and formation of_____ 70–72, 77–87
    relation to country rock_____ 31–32
    scope of descriptions of_____ 28–29
    size and structure of_____ 33–38
    vertical range of gold deposition in_____ 61–70

W

Western Cross claim, description of_____ 120, 121
Wheeler vein, location of_____ 132
Wilson, W. V., acknowledgments to_____ 5
Wonder prospect, description of_____ 102
Wyoming group, description of_____ 102

Y

Yellow Jacket mine, description of_____ 119–120
    map of_____ pl. 55
    quartz in, sketches showing character of_____ 78, 79
Young America mine. See Mugwump mine.
Yuba River, Middle Fork of, mines in drainage basin of__ 125–136
    outcrops of rock along, sketch map showing_____ 8
    section of Kanaka formation along_____ 12

This original publication came with additional plates and/or maps included in the document or the back pocket of the publication. Due to our distribution and printing process and to keep the cost down and make this publication affordable to everyone Miningbooks.com has digitally scanned and formatted these plates and/or maps and they are available for free as a download from our website. You may also choose to purchase these plates and or maps on CD Rom for a nominal fee in order to cover materials, labor, shipping and handling. Our website to download your maps for free or to purchase a CD ROM is www.miningbooks.com .

www.ingramcontent.com/pod-product-compliance
Lightning Source LLC
Chambersburg PA
CBHW08054722O326
41599CB00032B/6395